Lecture Notes in Computer Science 7469

Commenced Publication in 1973
Founding and Former Series Editors:
Gerhard Goos, Juris Hartmanis, and Jan van Leeuwen

Sergey Andreev   Sergey Balandin
Yevgeni Koucheryavy (Eds.)

Internet of Things, Smart Spaces, and Next Generation Networking

12th International Conference, NEW2AN 2012,
and 5th Conference, ruSMART 2012,
St. Petersburg, Russia, August 27-29, 2012
Proceedings

 Springer

Volume Editors

Sergey Andreev
Yevgeni Koucheryavy
Tampere University of Technology (TUT)
Department of Communications Engineering
Korkeakoulunkatu 1, 33720 Tampere, Finland
E-mail: sergey.andreev@tut.fi; yk@cs.tut.fi

Sergey Balandin
FRUCT Oy
Kissankellontie 20B, 00930, Helsinki, Finland
E-mail: sergey.balandin@fruct.org

ISSN 0302-9743   e-ISSN 1611-3349
ISBN 978-3-642-32685-1   e-ISBN 978-3-642-32686-8
DOI 10.1007/978-3-642-32686-8
Springer Heidelberg Dordrecht London New York

Library of Congress Control Number: 2012944210

CR Subject Classification (1998): C.2, B.8, C.4, D.2, K.6, I.2, H.3

LNCS Sublibrary: SL 5 – Computer Communication Networks and Telecommunications

Typesetting: Camera-ready by author, data conversion by Scientific Publishing Services, Chennai, India

Printed on acid-free paper

Springer is part of Springer Science+Business Media (www.springer.com)

Preface

We welcome you to the joint proceedings of the 12th NEW2AN (Next-Generation Teletraffic and Wired/Wireless Advanced Networking) and 5th Conference on Internet of Things and Smart Spaces ruSMART (Are You Smart) held in St. Petersburg, Russia, during August 27–29, 2012.

Originally, the NEW2AN conference was launched by ITC (International Teletraffic Congress) in St. Petersburg in June 1993 as an ITC-Sponsored Regional International Teletraffic Seminar. The first edition was entitled "Traffic Management and Routing in SDH Networks" and held by R&D LONIIS. In 2002, the event received its current name, the NEW2AN. In 2008, NEW2AN acquired a new counterpart in Smart Spaces, ruSMART, hence boosting interaction between researchers, practitioners, and engineers across different areas of ICT. Presently, NEW2AN and ruSMART are well-established conferences with a unique cross-disciplinary mixture of telecommunications-related research and science. NEW2AN/ruSMART is accompanied by outstanding keynotes from universities and companies from Europe, USA, and Russia.

The 12th NEW2AN technical program addressed the various aspects of next-generation data networks. This year, special attention was given to radio access networks and the related problems. The authors presented novel and innovative improvements for advanced signaling protocols, enhanced QoS mechanisms, cross-layer optimization solutions, and traffic characterization models. In particular, the issues of QoE in wireless and IP-based multiservice networks were studied, as well as some economical aspects of future networks. In addition, there was a traditional emphasis on wireless technologies, including, but not limited to, cellular, mesh, ad hoc, and sensor networks.

The 5th Conference on Internet of Things and Smart Spaces, ruSMART 2012, provided a forum for academic and industrial researchers to discuss new ideas and trends in the emerging areas of Internet of things and smart spaces that create new opportunities for fully customized applications and services. The conference brought together leading experts from top affiliations around the world. This year, there was active participation by industrial world-leader companies and particularly strong interest from attendees representing Russian R&D centers, which have a good reputation for high-quality research and business in innovative service creation and applications development.

This year, the Technical Program of NEW2AN/ruSMART benefited from joint keynote speakers from European and Russian universities and companies.

We would like to thank the Technical Program Committee members of both conferences, as well as the associated reviewers, for their hard work and important contribution to the conference.

The conferences were organized in cooperation with Open Innovations Association FRUCT, ITC (International Teletraffic Congress), IEEE, Tampere University of Technology, St. Petersburg State University of Telecommunications, and Popov Society. The support of these organizations is gratefully acknowledged.

We also wish to thank all those who contributed to the organization of the conferences. In particular, we are grateful to Jakub Jakubiak for his substantial work on supporting the conference website and his excellent job on the compilation of camera-ready papers and interaction with Springer.

We believe that the 12th NEW2AN and 5th ruSMART conferences delivered an interesting, high-quality, and up-to-date scientific program. We also hope that participants enjoyed the technical and social conference programs, the Russian hospitality, and the beautiful city of St. Petersburg.

August 2012

Sergey Balandin
Sergey Andreev
Yevgeni Koucheryavy

Organization

NEW2AN International Advisory Committee

Nina Bhatti	Hewlett Packard, USA
Igor Faynberg	Alcatel Lucent, USA
Jarmo Harju	Tampere University of Technology, Finland
Andrey Koucheryavy	Giprosviaz, Russia
Villy B. Iversen	Technical University of Denmark, Denmark
Paul Kühn	University of Stuttgart, Germany
Kyu Ouk Lee	ETRI, Republic of Korea
Mohammad S. Obaidat	Monmouth University, USA
Michael Smirnov	Fraunhofer FOKUS, Germany
Manfred Sneps-Sneppe	Ventspils University College, Latvia
Ioannis Stavrakakis	University of Athens, Greece
Sergey Stepanov	Sistema Telecom, Russia
Phuoc Tran-Gia	University of Würzburg, Germany

NEW2AN Technical Program Commitee

Mari Carmen Aguayo-Torres	University of Malaga, Spain
Ozgur B. Akan	METU, Turkey
Khalid Al-Begain	University of Glamorgan, UK
Sergey Andreev	Tampere University of Technology, Finland (TPC Chair)
Tricha Anjali	Illinois Institute of Technology, USA
Konstantin Avrachenkov	INRIA, France
Francisco Barcelo	UPC, Spain
Sergey Balandin	FRUCT, Finland
Thomas M. Bohnert	SAP Research, Switzerland
Torsten Braun	University of Bern, Switzerland
Chrysostomos Chrysostomou	University of Cyprus, Cyprus
Nirbhay Chaubey	Institute of Science and Technology for Advanced Studies and Research (ISTAR), India
Ibrahim Develi	Erciyes University, Turkey
Roman Dunaytsev	Tampere University of Technology, Finland
Eylem Ekici	Ohio State University, USA
Sergey Gorinsky	Washington University in St. Louis, USA
Markus Fidler	NTNU Trondheim, Norway
Giovanni Giambene	University of Siena, Italy
Stefano Giordano	University of Pisa, Italy
Ivan Ganchev	University of Limerick, Ireland

NEW2AN Additional Reviewers

Bernardo Vitor
Biernacki Arkadiusz
Borges Vinicius
Chaudhry Fazal
Gerasimenko Mikhail
Jakubiak Jakub
Pereira Vasco

Podnar Zarko Ivana
Pyattaev Alexander
Sadkhan Sattar
Vukovic Marin
Wagenknecht Gerald
Wang Ning

ruSMART Executive Technical Program Committee

Sergey Boldyrev	Nokia, Helsinki, Finland
Nikolai Nefedov	Nokia Research Center, Switzerland
Ian Oliver	Nokia, Helsinki, Finland
Alexander Smirnov	SPIIRAS, St. Petersburg, Russia
Vladimir Gorodetsky	SPIIRAS, St. Petersburg, Russia
Michael Lawo	Center for Computing Technologies (TZI), University of Bremen, Germany
Michael Smirnov	Fraunhofer FOKUS, Germany
Dieter Uckelmann	LogDynamics Lab, University of Bremen, Germany
Cornel Klein	Siemens Corporate Technology, Germany

ruSMART Technical Program Committee

Sergey Balandin	FRUCT, Finland
Michel Banâtre	IRISA, France
Mohamed Baqer	University of Bahrain, Bahrain
Sergei Bogomolov	LGERP R&D Lab, Russia
Gianpaolo Cugola	Politecnico di Milano, Italy
Alexey Dudkov	NRPL Group, Finland
Kim Geun-Hyung	Dong Eui University, Republic of Korea
Didem Gozupek	Bogazici University, Turkey
Victor Govindaswamy	Texas A&M University, USA
Prem Jayaraman	Monash University, Australia
Jukka Honkola	Innorange Oy, Finland
Dimitri Konstantas	University of Geneva, Switzerland
Reto Krummenacher	STI Innsbruck, Austria
Alexey Kashevnik	SPIIRAS, Russia
Dmitry Korzun	Petrozavodsk State University, Russia
Kirill Krinkin	Academic University of Russian Academy of Science, Russia
Juha Laurila	Nokia Research Center, Switzerland
Pedro Merino	University of Malaga, Spain
Aaron J. Quigley	University College Dublin, Ireland

Luca Roffia   University of Bologna, Italy
Bilhanan Silverajan   Tampere University of Technology, Finland
Markus Taumberger   VTT, Finland

ruSMART Additional Reviewers

D'Elia Alfredo   Paramonov Ilya
Gurtov Andrei   Petrov Vitaly
Jakubiak Jakub   Pyattaev Alexander
Koucheryavy Yevgeni   Ukhanova Anna
Luukkala Vesa
Muromtsev Dmitry

Table of Contents

Future Services II

Smart Space Governing through Service Mashups

Part II: NEW2AN

Wireless Cellular Networks I

Wireless Cellular Networks II

Ad-Hoc, Mesh, and Delay-Tolerant Networks

Scalability, Cognition, and Self-organization

Traffic and Internet Applications

Wireless Sensor Networks

Selected Papers from NEW2AN 2012 Winter Session

Defining an Internet-of-Things Ecosystem

Oleksiy Mazhelis[1], Eetu Luoma[1], and Henna Warma[2]

[1] University of Jyväskylä, Jyväskylä, Finland
[2] Aalto University, Helsinki, Finland
{mazhelis,eetu.luoma}@jyu.fi, henna.warma@aalto.fi

Abstract. By bringing the Internet connectivity to the things, the Internet-of-Things (IoT) promises a number of benefits to its customers, varying from faster and more accurate sensing of our environment, to more cost-efficient tracking of industrial processes. Likewise, from the business perspective, the wide adoption of IoT is expected to generate significant revenues to the providers of IoT application and services. The IoT adoption depends on whether the ecosystems of the companies focusing on IoT technologies would emerge and succeed in delivering to the market the solutions attractive to the customers. Therefore, it is of utmost importance to understand the essence and constituents of IoT ecosystems. This paper is aimed at defining an IoT ecosystem from the business perspective. Based on a literature survey, the technologies that could form a core of an IoT ecosystem are summarized, and the roles of the firms which may comprise such an ecosystem are identified.

Keywords: Internet-of-things, business ecosystem, roles of firms.

1 Introduction

Internet-of-Things (IoT) represents a "global network and service infrastructure of variable density and connectivity with self-configuring capabilities based on standard and interoperable protocols and formats [which] consists of heterogeneous things that have identities, physical and virtual attributes, and are seamlessly and securely integrated into the Internet" [1]. Thus, IoT follows the "anything connected" vision by ITU [2] and assumes that any physical or virtual thing which could benefit from a connection to the Internet will eventually be connected.

IoT research has its roots in several domains, where different IoT aspects and challenges were being addressed. These research domains include the radio-frequency identification (RFID), machine-to-machine (M2M) communication and machine-type communication (MTC), wireless sensor and actuator networks (WSAN), ubiquitous computing, web-of-things (WoT), to name a few. Furthermore, these technologies have been applied in many vertical application domains, varying from automotive and machinery to home automation and consumer electronics. Thus, what is known today as the IoT represents a convergence of multiple domains, and can be seen as an umbrella term uniting the related visions and underlying technologies [3].

S. Andreev et al. (Eds.): NEW2AN/ruSMART 2012, LNCS 7469, pp. 1–14, 2012.

Business-wise, IoT represents a tremendous opportunity for various types of firms, including the telecom operators, application and service providers, as well as the platform providers and integrators. According to some estimates, M2M communications alone will generate circa 714 billion euros in revenues by 2020 [4]. Meanwhile, at present the market is in the very early stage, with fragmented solutions targeting specific vertical domains and/or specific types of applications. The current solutions are also characterized by a variety of proprietary platforms, protocols, and interfaces, making the components of solutions by different vendors barely compatible, while keeping the prices of the components high. Some of the available technologies could be seen as de facto standards, but no fully open standards have succeeded yet in the domain of sensor networking. Standard protocols and interfaces are also either available or being developed (e.g. by Zigbee alliance, IPSO alliance), but no single dominating set of standard protocols, interfaces, platforms has emerged yet. The lack of a generally accepted dominant design and the resulting high costs of the solutions, along with the lack of reference architectures and the lack of the vendor-independent guidelines on how to choose among the solutions or their components inhibit the wider adoption of the IoT technologies [5].

Thus, the expected rapid growth of the IoT market is contingent on the emergence of IoT ecosystems forming around common/dominant standards, platforms, and interfaces. At present, IoT ecosystems are formed around technological innovations focusing on a specific application domain, such as RFID solutions in retail, mobile M2M communications in remote automated meter reading (AMR), or ZigBee communications in smart home. These ecosystems are mainly in the formation stage, where no single firm could be identified as the leader to play the role of a keystone or a dominator in the ecosystem.

Within the last decade, notable research efforts have been devoted to studying the ecosystems' business entities and their roles in the domain of telecommunications [6,7] and in the IoT domain [8]. These academic studies, as well as numerous industrial reports, focus on a specific aspect of the IoT ecosystem, such as the role of a telecommunications operator [6,7], M2M value chain scenarios [9,10], or the role of a broker [7,8]. As a result, only a subset of the IoT roles has been emphasized in these studies; furthermore, the term of the IoT ecosystem has been left undefined.

This paper is aimed at deepening our understanding of current and potential future IoT ecosystems. In particular, this paper based on the available literature sources and the discussions with domain experts, strives at defining an IoT ecosystem from the business perspective, considers the potential core technological assets around which such ecosystems could form, and provides a list of potential IoT ecosystem roles.

The remainder of the paper is organized as follows. The next section is defining the IoT business ecosystem(s) of companies sharing certain assets in their products and services. Section 3 identifies the players of the ecosystem and their roles. The likely future developments in the IoT ecosystems are discussed in Section 4. Finally, Section 5 provides concluding remarks and outlines the direction for further work.

2   Internet-of-Things Ecosystem

In this section, we consider the business ecosystem concept as a metaphor adopted from the biological studies, and define the IoT ecosystem both from the technical and business perspectives.

2.1   Natural Life Ecosystem versus Business Ecosystem

The concept of an ecosystem and ecosystem modeling has been brought to business context by James F. Moore from the biological studies, where a *natural life ecosystem* is defined as a biological community of interacting organisms plus their physical environment with which they also interact [11].

By analogy with natural life ecosystem, Moore has defined *a business ecosystem* as "the network of buyers, suppliers and makers of related products or services" plus the socio-economic environment, including the institutional and regulatory framework. Interacting organizations and individuals represent the 'organisms of the business world' and form the foundation of the economic community delivering goods and services to customers – as well the members of the business ecosystem [11].

Besides the ecosystem's core business consisting of the firms delivering the goods/services along with their customers, market intermediaries, and suppliers, the business ecosystem includes the owners and other stakeholders of the core, as well as regulatory bodies and competing organizations, as shown in Fig. 1 below.

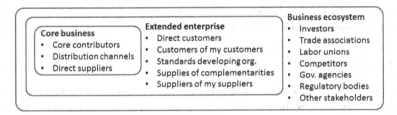

Fig. 1. Generic actors in business ecosystem [11]

Similarly to the organisms in the biological ecosystems, the firms in the business ecosystem co-evolve their capabilities around specific innovation(s), by both competing and cooperating with each other. However, the pace of evolutionary and ecological changes is different. In biological systems, evolutionary changes span several generations while multiple ecological changes are likely to occur within a lifetime of an organism. In business ecosystem, both types of changes are co-occurring, owing to the firms' ability to guide their own evolution, also proactively.

According to Moore [11], firm's capabilities in a business ecosystem co-evolve around innovations (compared to species' evolutionary paths). Similarly, Talvitie [12] following Iansiti and Levien [13] argues that the business ecosystems are formed around a specific *core*, i.e. a set of assets shared and commonly utilized by the firms forming the ecosystem. The core can be in a form of platforms, technologies, processes, standards or other assets common to and used by the members of the eco-system in their businesses. Making these assets available to the ecosystem members helps them in product and service creation by enabling higher levels of productivity, stability, and innovativeness, while also creating the positive network effect.

From the technical perspective, the platform forming the core of a business ecosystem, also represents a core of the corresponding *technical ecosystem* that can be defined as the collection of the platform and the modules specific to it [14]. Often, the platform is im-plemented as a software platform, i.e. as the "extensible codebase of a software-based system that provides core functionality shared by the modules that interoperate with it and the interfaces through which they interoperate" [15]. In turn, the modules represent software subsystems connecting to and adding functionality to the platform [16].

Topologically, the ecosystems can generally have either hub-centered star structure or a flat mesh-like structure. The star structure (often hierarchical) can be exemplified with the so-called *keystone model* matching the typical structures in the US. This model, as described by Moore [11] and elaborated by Iansiti and Levien [13], assumes that the ecosystem is dominated by a large hub firm interacting with a large number of small suppliers. On the other extreme, the *flat model* of the business ecosystem more typical for Europe is composed of mainly small and medium firms, accommodating also large ones [17].

According to Iansiti and Levien [13], the presence of the hubs makes the network robust to the removal of individual nodes, as soon as the hubs are intact. On the other hand, a removal of a hub often results in a collapse of the whole network. Following the biological metaphor, Iansiti and Levien [13] suggest that the *roles* in the biologi-cal ecosystem correspond to the *strategies* of the firms in the business ecosystem, with the most critical roles in the business ecosystem being the roles of the so-called keystone, dominator, and niche player.

2.2   Defining an IoT Ecosystem

As discussed above, an ecosystem emerges around a core, which represents some assets commonly used by the ecosystem members. Since the essence of the IoT is the interconnection of the physical world of things with the virtual world of Internet, the software and hardware platforms as well as the standards commonly used for enabling such interconnection may become a core of an IoT ecosystem. More specifically, such core may focus:

— On the connected devices and gateways, including both hardware platforms (Ar-duino prototyping platform [18], T-Mote Sky, Zolertia Z1, and other platforms based on Texas Instruments MSP430) and software platforms (TinyOS, Contiki OS), as well as the related standards (such as the gateway specifications by Home Gateway Initiative [19]);
— On the connectivity between the devices and the Internet, that may be implemented e.g. through a mobile wireless modem or a Wi-Fi router, or through a WPAN ga-teway device; namely, on hardware platforms (e.g. single-chip modems by RMC [20]), on the standards and protocols governing the communication (e.g. IETF 6LoWPAN, ROLL, and CoAP protocols promoted by IPSO Alliance [21], WPAN

standards by ZigBee Alliance [22]), or on the software platforms to support the connectivity (e.g. Californium Java CoAP framework [23], Erbium CoAP framework for Contiki [24]);

— On the application services built on top of this connectivity with the help of common software platforms (e.g. Pachube [25]) and standards governing the service composition and data format compatibility (EPC, JSON, SOA);

— On the supporting services that are needed for the provisioning, assurance, and billing of the application services (e.g. NSN M2M software suite [26], Ericsson Device Connection Platform [27], as well as M2M optimized network elements, e.g. the GGSN enabling network initiated PDP context) and related standards (e.g. the standards developed by ETSI M2M technical committee [28]).

These common assets with a potential to serve as a core for an IoT ecosystem are shown in Table 1, where they are categorized as hardware platforms, software platforms, or standards. This categorization is not fully exclusive, since e.g. the standards for the application services are likely to concern the connected devices.

Table 1. Examples of IoT ecosystem cores

Core	Hardware platform	Software platform	Standards
Connected device	Arduino, T-Mote Sky	TinyOS, Processing, Contiki OS	HGI
Connectivity	Wi-Fi or ZigBee systems-on-chip	Californium, Erbium	IPSO Alliance, ZigBee Alliance
Application services	Cloud infrastructure	Pachube	SOA, JSON, EPC
Supporting services	M2M optimized GGSN	NSN M2M suite, EDCP	ETSI M2M TC

Thus, from the business perspective, deriving from the definitions of Moore [11], Iansiti and Levien [13], and Talvitie [12], an *IoT business ecosystem* can be defined as a special type of business ecosystem which is comprised of the community of interacting companies and individuals along with their socio-economic environment, where the companies are competing and cooperating by utilizing a common set of core assets related to the interconnection of the physical world of things with the virtual world of Internet. These assets may be in a form of hardware and software products, platforms or standards that focus on the connected devices, on the connectivity thereof, on the application services built on top of this connectivity, or on the supporting services needed for the provisioning, assurance, and billing of the application services.

3   Roles in an Internet-of-Things Ecosystem

As described in the preceding section, the IoT ecosystems will form around commonly used IoT hardware and software platforms and standards. Therefore, the roles of the companies in an IoT ecosystem can be derived from the technical domains comprising the IoT solutions. As reflected in the end-to-end architectural vision of ETSI M2M TC [28] shown in Fig. 2, the three key IoT technical domains are the *device* domain including the identification and sensing/actuating technologies, the *connectivity* domain providing the access and core network connectivity and other service capabilities, and the application *service* domain.

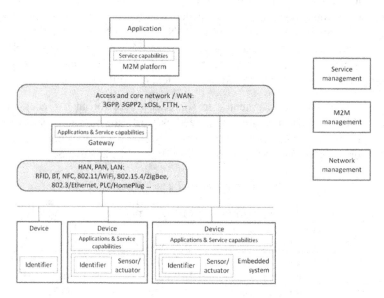

Fig. 2. A generic high-level end-to-end architecture based on ETSI M2M TC [28]

Within the last decade, the IoT related business entities and their roles have been studied in the domain of telecommunications [6,7] and also in the IoT domain [8]. Some of the relevant roles are described in industrial reports on M2M communications [29,30] and architectural frameworks of relevant standardization organizations [31,28]. The typical roles in telecommunications business and their relationships were identified by the ECOSYS project [6]. In addition to the telecommunications-specific ECOSYS roles, the study by the C-CAST project has identified the role of a broker aggregating and mediating the access to the contextual information about the user [7]. The importance of the broker role in the IoT domain is considered by the SENSEI project, where also the roles of wireless sensor and actuator network (WSAN) operator, WSAN service provider, and Sensor/Actuator Service Broker are differentiated [8].

As compared with the studies above, the report by ABI Research [29] dealing with M2M value chain adds the details on the roles visible in the product chain (chip and module manufacturer, as well as OEM/ODM) who help the ASPs in designing and pre-certifying with MNO the radio-subcomponents of M2M products.

Finally, the above studies on M2M and WSAN-related business roles assume that the available core network and transmission network infrastructure, including routers, domain-name servers, etc. are sufficient for the IoT applications. Whereas this may be the case with the majority of the M2M and WSAN-based applications, the applications relying on RFID tags are likely to require further infrastructure components, such as discovery services and object name services (ONS) [32]. The presence of these additional components may require the introduction of new business roles – for instance, the role of the ONS provider.

Fig. 3 aggregates the roles identified in various IoT-related works into a generic map of IoT ecosystem roles; the definitions for these roles are given in Table 2. The roles in the figure are mainly organized along the service delivery dimension, where several groups of related roles are identified, including the device, connectivity, and service related groups. The figure also shows the dimension of product/service life-cycle, consisting of development, distribution, provisioning, assurance, and billing roles. Other essential roles shown in the figure include the legislative, regulatory, standardization, and other bodies that directly or indirectly affect the IoT domain, as well as the roles responsible for domain specific auxiliary (e.g. infrastructure) technologies.

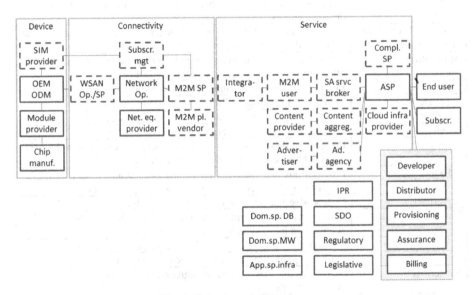

Fig. 3. Roles in an IoT ecosystem

Table 2. Definitions for the IoT ecosystem roles

Role	Description
Chip manufacturer	Designs and manufactures integrated circuits for module and device manufacturers.
Module provider	Manufactures components such as sensors of modems and deliver them to OEM/ODM.
OEM/ODM	Integrates components to produce the device or another piece of equipment.
SIM provider	Manufacturers SIM cards for network operators.

Table 2. (*Continued*)

Role	Description
WSAN operator and service provider (SP)	Operates and delivers services/information from WSANs under its responsibility.
Network operator	Provides connectivity between WSAN and the IoT applications; it may encompass access (mobile or landline) network, the core network, and the transmission network.
Network equipment provider	Manufactures network elements and related services and offers them to network operators.
Subscription management	A third party that manages SIMs and contracts on behalf of M2M user; is responsible for roaming and switching of networks (similar to MVNO).
M2M service provider	Manages the M2M service platform.
M2M platform vendor	Produces the M2M service platform which handles device specific tasks, including fault detection, management of SIM cards, etc.
Integrator	Ensures seamless inter-operation between the devices and the M2M platform.
M2M user	Is an organization that is formally in charge of the sensor and actuator devices/network.
Sensor and actuation service broker	Acts as a broker between the providers and the consumers of the sensor and actuator services.
Application SP	Builds the application/service from the components (own or made by other service providers) and deliver it to the user.
Complementary SP	Provides the services complementary to the one(s) of ASP.
(App) Developer	Designs and develops IoT applications and services.
Distributor	Is a retailer of physical or digital goods and services.
Provisioning SP	Deploys the application/service.
Billing SP	Provides billing services to a service operator, serving as a financial intermediary between them and their customers.
Ad agency	Provides ads and manages ad campaigns for advertisers, acting as intermediary between advertiser and a service provider.
Advertiser	Orders advertisements (individual or campaigns).
Content aggregator	Distributes content from different content providers to different SPs, acting as an intermediary between them.
Content provider	Provides user-generated or professionally created content.
User	Uses the application/service provided by the ASP.
Subscriber	Negotiates and commits to the agreement with the ASP on the service and its qualities.
Intellectual property rights (IPR) holder	Possesses exclusive rights for some intangible assets, and is entitled to giving the others an authorization for exploiting these assets
Standard development org. (SDO)	Official organizations, industrial alliances, special interests groups focusing on standard development.
Regulatory body	Controls the processes, as mandated by a legislative body
Legislative body	Makes, amends, or repeal laws.

Depending on the application domains, some of the roles in the table may be unnecessary. For instance, in case of a private WSAN solution, SIM provider may be unnecessary. Such optional roles are shown in the figure using dashed boxes.

Fig. 4 illustrates an example of IoT ecosystem forming around a standard Wi-Fi connectivity. In this rather simple flat ecosystem, the sensor devices, e.g. Wi-Fi enabled temperature sensors by Temperature@lert[1], are directly connected to Internet over a standard Wi-Fi router, without the need for SIMs, subscription management, M2M or other platforms and integration, and many other roles, making the ecosystem rather simplistic. Amazon, when it introduced the third generation of Kindle devices (without 3G) could be also seen as a member of this ecosystem.

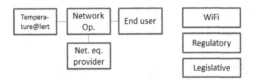

Fig. 4. A simple ecosystem formed around Wi-Fi connectivity

In a similar fashion, successful horizontalization of different IoT systems, protocols, interfaces, and data formats would enable a variety IoT services with a relatively simple ecosystem. For example, in future scenarios, the end user could acquire various IoT services through a home gateway that supports several technologies. Automated control of lightning, heating and security but also entertainment services could be provisioned through this gateway. With the interoperability issues diminishing, the end user could separately create contracts with its network operator and the application service providers, such as a utility company or a content provider. This scenario, which is presented in Fig. 5, resembles the contemporary Internet service provisioning model.

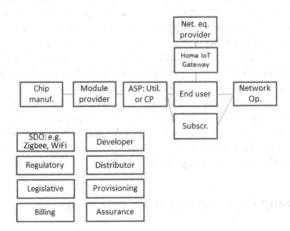

Fig. 5. Smart home ecosystem scenario

---

[1] See http://www.temperaturealert.com/Wireless-Temperature-Store/Temperature-Alert-WiFi-Sensor.aspx

Often, however, an ecosystem is more complex, and includes significantly larger number of roles. The example of Amazon Kindle, for instance, would require also the role of an ASP (Amazon), the role of complementary service providers (publishers), the role of the content providers (authors of the books), etc. Fig. 6 depicts another example of a more elaborate ecosystem – namely, the OnStar ecosystem formed around a vehicle telematics platform focusing on in-vehicle security, diagnostics, and other services[2]. As shown in the figure, this ecosystem involves, in addition to OEM (vehicle manufacturer) and MNO (Verizon Wireless), also system development (EDC), platform vendor (Hughes, LG), ASP (OnStar), and numerous complementary service providers (emergency call centers, roadside assistance, insurance companies), as well as SDOs and governmental agencies dealing with automotive telematics and positioning.

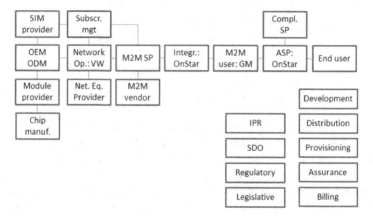

Fig. 6. On-star business ecosystem

The simple Wi-Fi centered ecosystem above may be sufficient when the service offered to the end users is simple and when both the provisioning and integration is straightforward (e.g. do-it-yourself approach) and no customer support is expected. However, for more complex scenarios, where several services (both core and complementary) with stringent security, safety, quality of service, and other requirements are to be provided, the business ecosystem usually demands a collaboration of a larger number of players. For instance, the need for mobility in case of OnStar necessitates the involvement of an MNO and a SIM provider, whereas the need for efficient subscription management necessitates the involvement of an M2M platform vendor and service provider. Note also that a player may take more than one role: for instance, OnStar combines the roles of the integrator and ASP, while Verizon Wireless combines the roles of subscription management and MNO.

4   Envisioning the Future of IoT Ecosystems

So far, there has been little discussion in the paper on the importance and weight of different roles in the development and sustainability of the IoT ecosystem. The general

---

[2] See www.onstar.com

theoretical treatment of ecosystems suggests that a keystone player would likely be maintaining the health and defining the future direction of an ecosystem [13]. Accordingly, the future structure and interactions between different roles in the ecosystem can be anticipated by identifying the potential keystone players.

IoT ecosystems are currently at their early development phases [33], and hence identifying keystones is naturally uncertain and challenging. Nevertheless, some useful insights into the likely future of the IoT ecosystems could be gained by comparing them to the structure of and the developments within the business ecosystems present in the domain of cloud computing. In cloud computing ecosystems, the vendors are producing services combining computing capacity, infrastructure software platforms and application, which all are delivered over the Internet [34]. To this end, we see a similarity in the general structure of the cloud and IoT ecosystems, as both exhibit roles related to either devices/hardware, connectivity and services.

Owing perhaps to the maturity of the hardware and software technologies compared to IoT, the cloud ecosystems have been forming relatively rapidly. Marston et al. [35] recognizes Amazon, Apache, Google, IBM, Microsoft and Salesforce.com as key players of the industry that have succeeded or are striving to create an ecosystem around their platform offerings. We observe that these platform providers create and capture most of the value of cloud computing. In addition, the platform providers are driving the growth of cloud computing market. Projecting this trend to the IoT domain suggests that the platform providers will eventually be creating most of the end-user value also in IoT ecosystems [10], and that these providers are the likely candidates for the keystone players.

Another similarity between the cloud and IoT ecosystems is their dependency on the connectivity. In cloud computing, the capacity of hardware and application services are usually provided to end-users as a bundle, but the bandwidth needs to be bought separately. We see a possibility for similar interaction from the end-users perspective also in the IoT; the end-user is required to contract with the two suppliers. The connectivity, to Internet, is nevertheless fairly standardized offering, and, as the communication service providers' capabilities in producing new service have generally declined, we find that they are less likely to drive the cloud or the IoT ecosystem developments.

There are some aspects, however, in which the IoT ecosystems are notably different from their counterparts in the cloud computing domain. In particular, the cloud computing ecosystems are relatively horizontal, i.e. different layers of the cloud computing solutions are provided by different vendors; furthermore, as the examples of the Amazon, Salesforce.com, and other key players evidence, the same elements of a cloud computing solution are often used "as is" in different vertical applications. This is not the case, yet, in the domain of the IoT, dominated currently by the integrated solutions targeting certain verticals or application domains. This can be attributed to the specific physical constraints and other requirements present in different vertical application domains, such as healthcare devices, in-vehicle telematics solutions, home automation products, etc., wherein the need for reliability, power efficiency, throughput of the communication link, and other characteristics may differ significantly. As opposite to the standardized hardware applied in cloud computing domain, this heterogeneity of requirements in the IoT domain results in the heterogeneity of application-specific hardware platforms and interfaces, multitude of communication protocols and data formats [36], and consequently a general lack of a software plat-forms applicable across different verticals [5]. Whereas, in the long term, the generic platforms with standardized interfaces driven by large keystones are likely to emerge, in the short term perspective, this heterogeneity creates business opportunities for small vendors who are able to enter the market by targeting an application domain with specific requirements and by offering products and complementary services to meet these specific requirements.

5   Concluding Remarks

The IoT technologies connecting the physical world with the world of Internet prom-ises a number of benefits both to its potential customers and to the vendors of the solutions. Despite these benefits, the adoption of these technologies is relatively mod-est, and the expected rapid adoption of the IoT technologies depends on the stability, diversity, and productivity of the business ecosystem that are being formed around these technologies.

In this paper, with the aim to improve our understanding of the IoT domain from the business perspective, an initial attempt has been made to define the IoT business ecosystem and the roles of the firms comprising it. The IoT business ecosystem has been defined in the paper as a community of interacting companies and individuals along their socio-economic environment, where the companies are competing and cooperating by utilizing a common share of core assets. These common assets can be in a form of hardware and software products, platforms or standards that focus on the connected devices, on the connectivity thereof, on the application services on top of this connectivity, or on the supporting services. Based on the literature survey, the paper has accumulated a list of potential IoT ecosystem roles, and illustrated them with some real-world examples.

In our future research, this work shall be expanded in several directions. First, the topologies of the emerging IoT ecosystems shall be analyzed, and the key hub-like roles shall be identified. Second, the roles in the IoT ecosystems serving different application domains, such as automotive, healthcare, connected home, etc., shall be compared, in order to identify common roles serving as horizontal layers in the IoT domain. Finally, the value of or the necessity for the individual roles shall be critically assessed, with the aim to identify the minimal set of roles whose presence is essential for the survival of the IoT ecosystems.

References

1. Tarkoma, S., Katasonov, A.: Internet of Things Strategic Research Agenda. Finnish Stra-tegic Centre for Science, Technology and Innovation (2011),
   http://www.internetofthings.fi/
2. ITU: The Internet of Things. Executive Summary. International Telecommunication Union (2005),
   http://www.itu.int/osg/spu/publications/internetofthings/

3. Atzori, L., Iera, A., Morabito, G.: The Internet of Things: A survey. Computer Networks 54, 2787–2805 (2010)
4. Machina Research: Machine-to-Machine connections to hit 12 billion in 2020, generating EUR714 billion revenue. Machina Research press release (2010),
   `http://www.machinaresearch.com/sitebuildercontent/sitebuilderfiles/machina_research_press_release_m2m_global_forecast_analysis_2010_20.pdf`
5. Batten, C., Wills-Sandford, T.: The Connected Home: a reality. Intellect report, Information Technology Telecommunications and Electronics Association (2011),
   `http://www.intellectuk.org/publications/intellect-reports/7743-the-connected-home-a-reality-report`
6. Kaleelazhicathu, R.K.: Business Models in Telecommunication. ECOSYS Deliverable 3 (2004), `http://ecosys.optcomm.di.uoa.gr`
7. Stanoevska-Slabeva, K.: Business Models - Opportunities and Barriers. C-CAST deliverable D15 (2010), `http://www.ict-ccast.eu`
8. Eurich, M., Lavoisy, O., Forest, F., Ytterstad, P., Akselsen, S., Tierno, A., Höller, J., Tsiatsis, V., Baugé, T., Villalonga, C.: Business Models and Value Creation. SENSEI Deliverable report D1.4 (2011), `http://www.ict-sensei.org/`
9. Jumira, O., Wolhuter, R.: Value Chain scenarios for M2M Ecosystem. In: Proc. IEEE Int. Workshop on Machine-to-Machine Communications, pp. 410–415 (2011)
10. Schlautmann, A., Levy, D., Keeping, S., Pankert, G.: Wanted: Smart market-makers for the "Internet of Things". Prism 2, 35–47 (2011)
11. Moore, J.F.: Death of competition. John Wiley & Sons (1996)
12. Talvitie, J.: Business Ecosystem Creation – Supporting collaborative business concept development. Tivit Business Forum (2011)
13. Iansiti, M., Levien, R.: The Keystone Advantage: What the New Dynamics of Business Ecosystems Mean for Strategy, Innovation, and Sustainability. Harvard Business Press (2004)
14. Cusumano, M., Gawer, A.: The elements of platform leadership. Sloan Management Rev. 43(3), 51–58 (2002)
15. Baldwin, C., Woodard, J.: Platforms, markets and innovation. In: Gawer, A. (ed.) The Architecture of Platforms: A Unified View, pp. 19–44. Edward Elgar, Cheltenham (2009)
16. Tiwana, A., Konsynski, B., Bush, A.A.: Platform Evolution: Coevolution of Platform Architecture, Governance, and Environmental Dynamics. Information Systems Research 21(4), 675–687 (2010)
17. Corallo, A., Passiante, G., Prencipe, A.: Digital Business Ecosystems. Edward Elgar Publishing, Cheltenham (2007)
18. Arduino open-source electronics prototyping platform, `http://www.arduino.cc/`
19. Home Gateway Initiative publications,
   `http://www.homegatewayinitiative.org/documents/publications.asp`
20. Wireless Modem Chipsets, Renesas Mobile Corporation,
   `http://renesasmobile.com/products/lte-modem.html`
21. IPSO Alliance, `http://www.ipso-alliance.org/`
22. Zigbee Alliance, `http://www.zigbee.org/Home.aspx`
23. Kovatsch, M., Mayer, S., Ostermaier, B.: Moving Application Logic from the Firmware to the Cloud: Towards the Thin Server Architecture for the Internet of Things. In: Proceedings of the 6th International Conference on Innovative Mobile and Internet Services in Ubiquitous Computing (IMIS 2012), Palermo, Italy (2012)

24. Kovatsch, M., Duquennoy, S., Dunkels, A.: A Low-Power CoAP for Contiki. In: Proceedings of the 8th IEEE International Conference on Mobile Ad-hoc and Sensor Systems (MASS 2011), Valencia, Spain, pp. 855–860 (2011)
25. Pachube real-time open data web service for the IoT, https://pachube.com/
26. Harjula, J.: Nokia Siemens Networks promotes GSM for Machine to Machine applications, Press release,
   http://www.nokiasiemensnetworks.com/news-events/press-room/press-releases/nokia-siemens-networks-promotes-gsm-for-machine-to-machine-applications
27. Blockstrand, M., Holm, T., Kling, L.-Ö., Skog, R., Wallin, B.: Operator opportunities in the internet of things. Ericsson Review 1 (2011)
28. ETSI Technical Committee for Machine to Machine Communications,
   http://www.etsi.org/Website/Technologies/M2M.aspx
29. Lucero, S.: Maximizing Mobile Operator Opportunities in M2M: The Benefits of an M2M-Optimized Network. ABI Research (2010)
30. OECD: Machine-to-Machine Communications: Connecting Billions of Devices. OECD Digital Economy Papers, No. 192. OECD Publishing (2012),
   http://dx.doi.org/10.1787/5k9gsh2gp043-en
31. The EPCglobal Network: Overview of Design, Benefits & Security. EPCglobal Inc. (2004),
   http://gs1nz.org/documents/TheEPCglobalNetworkfromepcglobalinc_001.pdf
32. Shih, D.H., Sun, P.L., Lin, B.: Securing industry-wide EPCglobal Network with WS-Security. Industrial Management & Data Systems 105(7), 972–996 (2005)
33. Sundmaeker, H., Guillemin, P., Friess, P., Woelffle, S. (eds.): Vision and Challenges for Realizing the Internet of Things. European Commission (2010)
34. Mell, P., Grance, T.: The NIST Definition of Cloud Computing, NIST Special Publication 800-145. National Institute of Standards and Technology, Information Technology Laboratory (2011)
35. Marston, S., Li, Z., Bandyopadhyay, S., Zhang, J., Ghalsasi, A.: Cloud computing – The business perspective. Decision Support Systems 51, 176–189 (2011)
36. Tschofenig, H., Arkko, J.: Report from the Smart Object Workshop, Internet Architecture Board (IAB), Request for Comments: 6574 (2012),
   http://www.rfc-editor.org/info/rfc6574

Towards IOT Ecosystems and Business Models

Seppo Leminen[1,2,*], Mika Westerlund[2,3], Mervi Rajahonka[4], and Riikka Siuruainen[1]

[1] Laurea University of Applied Sciences, Espoo, Finland
{seppo.leminen,riikka.siuruainen}@laurea.fi
[2] Aalto University School of Economics, Department of Marketing, Helsinki, Finland
{seppo.leminen,mika.westerlund}@aalto.fi
[3] University of California Berkeley, Haas School of Business
{mika.westerlund}@berkeley.edu
[4] Aalto University School of Economics, Department of Information and Service Economy
{mervi.rajahonka}@aalto.fi

Abstract. The recent discourse on IOT (Internet of Things) has emphasized technology and different technology layers. Currently, there is a pressing need for research into the emerging IOT ecosystems from the business perspective. This is due to their promise to create new business models, improve business processes, and reduce costs and risks. However, the existing literature lacks understanding and empirical research on what IOT business models are and how they are connected to the underlying ecosystem. We focus on this critical research gap by studying business models in the IOT ecosystem context. In the study, we construct a framework for analyzing different types of IOT business models. Furthermore, we use case examples from the automotive industry in explaining how these models are employed in connection with ecosystems. The study concludes by giving suggestions to business model developers and researchers.

Keywords: Internet of things, business models, ecosystems.

1 Introduction

More devices are becoming embedded with sensors and gaining the ability to communicate. A plethora of terms describe this emerging shift in information processing, such as 'ubiquitous computing', 'pervasive computing', 'things that think', 'ambient intelligence', and 'silent commerce' [1]. Albeit each of the terms describes a significant research avenue, they are all characterized by the advent of everyday physical objects equipped with digital logic, sensors, and networking capabilities [1], which together form the Internet of Things (IOT). Haller et al. [2] define IOT as "a world where physical objects are seamlessly integrated into the information network, and where the physical objects can become active participants in business processes."

The development of IOT is enabled by combining the Internet, near-field communications, hardware, and embedded sensors with real-time localization. The decline of

---

* Corresponding author.

S. Andreev et al. (Eds.): NEW2AN/ruSMART 2012, LNCS 7469, pp. 15–26, 2012.

sensors´ size, price, and energy consumption as well as increased performance boost the market from a limited number of communication devices to thousands or even millions of devices [1],[3-7]. Coupling tiny networked sensors together allows physical world things and places to generate data. Consequently, ranging and managing the vast amount of data lead to novel business opportunities. However, profiting from the IOT ecosystem's business potential depends on the progression of vertical domains such as consumer electronics, factory automation, smart home, smart grid, telecommunication, and logistics. Their expansion will create economy of scales and thus critical mass on the markets.

The central elements of IOT include the concepts of 'the ecosystem', 'the ecosystem core', and 'the business model'. Originally presented by James F. Moore in 1993 [8], the concept of business ecosystem stems from the insight that innovative businesses cannot evolve in a vacuum; rather, they rely on different resources. They need capital, partners, suppliers, and customers with which they create cooperative networks. Platforms, technologies, processes, and standards form the ecosystem core, and the members of the ecosystem create business models. These members include companies, public institutions, and individuals. The benefits for companies and public institutions to join a business ecosystem include (i) market creation, (ii) market expansion, (iii) market access, and (iv) the access to complementary competences and business models [9]. IOT allows consumer's choice of providers and leads to the adoption of ecosystems. Nevertheless, prior research remains silent on the types of IOT business models and their links to the underlying ecosystems.

In this paper, we aim to build a framework for analyzing different IOT business models. The objective is to bring forth the existing and potential models by discussing them in connection with the underlying ecosystems in designated domains. Business models and their elements are widely conferred in the academic literature [10-14] and previous research suggests that they are industry and context-dependent. Therefore, they should be concerned in the business domain at hand. This study contributes to the recent discussion on business models with a robust framework and a view from the IOT ecosystem perspective, which advises corporate business models based on connected devices. Specifically, we intend to:

- Discuss the links between the business models and IOT ecosystems.
- Categorize the IOT business models of contemporary companies.
- Describe the challenges of IOT business model development.

The study is divided as follows. After this brief introduction, we review the theoretical foundations of business models. Then, we construct and describe our theoretical framework for categorizing different IOT business models. Next, we utilize case examples from the automotive industry to discuss these models in connection with their underlying ecosystems. Finally, we review our key findings and conclude on the challenges of IOT business model development.

2   Business Models

2.1   Background

Research on business models has boomed since the 1990s among the scientific community. As a result of intense study, many characterizations, perspectives, and identifiable models have been suggested in the academic literature. However, no widely accepted definition for the term 'business model' has emerged [11-15]. The business model of a firm basically defines how the organization operates in the market and what the basis of its value creation is [16]. Osterwalder et al. [13] depict the evolution of the concept and suggest that business model studies are in the path towards applied models in practices. The conceptual progression of the business model can be divided into a number of distinct phases.

The early uses of business models define and classify them in e-business [17]. This approach was *de facto* prior to the Internet bubble burst in the beginning of the new millennium. The focus then moved on to identifying various components [10], which were seen as business model's building blocks. Modeling and testing business models provided diverse reference models and ontologies in the following phases [18]. Various management applications and conceptual tools were primary outcomes when applying business models in practice [13], [19]. For example, Tikkanen et al. [20] suggest a framework to design business models by focusing on the beliefs of industrial logic, social constructions, competitive analysis of a company and competitors, as well as cognitive representation of product and service or their usage.

Zott and Amit [21] continue the design view by suggesting a fit between the firm's market strategy and its business model. They [21] identify novelty-centered and efficiency-centered business strategies. This perspective also suggests a link between business model and ecosystems. Recently, Zott and Amit [22] defined business model as a system of activities dependent of each other through the focal company and the surrounding network. Furthermore, Westerlund et al. [16] studied business model management and proposed that by linking business models with a firm's relationships, researchers can identify various business models and their differences in the ecosystem context. Business model management is important when targeting to robust and profitable models in the emerging markets and domains such as the IOT [16].

2.2   IOT Business Models

Smart objects in IOT facilitate novel applications and business models [23]. However, designing feasible business models requires sufficient data, because the variety of data collected automatically from devices' information exchange helps to solve problems and enables the development of embedded services and revenue models. Despite the importance of developing IOT business models has been widely identified [3-4],[6-7], the mainstream of the IOT research has focused on technology and different technology layers. There only have been sparse scholarly attempts to increase understanding of the emerging IOT business models and ecosystems by applying:

- *Structural approaches*, including identifying value chains in ubiquitous computing environments [24], IOT value drivers' identification [4], IOT value chain analysis [25], and digital business ecosystems discussion [26].
- *Methodology approaches*, including business model development methodology in ubiquitous computing environments [25] and multipath deployment scenarios [28].
- *Design approaches*, including networked business model itemization for emerging technology-based services in ubiquitous computing [29] and the IOT business model canvas framework [30].

IOT enables the pooling of resources in networks that include multiple nodes and links between the nodes. It means that there are almost endless ways to utilize information and IOT structures. Bucherer and Uckelmann [30] emphasize that a networked infrastructure of IOT enables both incremental and radical business changes, but so far the full potential has not been leveraged. They [30] stress that information exchange between the nodes in IOT network and the involvement of all stakeholders in the "win-win" information exchange are key issues in designing IOT business models. Therefore, to unleash the business potential of IOT, the cost-centric approach has to be replaced by a value-focused perspective. For this purpose, Bucherer and Uckelmann [30] suggest a triangle of information exchange that consists of a thing, a business, and consumer between and within these.

Although IOT business model design requires much information, there are lacks of integration of information exchange [31]. Kourouthanassis et al. [32] propose that the needed data types are context specific, and a supporting infrastructure to manage the complexity between devices, information, and users is needed. Platforms underlie the leading global businesses where information is changed. In principle, there are one-sided platforms and two-sided platforms, in which competition takes place. That is, competition focuses on how to substitute and charge parties in different sides of platforms [33-35]. Modularity may help in winning competition, as it is possible to reduce complexity and to increase flexibility of a system with modules [36-37].

Fleisch et al. [1] stress that IOT bundles diverse technologies and systems together by combining technologies and functionalities. A complex product or service from smaller subsystems can he designed and built independently. Furthermore, there are two major paradigms behind business value potential of IOT: real-world visibility and business process decomposition [2]. The latter leads to a shift of power towards the edges of the network, as business processes can be decomposed into process steps that can be executed at the network edge. Therefore, IOT enhances the modularization of business processes, which in turn increases scalability and performance of the system. Business process decomposition also allows local decision-making in a decentralized manner [2]. Recent research applies modularity principles in the service context [38], because in the extreme case of IOT all devices would offer their functionality as a web service, and device integration would mean service integration.

Mejtoft [39] depicts three layers of value creation in IOT: manufacturing, supporting, and value co-creation. The first layer means that manufacturers or retailers benefit from the simple possibility of tracking single items. The second layer creates value as the collected data can be used in both industry and customer driven value creation processes. The third layer uses IOT as a co-creative partner, because the network of things can think for itself. We draw on these views on IOT and utilize the business model design approach discussed previously, because together they give us pieces to establish a framework for identifying and analyzing IOT business models.

3 Framework for the IOT Business Model Analysis

We adopt Osterwalder's [40] business model canvas thinking in order to build a conceptual framework for analyzing different types of IOT business models. We consider business model canvas thinking useful for the endeavor, because it suggests value proposition, customer, infrastructure, and financial perspectives as the basic elements of business models. These elements are among the regularly advocated key components of business models in the academic literature on business models [16].

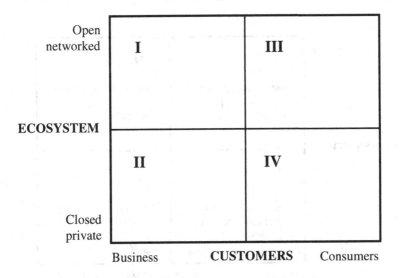

Fig. 1. Framework for analyzing diverse IOT business models

We use 'ecosystem' and 'customers' as the foundational dimensions in our framework, which helps to visualize a variety of current and potential IOT business models. Our framework describes diverse aspects of IOT on a 2*2 matrix, in which one dimension considers the type of the ecosystem (closed private or open networked) and the other employs the type of customers (business or consumers). Using these bipolar dimensions as the principal axes, we distinguish four different types of IOT business models. We expect that by further analyzing these distinctions it is possible to capture most of the differences between the models. The framework identifies diverse IOT business models using consecutive numbers from I to IV (see Figure 1).

Open, scalable, secure, and standardized infrastructures are needed for IOT, although they do not necessarily exist today [41]. Nevertheless, there is a clear trend from closed private ecosystems (the lower-end models II and IV) towards open networked ecosystems and business models (the upper-end models I and III). At the same time, more and more business-to-consumer (B2C) solutions (the right-end models III and IV) are expected to emerge besides business-to-business (B2B) solutions (the left-end models I and II) that dominate the IOT solution markets today.

3.1   Case Examples from the Automotive Industry

There are numerous examples of using IOT –based solutions in automotive manufacturing and logistics. In addition, there is a constant and explicit shift in the industry towards standardized systems that allow extensively networked ecosystems [41]. We put forward RFID, car sharing, intelligent manufacturing, and traffic safety services as virtuous illustrations of IOT business models in our framework (see Figure 2).

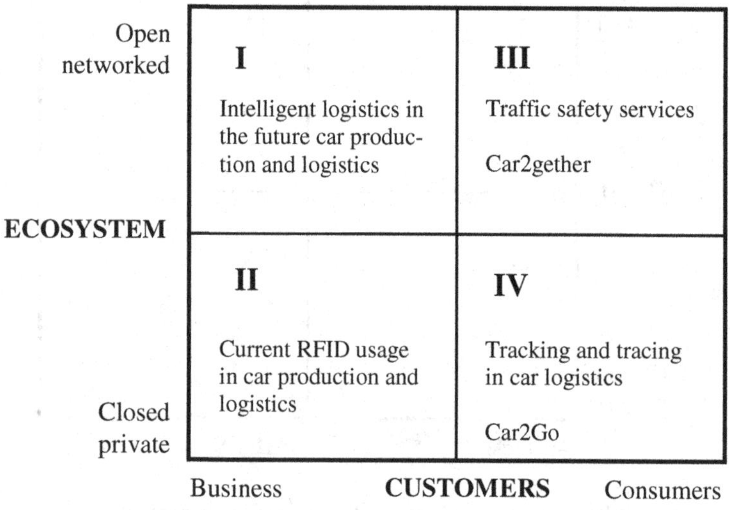

Open networked	**I** Intelligent logistics in the future car production and logistics	**III** Traffic safety services Car2gether
ECOSYSTEM		
Closed private	**II** Current RFID usage in car production and logistics	**IV** Tracking and tracing in car logistics Car2Go
	Business **CUSTOMERS** Consumers	

Fig. 2. IOT business model examples in the automotive industry

Intelligent logistics provides an example of model I in our framework. Approximately 80 percent of all car parts are today produced by external suppliers. For instance, Geis Group has been providing logistics services to the automotive industry and its suppliers for decades. As an area contract freight forwarder, this German company ensures that car parts are transported securely and on time to the respective manufacturer's production facility. IOT technology will be increasingly used in the future to help the company in processing its customer orders as economically and as environmentally-friendly as possible, whilst also taking into account the time and quality parameters as required by the just-in-time thinking in procurement [42].

Business models that lean on closed private infrastructures (models II and IV), such as wireless RFID installations or current machine-to-machine (M2M) solutions that are used in automotive production or logistics, can be described as Intranet or Extranet of Things solutions and may not strictly be based on the Internet of Things. They should, however, be extended to support open Internet architectures [41]. Therefore, they have the potential to become integral parts of open networked models, as the following example of Daimler AG shows.

Daimler AG launched 'Car2Go', a new mobility concept, in Germany in 2008. It is an innovative car share program that places an emphasis on making things easier for customers. This PaaS (product as a service) concept, which provides a simple, flexible, and value-for-money approach to mobility with environmentally-friendly vehicles, has spread into 13 cities in Europe and USA by 2012 [48]. It is offered to consumers through a closed ecosystem that relies mostly on the company-provided services (model IV in our framework). Small two-seater city cars can be hired at any time of day or night after a membership registration. Their use is charged on a minute-to-minute basis, including tax, insurance, mileage, and fuel. On registering for Car2Go, the customer's driving license is provided with an electronic chip that enables the car to be unlocked. The customer can unlock a car by holding his/her membership card against a reader in the windscreen. After the ride, the customer can return the car by parking it in any public parking space in the city [43].

Since 2010, Daimler AG has also tested a ride-share community 'Car2gether' in two German regions, namely Ulm and Aachen. After renting a Car2Go vehicle, the customer can offer a free seat via Car2gether for other members (model III in the framework). Available vehicles or seats can be located via Internet in both Car2Go and Car2together concepts. In the Aachen region, the Car2gether system is linked with the local public transport system making the service ecosystem more networked, so that also suitable public transport connections are suggested as transportation alternatives for the customer. [49]

The "Volvo on Call" system consists of an integrated mobile phone and a built-in GPS satellite unit. It automatically calls emergency services, if an airbag is deployed in an accident. By pressing SOS button the driver can contact emergency services and "Volvo on Call" button puts the driver through to a Volvo operator if the car breaks down. Emergency aid can be used automatically or manually. Automatic emergency aid starts when the car's airbag inflates in an accident. Emergency aid will then send a message and opens the audio channel to the service center. If the driver is injured and unable to speak, the central contacts emergency service and informs the car's location as vehicle tracking is available via satellite. The system also informs the operator when the car is stolen or the alarm is otherwise activated. Central contacts the owner and makes sure it is not a false alarm and on owner's confirmation of a crime, the system begins tracking the stolen car and reports its location to the police [44].

There are currently several ongoing projects aiming to enhance traffic data services. Vehicle-to-vehicle (V2V) and vehicle-to-infrastructure (V2I) communication create new opportunities to increase road safety [45]. Such emergency service systems require open networked ecosystems with multiple actors such as car manufacturers and public sector actors including police, hospitals, and other authorities and public bodies. We categorize these traffic safety services as model III in our framework.

4   Discussion

4.1   Findings

This study focused on understanding what IOT business models are, which types of business models there are, and how they are connected to the underlying ecosystem. The success of IOT depends on the right technology, business models, and acceptability to users. This justifies techno-economic and human centric studies of adoption, value networks, and ecosystems creation. IOT is about a large number of ever smaller and more specialized "things", i.e. devices and sensors connected (often wirelessly) to each other and to Internet. These "things" expand existing Internet applications and services and enable new ones. The new functionality creates and requires new roles and technical components. It also enables the configuration of new business models in the ecosystem. IOT increases the complexity of communications and encourages designers to prepare for increasingly adaptive technical solutions.

In the study, we established a framework to identify and analyze different IOT business models. We first reckoned that the previous studies on IOT business ecosystems or business models propose three divergent approaches: structural, methodology, and design. We concluded that the design approach is helpful for our purpose, as it provides specific items to use in our business model framework. Consequently, our framework was based on two dimensions: ecosystem (closed private vs. open networked) and customers (business vs. consumer). As a result, we identified four distinct IOT business models and explained those models using examples from the automotive industry. We found that there is an ongoing trend in the industry from closed to open ecosystems, which follows the worldwide phenomenon of open innovation. Many closed private models are increasingly connected to new open ecosystems and become a part of open networked business models. Furthermore, we found that there are more and more B2C solutions among IOT technology in the automotive industry.

There are some points to discuss about the future of IOT. We anticipate that large companies drive the development of IOT due to their plentiful resources and key roles in technology provision. However, user-centered approaches, which mean opening up innovation and ecosystems, should be utilized when creating IOT business models. It means that users share their own expertise and knowhow and become producing actors [4-5], [46]. The problem is the lack of viable business models and the ways needed to create them in relation to IOT ecosystems. Moreover, we share the view of Haller et al. [2], who maintain that the lack of trustworthy service intermediaries as well as challenges in user-service interaction and measuring quality of information still prevent the seamless and agile interoperation of smart items in IOT. They [2] state that there is a need for open global standards and new business models have to be developed [2]. We look forward to innovative IOT business models that emerge in different industries.

4.2   Needs for Future Research

There are ample research gaps related to IOT business models, because (i) there is little research on IOT business models as such, (ii) generic business model research frameworks have challenges as IOT applications are often context specific, (iii) current literature describes business models within a single company and not across the networks of companies, (iv) IOT is modular since it bundles not only technologies and systems together, but also diverse industrial applications and business elements, and (v) there are almost endless possibilities to connect a thing, a business, and a consumer together, which makes it virtually impossible to invent a particular "killer" business model.

The IOT evolves stepwise, so relevant ecosystems and companies need to adopt "trial and error" methods in creating novel business models. This is a challenge for future research, which struggles with the contextual problems. Networked business models are context-specific and should be studied with an emphasis on both generic business model configurations [47] and the specific modules of business models [29].

To sum up, there is a need for research that reveals the embedded structures and creates comprehensive understanding of the networked IOT business models, as well as depicts the roles of diverse IOT actors and the dynamics of mega-ecosystems, in which different industries and clusters are integrated into a large ecosystem. The framework presented in this paper helps to analyze networked IOT business models based on their design factors, both in infrastructure and customer relations.

5   Conclusion

The preliminary results of our ecosystems and business models study will support the bootstrapping of new IOT services in service and manufacturing industries. The emerging business models are assumed to be the main vehicles of an IOT ecosystem creation. Future research and development should stress studies and projects in which the core business partnerships in the emerging IOT ecosystem are simultaneously developed from the perspectives of business, technology, and users, thus increasing trust that is necessary for creating a working collaborative ecosystem.

References

1. Fleisch, E., Sarma, S., Thiesse, F.: Preface to the focus theme sections: 'Internet of things'. Elektron Markets 19, 99–102 (2009)
2. Haller, S., Karnouskos, S., Schroth, C.: The Internet of Things in an Enterprise Context. In: Domingue, J., Fensel, D., Traverso, P. (eds.) FIS 2008. LNCS, vol. 5468, pp. 14–28. Springer, Heidelberg (2009)

3. ITU (2005). ITU Internet Reports 2005: The Internet of Things, 23 p. (2005)
4. Fleisch, E.: What is the Internet of Things? An Economic Perspective. Economics, Management, and Financial Markets 5(2), 125–157 (2010)
5. Kortuem, G., Kawsar, F., Fitton, D., Sundramoorthy, V.: Smart objects as building blocks for the Internet of things. IEEE Internet Computing, 44–51 (2010) ISSN: 1089-7801
6. Internet of Things (2011) Internet of Things (IoT) and Machine to Machine Communications (M2M): Challenges and opportunities Interim positioning paper, 29 p. (September 2011)
7. OECD 2012, Machine-to-Machine Communications: Connecting Billions of Devices. OECD Digital Economy Papers, No. 192. OECD Publishing (2012), http://dx.doi.org/10.1787/5k9gsh2gp043-en
8. Moore, J.F.: Predators and Prey: A New Ecology of Competition. Harvard Business Review 71(3), 75–83 (1993)
9. Talvitie, J.: Business ecosystem creation, supporting collaborative business concept development. Tivit Business Forum (April 12, 2011)
10. Amit, R., Zott, C.: Value creation in E-business. Strategic Management Journal 22(6-7), 493–520 (2001)
11. Magretta, J.: Why Business Models Matter. Harvard Business Review 80(5), 86–92 (2002)
12. Leminen, S., Anttila, M., Tinnilä, M., Miikkulainen, K.: Strategic Pricing in Business Relationships - Do Not Miss the Opportunity to Create Value for the Customers. In: The Proceedings of the Annual ANZMAC Conference, Brisbane (December 2006), http://smib.vuw.ac.nz:8081/WWW/ANZMAC2006/program.html
13. Osterwalder, A., Pigneur, Y., Tucci, C.L.: Clarifying business models: origins, present and future of the concept. Communications for AIS 16(1) (2005)
14. Amit, R., Zott, C.: Business Model Design: An Activity System Perspective. Long Range Planning 43, 216–226 (2010)
15. Casadesus-Masanell, R., Ricart, J.E.: From strategy to business model and onto tactics. Long Range Planning 43, 195–215 (2010)
16. Westerlund, M., Rajala, R., Leminen, S.: Insights into the dynamics of business models in the media industry, 42 p. Laurea Publications A74, Vantaa (2011) ISBN 978-951-799-228-2
17. Timmers, P.: Business Models for Electronic Markets. Electronic Markets 8(2), 3–8 (1998)
18. Osterwalder, A., Pigneur, Y.: An e-Business Model Ontology for Modeling e-Business. In: 15th Bled Electronic Commerce Conference e-Reality: Constructing the e-Economy Bled, Slovenia, June 17-19, 11 p. (2002)
19. Sosna, M., Trevinyo-Rodríguez, R.N., Velamuri, S.R.: Business Model Innovation through Trial-and –Error Learning: The Naturhouse Case. Long Range Planning 43(2/3), 383–407 (2010)
20. Tikkanen, H., Lamberg, J.-A., Parvinen, P., Kallunki, J.: Managerial cognition, action, and the business model of the firm. Management Decision 43(6), 789–809 (2005)
21. Zott, C., Amit, R.: The fit between product market strategy and business model: implications for firm performance. Strategic Management Journal 29, 1–26 (2008)
22. Zott, C., Amit, R.: Designing your future business model: An activity system perspective. Long Range Planning 43, 216–226 (2010)
23. Bohn, J., Coroama, V., Langheinrich, M., Mattern, F., Rohs, M.: Social, Economic, and Ethical Implications of Ambient Intelligence and Ubiquitous Computing. In: Weber, W., Rabaey, J., Aarts, E. (eds.) Ambient Intelligence, pp. 5–29. Springer, Heidelberg (2005)
24. Lee, H.J., Leem, C.S.: A Study on Value Chain in a Ubiquitous Computing Environment. In: Gervasi, O., Gavrilova, M.L., Kumar, V., Laganá, A., Lee, H.P., Mun, Y., Taniar, D., Tan, C.J.K. (eds.) ICCSA 2005, Part IV. LNCS, vol. 3483, pp. 113–121. Springer, Heidelberg (2005)

25. Banniza, T.R., Biraghi, A.M., Correia, L.M., Goncalves, J., Kind, M., Monath, T., Salo, J., Sebastiao, D., Wuenstal, K.: Project-wide Evaluation of Business Use Cases. Project report. FP7-ICT-2007-1-216041-4WARD/D1.2 (2010)
26. Nashira, F., Nicolai, A., Dini, P., Louarn, M.L., Leon, L.R.: Digital Business Ecosystem, 214 p. (2010), http://www.digital-ecosystems.org/book/de-book2007.html
27. Leem, C.S., Jeon, N.J., Choi, J.H., Shin, H.G.: A Business Model (BM) Development Methodology in Ubiquitous Computing Environments. In: Gervasi, O., Gavrilova, M.L., Kumar, V., Laganá, A., Lee, H.P., Mun, Y., Taniar, D., Tan, C.J.K. (eds.) ICCSA 2005, Part IV. LNCS, vol. 3483, pp. 86–95. Springer, Heidelberg (2005)
28. Levä, T., Warma, H., Ford, A., Kostopoulos, A., Heinrich, B., Widera, R., Eardley, P.: Business Aspects of Multipath TCP Adoption. In: Tselentis, G., et al. (eds.) Towards the Future Internet: Emerging Trends from European Research, pp. 21–30. IOS Press (April 2010) ISBN 978-1-60750-538-9 (print) and ISBN 978-60750-539-6 (online)
29. Palo, T., Tähtinen, J.: A networked perspective on business models for emerging technology-based services. Journal of Business and Industrial Marketing 26(5), 377–388 (2011)
30. Bucherer, E., Uckelmann, D.: Business Models for the Internet of Things. In: Uckelmann, D., Michahelles, F., Harisson, M. (eds.) Architecting the Internet of Things. Springer, Berlin (2011) ISBN: 978-3-642-19156-5
31. Fleisch, E.: Business Impact of Pervasive Technologies: Opportunities and Risks. Human and Ecological Risk Assesment 10(5), 817–829 (2004)
32. Kourouthanassis, P., Giaglis, G., Karaiskos, D.: Journal of Information Technology 25, 273–287 (2010)
33. Rochet, C., Tirole, J.: Platform competition in two-sided markets. Journal of the European Economic Association (2003)
34. Parker, G., Van Alstyne, M.: Two-sided networks: A theory of information product design. Management Science (2005)
35. Eisenmann, T., Parker, G., Van Alstyne, M.: Strategies for two-sided markets. Harvard Business Review (2006)
36. Baldwin, C.Y., Clark, K.B.: Design Rules: The Power of Modularity, vol. 1. MIT Press, Cambridge (2000)
37. Schilling, M.A., Steensma, H.K.: The use of modular organizational forms: An industry-level analysis. Academy of Management Journal 44(6), 1149–1168 (2001)
38. Voss, C.A., Hsuan, J.: Service architecture and modularity. Decision Sciences 40(3), 541–569 (2009)
39. Mejtoft, T.: Internet of Things and co-creation of value. In: Xia, F., Chen, Z., Pan, G., Yang, L.T., Ma, J. (eds.) 2011 IEEE International Conferences on Internet of Things, and Cyber, Physical and Social Computing, pp. 672–677. Institute of Electrical and Electronics Engineers (2011)
40. Osterwalder, A., Pigneur, Y.: Business Model Generation. A Handbook for Visionaries, Game Changers, and Challengers (2009)
41. Uckelmann, D., Michahelles, F., Harisson, M.: An Architectural Approach Towards the Future Internet of Things. In: Uckelmann, D., Michahelles, F., Harisson, M. (eds.) Architecting the Internet of Things. Springer, Berlin (2011) ISBN: 978-3-642-19156-5
42. Geis Group: Intelligent procurement logistics for the automotive industry. Transfer No. 27, p. 9 (October 2010)
43. Smart "car2go edition", Training Video in English (accessed April 13, 2012), http://www.youtube.com/watch?v=gsx8kEv0DWc&list=UUvqZkPdzMt3tW-XbzBa8UaA&index=8&feature=plcp

44. Volvo Car UK Newsroom: C70 Technology overview (accessed April 13, 2012), https://www.media.volvocars.com/uk/enhanced/en-uk/Models/2011/C70/Model/Home.aspx

45. Martínez, F., Toh, C.K., Cano, J.C., Calafate, C., Manzoni, P.: Emergency Services in Future Intelligent Transportation Systems based on Vehicular Communication Networks. IEEE Intelligent Transportation Systems Magazine (2010)

46. Trappeniers, L., Roelands, M., Godon, M., Criel, J., Dobbelaere, P.: 13th International Conference on Intelligence in Next Generation Networks, ICIN 2009 (2009)

47. Nenonen, S., Storbacka, K.: Business model design: Conceptualizing networked value co-creation. Industrial Journal of Quality and Service Sciences 2(1), 43–59 (2011)

48. Daimler AG: Car2Go, http://www.car2go.com/

49. Daimler AG: Mobility concepts, http://www.daimler.com/

Open and Scalable IoT Platform and Its Applications for Real Time Access Line Monitoring and Alarm Correlation

Andrej Kos[1], Damijan Pristov[2], Urban Sedlar[1], Janez Sterle[1], Mojca Volk[1], Tomaž Vidonja[3], Marko Bajec[4], Drago Bokal[5], and Janez Bešter[1]

[1] University of Ljubljana, Faculty of Electrical Engineering,
Tržaška cesta 25, 1000 Ljubljana, Slovenia
andrej.kos@fe.uni-lj.si
http://www.ltfe.org
[2] IskraTEL d. o. o., Ljubljanska c. 24a, 4000 Kranj, Slovenia
pristov@iskratel.si
http://www.iskratel.si
[3] Technology Network ICT,
Tehnološki park 24 – Building E, 1000 Ljubljana, Slovenia
tomaz.vidonja@ict-slovenia.net
http://www.ict-slovenia.net
[4] University of Ljubljana, Faculty of Computer and Information Science,
Tržaška cesta 25, 1000 Ljubljana, Slovenia
marko.bajec@fri.uni-lj.si
http://www.fri.uni-lj.si
[5] Cosylab d. d., Teslova ulica 30, 1000 Ljubljana, Slovenia
drago.bokal@cosylab.com
http://www.cosylab.com

Abstract. We present an open intelligent communication platform that can be used to support various usage scenarios related to future internet of things. Its purpose is twofold: to support fusion of large amounts of data, irrespective of their source or structure and to provide users or devices with semantically analysed and enriched data according to their needs and context. Using the platform, users are able to access enriched data, receive warnings and notifications about events recognised by the system. We present applications for the real time access line monitoring and alarm correlation that respond to telecommunications operators' needs in the field of proactive line measurement in enabling the helpdesk and field technical teams to pinpoint the cause of service degradations.

Keywords: IoT platform, access line monitoring, event correlation, real-time.

1 Introduction

Internet of Things (IoT) offers a promising vision of the future where the virtual world integrates seamlessly with the real world of things. The IoT enables many applications with high social and business impact, and once the enabling

S. Andreev et al. (Eds.): NEW2AN/ruSMART 2012, LNCS 7469, pp. 27–38, 2012.

technologies are stable, it is expected new ones will emerge as well. Most important challenges in current research and development in the IoT domain are the definition of architectures, protocols and development of algorithms for efficient interconnection of smart objects, both between the objects themselves as well as with the future internet. Businesswise, the most important aspect is creation of value-added services, especially open and interoperable, that are enabled by interconnection of things, machines, and smart objects and can be integrated with current and future business development process [1].

Support of the future commercial success of IoT solutions requires scalable, expandable, and flexible infrastructure for real-time and non-real time aggregation, storage, processing, management, and sharing of data of various types, obtained from and delivered to the broad spectrum of different devices and applications. With ever increasing volume and value of information, new platforms, algorithms, methods and solutions for fast and efficient processing, correlating, reusing and understanding of gathered and aggregated data are becoming more and more important and influence both, the infrastructure and application design. Further research and development in this area, with data as input and cloud based visualized information as output [2, 3], will enable new innovative business models and proactive services in the fields of communications, smart-grid, energy, health, asset management, transport, personal data tracking, social relationships, and many more [4–10].

Based on the experience of using event driven architectures in the fields of merging different data sources on semantics [5], intelligent networked vehicles [7, 10], contextualization of personal health data, monitoring and visualization of IPTV system performance [4], asset management and network planning [9] and context-aware communications, we developed an advanced IoT multi-application platform that is able to support and fulfill the majority of the above mentioned features. The platform is cloud computing enabled, allows a plethora of different visualisations, and resembles some of the functionalities of other available platforms [11–15], while its main advantage is inherent real-time support and the ability to process large amounts and data.

Furthermore, we present a solution using this IoT platform in the domain specific field of telecommunications network monitoring and event correlation. The results partly came from the research and development project "Open Communication Platform for Integrated Services", with the company Iskratel as a partner and its customers as end users.

Today telecommunications operators offering xDSL access line services are confronted with the problem of SLA (Service Level Agreement) fulfillment due to FEXT (Far-End Crosstalk), which is the interference between two pairs of a cable measured at the other end of the cable from the transmitter. Increasing the number of subscribers on the same cable and increasing the speeds and mixture of different broadband technologies (e.g., ADSL and VDSL) causes degradation of service. Operators are failing to provide services that they have sold to users in the past. As the need for bandwidth is becoming larger, operators are forced to move DSLAMs (Digital Subscriber Line Access Multiplexer) closer to the users.

Therefore, the operators have to deploy efficient monitoring and correlation tools to keep the current network operation under control and proactively detect changes and trends in their network.

Telecommunications operators also reported a strong need for a tool for real-time proactive network measurement. Data collected and correlated enables commercial sectors to better plan, define and offer new packages, technical departments to better plan and upgrade networks, and helpdesk teams to pinpoint the cause of eventual service degradations, thus being able to respond faster and more accurately.

Such access line monitoring and alarm correlation tool is a combination of performance and alarm monitoring with addition of an event correlation and pattern matching expert tool.

Iskratel today uses two separate products for alarm monitoring, performance and quality analysis. These products currently lack the required expert tool. Based on the analysis and comparison of mayor commercial platforms and tools on the market, Iskratel with partners decided to help develop such an expert tool, based on the IoT platform.

Evaluation of major commercial platforms that offer the required functionality from big companies showed that these solutions proved not to be scalable enough and suitable (in terms of licensing and required servers) for Iskratel's specific needs to cover networks of different sizes (from very small to very big networks) with one common solution. Additionally, the Iskratel company policy is to use open source solutions if they are available and suitable.

2 IoT Platform

Our IoT platform is an open communication platform for data integration and development of data-driven and event-driven services to be used in various communication and service environments. It is open in two ways; by using open source software building blocks and by using open internal and external high level and low level interfaces.

The platform is designed universally and can be scaled in order to accommodate different scenarios (i.e. different fields of application), further enhancements, and modifications. The majority of platform components are generic and can be used in various use cases (where simple user scenarions apply), while domain specific components are kept to minimum (required for complex user scenarions).

One of the key functionalities of the platform is the possibility of merging data from different sources, use cases, and scenarios (Fig. 1). This prospect paves the way towards enhancing data with data from other domains. Consequently new use cases and new data integrations are possible. The platform architecture enables simple and straightforward inclusion and integration of new domain scenarios. Based on that and through open interfaces, development of new and innovative services is easy and straightforward.

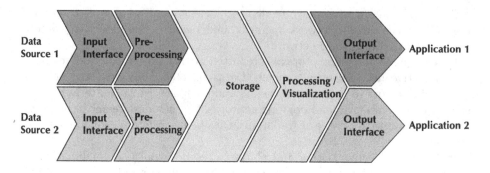

Fig. 1. Data fusion

2.1 Open Interfaces

The platform's internal open interfaces support abstract access to large volumes
of distributed data for information storage, analysis, correlation, and visualization. Interfaces provide high-level access to data without any knowledge on
low-level specifics on data stores and disk arrays where data is located. Due to
their internal nature, these interfaces provide access to original data over high-
performance interfaces, which allows for use of demanding algorithms with high
data rates and processing requirements that are able to flexibly employ virtualized hardware resources for parallel processing. Additionally, the use of internal
interfaces facilitates a formal separation of trusted procedures and applications
for data analysis from general outside users, the actions of which are prone to
detailed regulation through the access control system due to their increased trust
distribution.

In addition to internal open interfaces for high-performance data access, available to trusted data analysis mechanisms, the system incorporates a range of external open interfaces for data brokering and acquisition, and interfaces for data
source and destination control. Due to confidentiality of some types of data,
the external interfaces are required to perform authentication and authorization as well as flexible policy-based access control prior to granting any request
related to data insertion or acquisition or control operations. External interfaces are generic in design; this supports extensibility of distribution of various
data types and assures data marking by means of metadata, providing detailed
descriptions and context definitions.

2.2 Architecture

High-level platform architecture (Fig. 2) comprises event-driven components (i.e.
real time analysis, business activity monitoring, and alarming), data-driven components (i.e. persistence, interoperability, expert analysis, and business intelligence), sources and external systems.

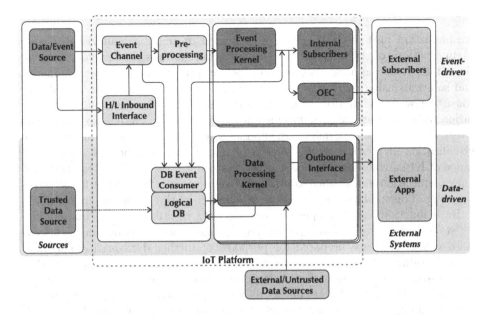

Fig. 2. IoT platform architecture

The inputs to the platform are data and event sources; sensors and devices forwarding measured data into the platform, i.e. applications based on service and event driven architectures and legacy systems, and trusted external data sources, i.e. databases.

Events are placed in the event channel. From there they are dispatched to the event processing kernel and the database. Event processing kernel provides for event filtering, identification of notifiable events, pattern matching, time series analysis, and event correlation. The outputs are the so called generated or output events, representing important information for various recipients, i.e. alarms and indicators. These events are placed into outbound queue, to make sure all concerned recipients are informed about particular events. The events are placed into two queues for the use of internal and external subscribers.

Processed events can be used internally or externally. Inside the platform, monitoring is possible with generic monitoring dashboard providing key performance indicators (KPI). A basic KPI is an individual event, that a user wants to monitor, while complex KPI consists of more (correlated) events. Users can select KPIs and define type and mode of visualization (graph type, refresh rate, labels, etc.). Special type of KPI, which can be forwarded to an email, short message service (SMS), or start an application, is an alarm.

Processing kernel provides for entity resolution, duplicate elimination, data enrichment, expert analysis, complex pattern matching, prediction, and forecasting.

External subscribers and web applications are notified about events via external output queue. These are, for example, external applications, SOA processes, mobile devices, etc.

2.3 Analytics and Visualization

An important part of the platform is dedicated to analytics and visualization. Main functions of the analytics and visualization modules are analysis, correlations, interpretation, presentation and export of processed data to environments and solutions using platforms services. Design of the basic analytic modules is consistent with the platform's universal design. Data processing and representation techniques are domain specific and adapted for concrete usage scenarios. Every identified use case defines requirements and specialized data analysis and visualization methods including variety of tools and techniques for the defined range of KPIs, consisting of raw and processed data. Additionally, methods for data preparation and export to other environments can be defined for selected scenarios. This enables upgrading the intelligence of the provided services.

Basic functional modules perform original data visualization, execution of basic statistical operations and data groupings into multi-dimensional matrices. As a result, data are prepared for purposes of further data mining and machine learning procedures, such as decision trees, clustering, neural networks, etc. This group of functionalities incorporates routines for semantic data analysis, and routines for end user context discovery and analysis of effects related to data handling and environmental context. An important data handling aspect is data anonymization in near-real-time, which is vital for personal information protection.

The modules for advanced data analysis and long-term pattern discovery utilize a selection of statistical, data mining and predictive techniques and algorithms. Aside basic statistics, text, data and flow mining techniques and graph discovery algorithms are implemented along with a subset of graph-based visual analytics tools. The selected methods represent a generic set of analytical functionalities useful for any kind of application domain where data correlation and graph analyses are of interest.

2.4 Usage Scenarios

The scope of the platform is to support user scenarios, which employ functionalities of the platform independently and at the same time allow for interdependent use and merging of data from different scenarios. This allows for greater added value and data enhancement and enrichment surpassing the borders of traditionally closed information systems.

The IoT platform is applicable to all usage scenarios with big amounts of events (data), especially if real-time or near-real-time response is required. Therefore, the platform can be used for many event-driven and data-driven scenarios (besides the scenario explained in this article), such as e-health, e-tolling, smart grids, smart traffic, smart city, smart logistics, real-time fraud detection, denial of service detection, and many more.

Adding new scenarios is straight-forward. When a new scenario is defined, an extension of the platform with domain specific layer is added as depicted in Fig. 3 for the scenario described in this paper.

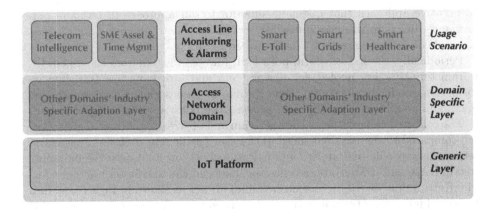

Fig. 3. IoT platform usage scenarios

3   Access Line Monitoring and Alarm Correlation

We designed and implemented a solution for real-time domain monitoring of access line and access network. It consists of DSLAMs with modems as network elements (NE), Iskratel management system (SI3000 MNS OpenMN) and the IoT platform (Fig. 4).

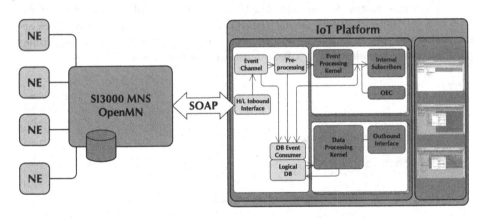

Fig. 4. Access line data collection and processing

The idea is to monitor the network (DSL part as well as aggregation and core part of the network) using an IoT approach. We use DSL system in combination with OpenMN as a sensor network. Each individual DSL line represents a sensor node that provides required information. Advantages of such approach are:

- it enables 100 % penetration (monitoring the entire network),
- monitoring the last operator network line means end-to-end operator network coverage.

Alternative approach that might be superior to our approach in terms of performance, but is in this case infeasible, would only be to monitor CPE devices. Example of such usage scenario is described in [4].

Periodically, the IoT platform polls DSL line parameters from the OpenMN. Parameters that are pooled, for downstream and upstream, are DSL line information (node, board, port), time, profile defined speed, actual speed, and maximum possible speed. The time between the polls is set to a needed value.

The collected values of actual downstream and upstream speeds on a given port are compared to the required speed of the user profile. For each user, a speed is set in a user profile (dsRateAs0 and usRateAs0, yellow line), also the minimum (dsMinRAte and usMinRate, red dashed line) and the maximum tolerance (dsMaxRate and usMaxRate, blue dashed line) are set as shown in Figs. 5 and 6. If the speed falls under the minimal required speed (if performance discrepancies in comparison to the predefined user profiles are detected), the alarm is triggered.

In that way, the operator is informed about the problem before most users notice it or call a call center. If the actual speed returns to the value that is within the limits defined in the user profile, the alarm is cleared. The solution also takes into account the state of a DSL line, which can be in active or in stand-by mode.

The maximum attainable rates are valuable information for operators and are observed very carefully. If a trend analysis shows that the value is declining, the reason for it may well be far-end crosstalk, which is a significant problem occurring in modern networks. The operator needs to be aware of the realistic speeds that can be offered to users on particular lines when providing new services or increasing the speed of existing service. The operator also has to optimize profile parameters or consider network reconfiguration in the future.

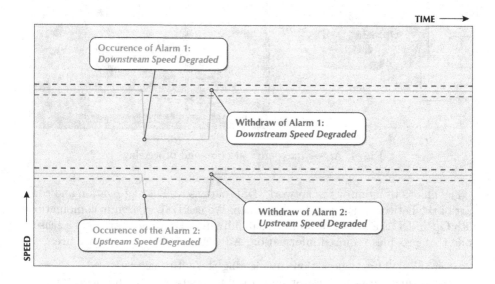

Fig. 5. DSL speeds and alarming – schematic view

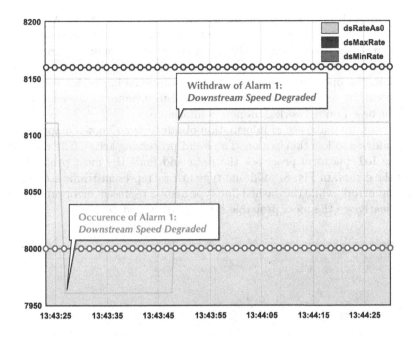

Fig. 6. Downstream speed degradation - dashboard view with alarm information added

Alarming is set for all cases, where the downstream or upstream speed is different from the speed defined in the user profile (as noted above, tolerance is defined in a user profile). The information and alarms are shown on different dashboards and alarm lists. Three of the views are shown in Figs. 5, 6 and 7. Figure 5 shows the schematic view of upstream and downstream DSL speeds with maximum and minimum speed tolerances, as well as alarm information. Figure 6 shows the dashboard view of downstream speed degradation. Alarm information is also added. In Fig. 7, an alarm dashboard view of downstream speed degradation is shown. Based on the alarm type, different actions are started (email, SMS, web pop-up, application trigger).

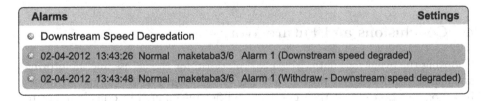

Fig. 7. Alarms – dashboard view

The second IoT enabled application is network error localization. As DSL lines are monitored, we are able to analyze and define the most probable location of network error. DSL lines with simultaneous errors share common parents in the tree (network elements that provide the service). Based on the network topology information, the operator can pinpoint the malfunctioning node. Topology information can be obtained in different ways; from inventory systems, manually entered, or based on network parameters information.

In our case, we make use of information obtained from known mapping of IP network address to location/region. The event processing kernel (the correlator) within the IoT platform processes the data and finds the most probable root cause of the errors. In Fig. 8, solid line type arrows represent traffic stream flowing without errors, while the dashed line type arrows represent error propagation. We can clearly see the most probable error location and device.

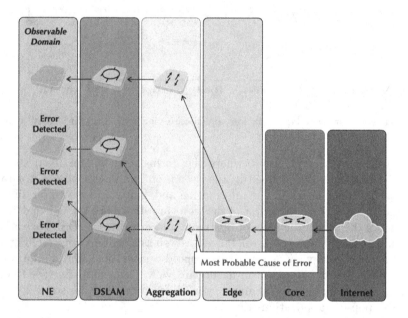

Fig. 8. Network error localization

4   Conclusions and Future Work

Choosing the event and data driven IoT platform for the access line monitoring and alarm correlation tool proved to be a good decision. The development process based on this platform is time effective and flexible enough to quickly adapt to new requirements that arose during the implementation. The team needed less than a month to add a new domain specific scenario.

Network field tests show that the solution is a very useful tool for operators. Next step is large-scale testing in real environment and making adaptations for big network implementation.

In the short run, we plan to further extend the solution to support not only the Iskratel DSLAMs and multi service access nodes (MSAN), but also access elements from other vendors. The solution will be integrated into existing Iskratel's alarm and performance monitoring products. Alarm correlation will be further extended to work on alarms triggered by DSLAMs/MSANs, which will give us even more precise picture of the problems in the network.

The current set of KPIs that are monitored by the platform will be extended by further set of KPIs that will allow for deeper insight into the access line quality. These include line attenuation monitoring, signal-to-noise ratio margins, modelling of line noise characteristics and others.

In the field of alarm cause correlation, we are already developing a solution to correlate lightning information with DSL data and parameters (alpha version is already running). The platform fetches the data from the trusted external server that collects and, over an API, exposes the information about lightings (location, location confidence range, time) [16, 17]. As the lightings are common cause of DSL problems, the application correlates them with network outages. The advantages are more-fold. Operator's control center and helpdesk center get instant information about the state of the lines and network as well as the state of environment, and can thus better respond to customers reports and more optimally plan the eventual repairs (pinpointing the most probable error location/part, timing, needed spare parts).

In the long run, we anticipate future internet of things to develop into the so-called collective adaptive systems that will consist of many units/nodes, with individual properties, objectives and actions. Decision-making will become more distributed and the interaction between units will lead to an increase in unexpected phenomena. The operating principles will go beyond existing monitoring, control and correlation tasks, taking into account the diversity of objectives within the system, and the need to reason in the presence of partial, noisy, out-of-date and inaccurate information.

Acknowledgements. The work was in part supported by the Ministry of Education, Science, Culture and Sport, the Slovenian Research Agency, and the Competence Center Open Communications Platform.

The authors would like to thank partners in Competence Center Open Communications Platform; Technology Network ICT, Alpineon d. o. o., Cosylab d. d., Innova IT d. o. o., Institut Jožef Stefan, Špica International d. o. o., Globtel d. o. o., University of Ljubljana, Faculty of Electrical Engineering, University of Ljubljana, Faculty of Computer and Information Science, Telekom Slovenije d. d., Amis d. o. o, and Iskratel d. o. o.

References

1. Future and Emerging Technologies: Information and Communication Technologies under FP7 Extract from Work Programme 2011-12 for ICT, http://cordis.europa.eu/fp7/ict/fet-proactive/docs/usef-48_en.pdf
2. Lee, G.M., Crespi, N.: Shaping Future Service Environments with the Cloud and Internet of Things: Networking Challenges and Service Evolution. In: Margaria, T., Steffen, B. (eds.) ISoLA 2010, Part I. LNCS, vol. 6415, pp. 399–410. Springer, Heidelberg (2010)
3. Keim, D.A., Mansmann, F., Schneidewind, J., Thomas, J., Ziegler, H.: Visual Analytics: Scope and Challenges. In: Simoff, S.J., Böhlen, M.H., Mazeika, A. (eds.) Visual Data Mining. LNCS, vol. 4404, pp. 76–90. Springer, Heidelberg (2008)
4. Sedlar, U., Volk, M., Sterle, J., Sernec, R., Kos, A.: Contextualized Monitoring and Root Cause Discovery in IPTV Systems Using Data Visualization. IEEE Netw. Mag., Special Issue - Computer Network Visualization, 1–9 (2012) (in review)
5. Šubelj, L., Jelenc, D., Zupančič, E., Lavbič, D., Trček, D., Krisper, M., Bajec, M.: Merging Data Sources based on Semantics, Contexts and Trust. IPSI BGD Trans. Internet Res. 7(1), 18–30 (2011)
6. Umberger, M., Lumbar, S., Humar, I.: Modeling the Influence of Network Delay on the User Experience in Distributed Home-Automation Networks. Information Systems Frontiers, 1–14–1 (August 14, 2010)
7. Moltchanov, D., Jakubiak, J., Koucheryavy, Y.: Infrastructure-Assisted Probabilistic Power Control for VANETs. In: Balandin, S., Koucheryavy, Y., Hu, H. (eds.) NEW2AN/ruSMART 2011. LNCS, vol. 6869, pp. 231–237. Springer, Heidelberg (2011)
8. Jakus, G., Sodnik, J., Tomažič, S.: The Architectural Design of a System for Interpreting Multilingual Web Documents in E-speranto. J. Univers. Comput. Sci. 17(3), 377–398 (2011)
9. Occapi platform, http://www.opcomm.eu/en/solutions/occapi-platform
10. Štern, A., Bešter, J.: Seamless Connectivity in a Networked Vehicle. In: Rijavec, R., Anžek, M., Hernavs, B., Meše, P., Štern, A., Gostiša, B., Kos, S., Petelin, S., Janša, S. (eds.) 20th International Symposium on Electronics in Transport (being) ISEP 2012, March 26-27. Linking people with ITS: proceedings, pp. 1–6. Electrotechnical Association of Slovenia, Ljubljana (2012)
11. ioBridge, http://iobridge.com/docs
12. Pachube, https://pachube.com/docs
13. Exosite, http://exosite.com
14. Libelium, http://www.libelium.com
15. Tendrilinc, http://www.tendrilinc.com
16. SCALAR System Network, http://www.scalar.si/en
17. Kosmač, J., Djurica, V., Babuder, M.: Automatic Fault Localization Based on Lighting Information. In: IEEE Power Engineering Society General Meeting, June 8-22, pp. 1–5. IEEE Xplore (2006)

Aligning Smart and Control Entities in the IoT[*]

Konstantinos Kotis, Artem Katasonov and Jarkko Leino

VTT, Technical Research Center of Finland
{Ext-konstantinos.kotis,artem.katasonov,jarkko.leino}@vtt.fi

Abstract. In our latest work, the problem of semantic interoperability for inter-connected and semantically coordinated smart/control entities in a Semantic Web of Things has been explicated and a framework for Semantic Smart Gateways (SSGF) has been proposed. This paper aims to report on a proof-of-concept implementation of the core component of this framework which is the automatic alignment of smart/control entities' semantic descriptions. More specific, the paper reports on a recent approach towards implementing a configurable, multilingual and synthesis-based ontology alignment tool that has been evaluated by the SEALS and OAEI initiatives.

Keywords: ontology alignment, semantic coordination, smart entity, smart gateway, semantic interoperability.

1 Introduction

One of the current trends in the Internet of Things (IoT) research domain is the integration of 'things' seamlessly with the existing Web infrastructure and the explosion of them uniformly as Web resources, shaping what is referred to as the Web of Things. Such an approach is a great facilitator of interoperability. However, for true interoperability within this setting, semantic interoperability is the key solver, i.e. the ability of the devices to unambiguously convey the meaning of data they communicate over Web protocols. Semantic Web technology can be used to extend WoT, shaping consequently what is sometimes referred to as the Semantic Web of Things.

The work reported in this paper is particularly motivated by our vision of an open and interoperable IoT, based also on the recent work of baseline solutions to cross-domain and cross-platform interoperability and information exchange approaches such as the Smart-M3. In this vision, the following four requirements must be satisfied:

- Ability to have gradually growing IoT environments, contrasted to installing and interconnecting all IoT devices and software at once.
- Ability to interconnect devices from different vendors.
- Ability of 3rd parties to develop software applications for IoT environments, contrasted to applications coming only from the devices' vendors.

---

[*] This work was carried out during the tenure of an ERCIM "Alain Bensoussan" Fellowship Programme. This Programme is supported by the Marie Curie Co-funding of Regional, National and International Programmes (COFUND) of the European Commission.

S. Andreev et al. (Eds.): NEW2AN/ruSMART 2012, LNCS 7469, pp. 39–50, 2012.

- Ability to develop applications that are generic in the sense of running on various IoT device sets (different vendors, same purpose), contrasted to developing applications for a very particular configuration of devices.

In contrast to the above vision, most IoT-related solutions today, except for UPnP/DLNA-based media sharing, fall into one the following categories:

- Buy all the devices from one vendor.
- Connect "smart" devices (phones, TVs) from different vendors through installing a particular software client (from one vendor) on each of them (limited list of supported platforms).
- Use a particular gateway box, then can connect devices from different vendors (from a limited list of supported by the gateway).

In all three cases, a single vendor is responsible for all of the "interoperability" issues.

In our recent work on a Semantic Smart Gateway Framework (SSGF), we introduce a high-level type of entities, which we refer to as 'smart entities'. They correspond directly to physical 'things' that are of some interest to humans, i.e. door, room, and package. In addition, they are equipped with other physical entities, the purpose of which is connecting the 'things' to IoT (e.g. sensors, actuators, RFIDs). Some smart entities, however, can be virtual and just equipped with datasets/data-streams exposed by Cloud services, e.g. Cosm (previously known as Pachube). Examples of smart entities are: a "smart door", a "smart room", a "smart package". Smart entities may also be controlled by other entities responsible for the execution of tasks, which we refer to as 'control entities'. More importantly, smart and control entities are equipped with a conceptual description (domain ontology) of the properties of the physical entities or functionalities they 'carry' and of the data they produce or consume.

The SSGF is designed to support the (semi-)automated translation process between smart or/and control entities' data, with minimum human involvement, by computing their ontology definitions' alignments. Specifically, the Semantic Smart Gateway Framework (SSGF) must provide (among other functionality), an ontology alignment component to support the discovery of similarities between smart/control entities' profiles and ontology definitions, in order to be able to i) support the search for similar smart entities in a common view, ii) support the clustering of smart entities into domain-specific clusters, iii) to support the merging of similar smart entities towards a more efficient network organization, iv) to support the interoperability of smart entities with control entities (e.g. smart applications) that also carry ontology definitions of their input/output parameters. Such an approach is a great facilitator of interoperability and will be able to enable realization of the following example use cases:

- Two temperature sensors are both delivering measurements over HTTP GET as JSON, but of different structure and with different object/property names.
- Two heater devices are accepting commands over HTTP PUT as JSON, but of different structure and with different object/property names.
- A motion detector and light switch control software (one integrated application) is communicating messages (accepting measurements data and delivering commands in XML) with the related devices (communicating in JSON format) using heterogeneous vocabularies (e.g. app:Motion and dev:Movement). The application must be able to 'understand' motion detection events and issue commands to the switch

actuator e.g. for switching on/off the attached device (lamp). The switch actuator must be able to 'understand' commands issued by the application. The process must be automatic or at least semi-automatic with minimal human involvement.

- A package with an RFID or a UCODE label attached to it sent by a post office in origin-A can be automatically managed and forwarded by the intermediate destination-B to the next destination-C, without the current requirement that all post-offices (origin and destinations) share the same database or even the same semantic repository.

The paper is structured as follows: section 2 summarizes the most relevant and recent work, section 3 presents the aim, requirements, design and implementation of our approach, section 4 reports on the evaluation approach and results, and finally section 5 concludes the paper with plans for future work.

2   Related Work

This paper reports current research work related to a (semi-)automated ontology alignment process in the context of smart environments and IoT. This work extends the one of a dynamic ontology linking process for the behavioral coordination of heterogeneous systems that is reported in [2], and complements the related work on ontology mapping in Smart-M3 smart space that has been recently reported in [3]. The proposed work is also motivated by the related work of semantic coordination of agents in P2P systems [4, 5] in respect to the automatic and self-organization of schema/ontology mappings towards 'healing' incorrect mappings of entities in smart spaces. These works however are emphasizing the decentralized computation of mappings, a goal that is out of the scope of this paper. Moreover, our work is also motivated by the latest effort within the SWISS-Experiment project, emphasizing the use of W3C XG SSN ontology in a large-scale federated sensor network for semantic data sensor search [6], which manually provides mappings of data to the reference SSN ontology using a custom mapping language. Our work focuses on the automation of this manual mapping process.

Since the main problem of the presented work is the computation of alignments between ontologies, we mainly consider the related work that has been heavily conducted and reported the past six years within the Ontology Alignment Evaluation Initiative (OAEI) contests [7]. In this context, a number of tools have been evaluated and reported [7], and most importantly, a number of challenges have been identified [8]. The work presented in this paper is an extension to author's work conducted for AUTOMS ontology alignment tool that has been participating in the OAEI 2006 contest, considering both these challenges and the specific requirements of the SSGF. Since it has been always acknowledged by OAEI contesters that there is not a single 'best tool' for ontology alignment [7, 8], and this has been driven due to the different application domains that tools must function, we do not compete with any of the related works. Instead, we focus on a comparative study based on their evaluation with common evaluation datasets.

3   Aligning Smart/Control Entities

In the SSGF, agents (software or human) must be able to register or search for smart or/and control entities (e.g. smart devices, smart applications) that match certain properties/criteria. Newly added heterogeneous entities, 'carrying' their vendors' ontology definitions (or simpler types of metadata) for describing their properties, should be able to be dynamically coordinated via the automatic alignment of their ontology definitions. Such alignment may be computed in a direct (point-to-point) way or via agreed and shared conceptualizations of a reference ontology. In the latter case, in order to align the ontology definitions of two temperature sensors of different vendors, a reference representation of the temperature domain must exist.

To demonstrate the point-to-point alignment process, let us assume we have 2 sensors (a temperature and a motion detection), 2 actuators (switches) and 2 smart applications, with the following heterogeneous schemas (Class:{property TYPE}):

— Sensor: {movement_detection BOOLEAN, timed DATETIME}   (Sensor-A)
— Temperature_Sensor: {value DOUBLE, time_stamp DATETIME}   (Sensor-B)
— AirCond_SwitchAct: {value INTEGER, time_stamp DATETIME}   (Actuator-A)
— Lights_SwitchAct: {value INTEGER, time_stamp DATETIME}   (Actuator-B)
— MotionDetectionApp: {motionDetected BOOLEAN, timed DATETIME}   (Application-A)
— Temperatur: {wert DOUBLE, zeit_stempel DATETIME}   (Application-B, in
  German)

Application-A utilizes temperature data to issue commands towards a switch actuator for switching on/off the air-conditioning appliance, and application-B utilizes motion detection data to issue commands towards a switch actuator for switching on/off the lights in a room. The first goal is to match the sensor and actuator schemas with the proper application (input/output requirements) i.e. Sensor-A and Actuator-A with application-A, and consequently Sensor B and Actuator-B with application-B. The second goal is to align the elements (classes/properties) of each sensor schema with the elements of the application schema in order to be able for both sides (device and application side) to understand the semantics of each other. In the following paragraphs we focus on the implementation and evaluation of the second goal.

As abovementioned, in case smart/control entities are not equipped with an ontology, a transformation step (e.g. JSON to OWL representation) must also be added. An overview of the architecture of a "smart proxy" that implements this transformation step and the ontology alignment of two individual entities (in this case, of a smart entity and of a control entity) is depicted in Figure 1. An Ontology Wizard component is responsible for transforming device's and application's messages that are exchanged between each other or via a gateway (e.g. ThereGate) from JSON or XML or URI format to sets of OWL classes and properties as well as to refine those sets using some heuristic rules (e.g. to handle structural issues). The two sets of ontology elements, one for the device and one for the application, are then processed by the Ontology Alignment component in order to obtain their similarities and compute alignments between them. These alignments (computed at the deployment time) are then used by a Message Translator component at run-time for a bi-directional translation of messages.

There are still some open issues that need to be treated in future work, i.e.:

- Where and how to place human involvement for the refinement of ontology definitions that cannot be fully automatically shaped from raw data (see Figure 1, 'domain expert' labeled grey face icon)
- Where and how to place human involvement (or other means of detecting and correcting erroneous alignment decisions e.g. wisdom of network) for the validation of the computed alignments between ontology definitions since a 100% accuracy of a fully automated method is not realistic (see Figure 1, 'validator' labeled grey face icon)

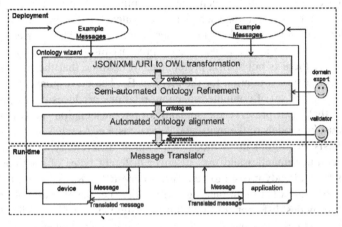

Fig. 1. 'Smart proxy' architecture for the (semi-)automated alignment of smart/control entities, providing syntactic and semantic interoperability of exchanged data between entities (in this case, device and application)

3.1   Requirements and Design

Based on OAEI reports and best practices [7], we have observed that ontology alignment methods which utilize reference ontologies or other external resources such as WordNet lexicon, produce more precise results than others. Although to depend on a reference ontology and external resources is not (generally) a very good practice (must ensure that they both exist and provide a quite complete coverage of the domain knowledge), we observed that in our experimental setting seems to have more benefits than obstacles, thus we decided to use WordNet-based ontology alignment methods where possible.

Regarding the size of the ontologies that may be developed for representing the smart/control entities, it has been also observed that in most cases they will not exceed the number of 1 class and a couple of properties, with the rare case to reach the size of 3 or 4 classes. Such observations have been made by experimenting with commercial products i.e. gateways and sensors-actuators for smart home (ThereGate[1]) and smart building (Niagara Framework[2]). Small sizes of ontologies for smart/control entities allow applying a wide range of alignment methods, even those that have high computational complexity. The approach we follow in this work is able to identify the size of ontologies and adapt the alignment process accordingly by selecting the most suitable alignment methods (a process called profiling), as explained in the following sections.

The problem of computing alignments between ontologies can be formally described as follows: Given two ontologies $O_1 = (S_1, A_1)$, $O_2 = (S_2, A_2)$ (where S_i denotes the signature and A_i the set of axioms that specify the intended meaning of terms in S_i) and an element (class or property) E_i^1 in the signature S_1 of O_1, locate a corresponding element E_j^2 in S_2, such that a mapping relation (E_i^1, E_j^2, r) holds between them. r can be any relation such as the equivalence ($\equiv$) or the subsumption ($\sqsubseteq$) axiom or any other semantic relation e.g. meronym. For any such correspondence a mapping method may relate a value γ that represents the preference to relating E_i^1 with E_j^2 via r. If there is not such a preference, we assume that the method equally prefers any such assessed relation for the element E_1. The correspondence is denoted by $(E_i^1, E_j^2, r, \gamma)$. The set of computed mapping relations produces the mapping function $f:S_1 \rightarrow S_2$ that must preserve the semantics of representation: i.e. all models of axioms A_2 must be models of the translated A_1 axioms: i.e. $A_2 \vDash f(A_1)$.

The synthesis of alignment methods that exploit different types of information (lexical, structural, semantic) and may discover different types of relations between elements has been already proved to be of great benefit [7, 8]. Based on the analysis of the characteristics of the input ontology definitions, i.e. the profiling of ontologies, our approach provides different configurations of alignment methods. The analysis of input ontologies is currently based on their size, the existence of individuals or not, the existence of class/properties annotations e.g. labels, and the existence of class names with an entry in WordNet lexicon. Part of the profiling is also a translation method that supports the translation of classes/properties annotations if these are given in a non-English language.

In the presented work we follow an advanced synthesis strategy, which performs composition of results at different levels (see Figure 2): the resulted alignments of individual methods are combined using specific operators, e.g. by taking the union or intersection of results, intersection of results or by combining the methods' different confidence values with weighing schemas. Given a set of k alignment methods (e.g. string-based, vector-based), each method computes different confidence values concerning any assessed relation (E_1, E_2, r). The synthesis of these k methods aims to compute an alignment of the input ontologies, with respect to the confidence values of the individual methods. Trimming of the resulted correspondences in terms of a threshold confidence value is then performed for optimization.

---

[1] http://therecorporation.com/
[2] http://www.tridium.com/

The alignment strategy followed in our work is outlined in the following steps:

— Step 0: Analyze ontology definitions to be aligned (profiling) and assign the correspondent configuration of alignment methods to be used (configuration). If needed, translate ontology into an English-language copy of it.
— Step 1: For each integrated alignment method k compute correspondence (E_i^1, E_j^2, r, γ) between elements of a peer ontology and a reference domain ontology definition that is specified in the IoT-ontology.
— Step 2: Apply synthesis of methods at different levels (using different aggregation operators) to the resulted set of alignments S_k.
— Step 3: Apply trimming process by allowing agents to change a variable threshold value for each alignments set S_k or for the alignments of a synthesized method.
— Step 4: Validate alignments (initially by human agents via ontology alignments' visualization interfaces) and allow for modifications.

The proposed ontology alignment approach considers most of the challenges in ontology alignment research [8, 9]) but emphasizes the a) alignment methods selection and synthesis, and b) user involvement. A general description of the ontology alignment process [8] extended with the profiling and configuration functionality is shown in Figure 2. Two alignment methods m and m', also called matchers, are selected based on the profiling and configuration method and used for aligning the input ontologies o and o'. In case of translation needed, this is performed before entering m and m' respectively. An initial alignment A is often used also as input to the alignment methods but in our case this is always empty. The resulting alignments are aggregated/merged in (a) using an aggregation operator (union is the most common one), resulting in another alignment (A''') which will be improved by another alignment method (m'') resulting to the final alignment (A'''').

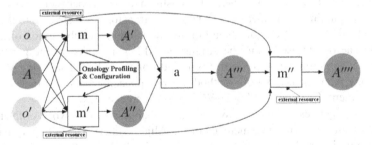

Fig. 2. General description of the ontology alignment process, extended with the profiling and configuration functionality

3.2   Implementation

AUTOMSv2 is an ontology alignment tool based on its first version in 2006 [10]. It computes 1:1 (one to one) alignments of two input domain ontologies, discovering equivalences between ontology elements, both classes and properties. The features that this new version integrates are summarized in the following points:

— It is implemented with the widely used open source Java Alignment API [11]
— It synthesizes alignment methods at various levels (lexical, structural, instance-based, vector-based, lexicon-based) with the possibility to aggregate their alignments using different aggregation operators (Union, Pythagorean means)
— It implements an alignment-methods' configuration strategy based on ontology profiling information (size, features, etc.)
— It integrates state-of-the-art alignment methods with standard Alignment API methods
— Implements a language translation method for non-English ontology elements

The tool has been developed from scratch, reusing some of the alignment methods already provided within the Alignment API. Other state-of-the-art methods such as the COCLU string-based and the LSA vector-based methods implemented in AUTOMS [10] using the AUTOMS-F API [12] have been re-implemented using the new API. The instance-based and structure-based alignment methods have been also implemented from scratch. The final experimental version of AUTOMSv2 (http://ai-lab-webserver.aegean.gr/kotis/AUTOMSv2) was delivered in 18th of March as a submission to the Ontology Alignment Evaluation Initiative 2011.5 Campaign (http://oaei.ontologymatching.org/2011.5/seals-eval.html), using the Semantic Evaluation At Large Scale (SEALS) platform. The detailed description of the alignment methods is out of the scope of this conference, hence of this paper also, since they have been adequately presented in previously published works [10, 12, 13].

The integrated string-based methods are used in two different synthesized methods and in one single method. All three methods use class and property names as input to their similarity distance metrics.

The first one synthesizes the alignments of two string-based similarity distance methods distributed with the Alignment API, namely, the 'smoaDistance' method and the 'levenshteinDistance'. A general Levenshtein distance between two strings is defined as the minimum number of edits needed to transform one string into the other, with the allowable edit operations being insertion, deletion, or substitution of a single character. The one re-used from the Alignment API is a version of the general distance metric, based on the Needleman Wunsch distance method. The String Matching for Ontology Alignment (SMOA) method utilizes a specialized string metric for ontology alignment, first published in ISWC 2005 conference [13].

The second synthesized method synthesizes the alignments of two WordNet-based string-based similarity distance methods of the Alignment API, namely, the 'basicSynonymySimilarity' and the 'cosynonymySimilarity'. The first computes the similarity of two terms based in their synonymic similarity, i.e. if they are synonyms in Word-Net lexicon (returns '1' if term-2 is a synonym of term-1, else returns a BasicString-Distance similarity score between term-1 and term-2), and the second computes the proportion of common synset between them, i.e. the proportion of common synonyms shared by both terms.

The third one is a single method that is implemented based on the state-of-the-art string similarity distance method COCLU, initially integrated in AUTOMS [10] and in other implementations using the AUTOMS-F API [12]. Since it is a complete re-implementation, in this version of the tool it is used in two different modes, i.e. in names-mode and in labels-mode, according to the type of input ontologies that the profiling method will return. COCLU is a partition-based clustering algorithm which divides data into clusters and searches the space of possible clusters using a greedy heuristic.

Regarding vector-based alignment methods, AUTOMSv2 integrates two LSA-based methods, versions of the original HCONE-merge alignment method implemented in AUTOMS [10]. The first version is based on LSA (Latent Semantic Analysis) and WordNet and the second just in LSA. In the first one, given two ontologies, the algorithm computes a morphism between each of these two ontologies and a "hidden intermediate" ontology. This morphism is computed by the Latent Semantic Indexing (LSI) technique and associates ontology concepts with WordNet senses. Latent Semantic Indexing (LSI) is a vector space technique originally proposed for information retrieval and indexing. It assumes that there is an underlying latent semantic space that it estimates by means of statistical techniques using an association matrix (n×m) of term-document data (WordNet senses in this case). The second version of this method is based on the same idea but instead of exploiting WordNet senses it builds the term-document matrix from the concepts' names/labels/comments and their vicinity (properties, direct superconcepts, direct subconcepts) of the input ontologies. The similarity between two vectors (each corresponding to class name and annotation as well as to its vicinity) is computed by means of the cosine similarity measure.

Finally, two more methods, a structure-based and an instance-based method, are integrated, based on the general principle that two classes can be considered similar if a percentage of their properties or their instances has been already considered to be similar. The similarity of properties and instances is computed using a simple string-matching method (Levenshtein). Although it seems that structure and instances are not present in the ontology definitions of smart/control entities, their integration in AUTOMSv2 does not influence its performance since, as already stated, the profiling analysis automatically detects the features of the input ontologies and exclude these methods from computing alignments (i.e. are not included in the synthesis configuration for the smart/control entities' ontology definitions).

The different configurations regarding the way the above methods were synthesized, i.e. computing and merging alignments, is based on the profiling information gathered after the analysis of the input ontologies. Both input ontologies (since our problem concerns the alignment of two ontologies), are examined using currently four different analysis methods:

1. Based on the size of the ontologies, i.e. the number of classes that ontologies have, if one of them has more than a specific number of classes (this number is currently set to 100), then this pair of ontologies is not provided as input to alignment methods with heavy computations since it will compromise the overall execution time of the tool. Such methods are the vector-based, WordNet-based and structure-based ones.
2. If an ontology pair has an ontology with no instances at all, then this pair is not provided as input to any instance-based alignment method (the explanation for this is straight forward).

3. If an ontology pair has two ontologies that half of their classes have no names with an entry in the WordNet, but they have label annotation(s), then provide this pair as input to alignment methods that a) do not consider WordNet as an external resource and b) consider labels matching instead of class names.

4. If an ontology pair has two ontologies that half of their classes have no names with an entry in the WordNet, and they also have no labels, then provide this pair as input to alignment methods that a) do not consider WordNet as an external resource and b) do not consider labels matching. In fact, this is the hardest case, and we have applied an experimental configuration using either instance-based methods if instances exist, or COCLU alignment method for class names string matching.

AUTOMSv2 is re-using a free Java API named WebTranslator (http://webtranslator.sourceforge.net/) in order to solve the multi-language problem. AUTOMSv2 translation method is converting the labels of classes and properties that are found to be in a non-English language (any language that WebTranslator supports, including Chinese and Russian) and creates a copy of an English-labeled ontology file for each non-English ontology. This process is performed before AUTOMSv2 profiling, configuration and matching methods are executed, so their input will consider only English-labeled copies of ontologies.

4   Evaluation

We have evaluated our tool with OAEI 2011 and 2011.5 contests datasets, measuring precision and recall. Precision is the ratio between true positive and all aligned objects (the correct alignments among the retrieved alignments have to be determined); Recall is the ratio between the true positive and all the correspondences that should have been found.

We present and compare results using datasets of OAEI 2011 contest, since the latest results of 2011.5 contests were not published until the submission of this paper. The dataset, namely 'Benchmarks2', concerns new systematically generated benchmarks, based on other ontologies than the bibliographic one (Benchmark1). It is made of a set of 103 pairs of ontologies related to the conference publications domain. The Ekaw ontology, one of the ontologies from the conference track, has been used as reference ontology for generating the dataset. It contains 74 classes and 33 object properties. Figure 3 presents the mean results of precision (Prec) and recall (Rec) of all tools that have been evaluated with this dataset.

We have computed three different average types, i.e. *h-mean*, *mean* and *mean+* in order to better understand the performance of our tool. H-mean (harmonic mean of non-zero values) and mean+ (simple average of non-zero values) implement averages that do not consider zero values in the computations. Zero values occur only for some ontology pairs (15 out of the 103) that organizers have completely changed by scrambling all possible elements' strings (names, labels, comments) using randomly generated meaningless strings. We conjecture that such ontologies are not much realistic and rare to find in real smart space environments.

As it is depicted in Figure 3, based on the different average computation types, AUTOMSv2 performs better regarding precision (from 0.83 to 1.00) than recall (from 0.49 to 0.62). Also, from the 16 tools compared, AUTOMS outperforms 6 of them in terms of MEAN precision and in terms of H-Mean and Mean+ it performs with a near to 100% accuracy. Also, only 6 out of the 16 tools achieved a better Mean recall than AUTOMSv2 did, however it outperforms almost all other tools in terms of H-mean recall.

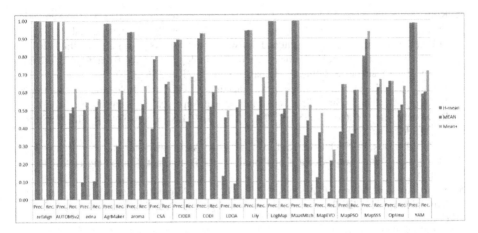

Fig. 3. OAEI 2011 Benchmark mean results compared to AUTOMSv2

It is our aim to create a new custom evaluation dataset that is focused on data gathered from the smart environments domain. The aim is to develop experimental OWL ontology definitions that represent such knowledge and evaluate our tool with ontology pairs that combine different variations of them.

5   Conclusion and Future Work

This paper presents an implementation of SSGF core component, i.e. the automatic alignment of smart/control entities' semantic descriptions. More specifically, it presents the implementation of AUTOMSv2 ontology alignment tool that has been designed in order to meet the alignment requirements of smart/control entities in the SSGF. The focus of the presented implementation is on a) the alignment-methods' configuration strategy based on ontology profiling information, the synthesis and integration of state-of-the-art ontology alignment methods, and c) the support of multilingualism. The presented tool is part of a proof-of-concept SSGF implementation that is called 'smart-proxy', for the (semi-)automated alignment of smart/control entities, providing syntactic and semantic interoperability of exchanged data between these entities. Future work concerns issues related to the degree and nature of human involvement during the ontology refinement and ontology alignments' validation process. Finally, the paper presents the overall performance of the implemented tool and compares it with other tools using the latest OAEI datasets.

Acknowledgements. Authors acknowledge that part of the work presented in this paper has been driven by related work in the following projects: Internet of Things (Finnish national TEKES project), iCORE (287708 - FP7 Integrated Project). We also thank the anonymous reviewers for their valuable comments.

References

1. Kotis, K., Katasonov, A.: Semantic Interoperability on the Web of Things: The Semantic Smart Gateway Framework. To appear in: Proceedings of the 6th International Workshop of Frontiers in Complex, Intelligent and Software Intensive Systems, Palermo, Italy (2012)
2. Katasonov, A., Terziyan, V.: Using Semantic Technology to Enable Behavioral Coordination of Heterogeneous Systems. In: Semantic Web, pp. 135–156 (2010)
3. Smirnov, A., Kashevnik, A., Shilov, N., Balandin, S., Oliver, I., Boldyrev, S.: On-the-Fly Ontology Matching in Smart Spaces: A Multi-model Approach. In: Balandin, S., Dunaytsev, R., Koucheryavy, Y. (eds.) ruSMART/NEW2AN 2010. LNCS, vol. 6294, pp. 72–83. Springer, Heidelberg (2010)
4. Cudré-Mauroux, P., Suchit, A., Karl, A.: GridVine: An Infrastructure for Peer Information Management and Large-Scale Collaboration. IEEE Internet Computing 11(5), 36–44 (2007)
5. Spiliopoulos, V., Vouros, G.A.: Synthesizing Ontology Alignment Methods Using the Max-Sum Algorithm. IEEE Transactions on Knowledge and Data Engineering 24(5), 940–951 (2012)
6. Calbimonte, J., Jeung, H., Corcho, O., Aberer, K.: Semantic Sensor Data Search in a Large-Scale Federated Sensor Network. In: Semantic Sensor Network Workshop, ISWC. CEUR-WS proceedings, pp. 23–38 (2011), http://CEUR-WS.org
7. Euzenat, J., Meilicke, C., Stuckenschmidt, H., Shvaiko, P., Trojahn, C.: Ontology Alignment Evaluation Initiative: Six Years of Experience. In: Spaccapietra, S. (ed.) Journal on Data Semantics XV. LNCS, vol. 6720, pp. 158–192. Springer, Heidelberg (2011)
8. Shvaiko, P., Euzenat, J.: Ontology matching: state of the art and future challenges. IEEE Transactions on Knowledge and Data Engineering (December 08, 2011), http://doi.ieeecomputersociety.org/10.1109/TKDE.2011.253
9. Kotis, K., Lanzenberger, M.: Ontology Matching: Current Status, Dilemmas and Future Challenges. In: International Conference of Complex, Intelligent and Software Intensive Systems, pp. 924–927 (2008)
10. Kotis, K., Valarakos, A., Vouros, G.A.: AUTOMS: Automating Ontology Mapping through Synthesis of Methods. In: International Semantic Web Conference, Ontology Matching International Workshop, Atlanta, USA (2006)
11. David, J., Euzenat, J., Scharffe, F., Trojahn dos Santos, C.: The Alignment API 4.0. Semantic Web - Interoperability, Usability, Applicability 2(1), 3–10 (2011)
12. Valarakos, A., Spiliopoulos, V., Kotis, K., Vouros, G.A.: AUTOMS-F: A Java Framework for Synthesizing Ontology Mapping Methods. In: International Conference i-Know, Graz, Austria (2007)
13. Stoilos, G., Stamou, G., Kollias, S.: A String Metric for Ontology Alignment. In: Gil, Y., Motta, E., Benjamins, V.R., Musen, M.A. (eds.) ISWC 2005. LNCS, vol. 3729, pp. 624–637. Springer, Heidelberg (2005)

Where Have You Been? Using Location Clustering and Context Awareness to Understand Places of Interest[*]

Andrey Boytsov[1,2], Arkady Zaslavsky[1,3], and Zahraa Abdallah[2]

[1] Department of Computer Science, Space and Electrical Engineering,
Luleå University of Technology,
SE-971 87 Luleå, Sweden
{andrey.boytsov,arkady.zaslavsky}@ltu.se
[2] Caulfield School of IT,
Monash University,
Melbourne, Australia
zahraa.said.abdallah@monash.edu
[3] CSIRO, Canberra, Australia
arkady.zaslavsky@csiro.au

Abstract. Mobile devices have access to multiple sources of location data, but at any particular time often only a fraction of the location information sources is available. Fusion of location information can provide reliable real-time location awareness on the mobile phone. In this paper we propose and evaluate a novel approach to detecting the places of interest based on density-based clustering. We address both extracting the information about relevant places from the combined location information, and detecting the visits to known places in the real time. In this paper we also propose and evaluate ContReMAR application – an application for mobile context and location awareness. We use Nokia MDC dataset to evaluate our findings, find the proper configuration of clustering algorithm and refine various aspects of place detection.

Keywords: context awareness, contextual reasoning, location awareness, sensor fusion.

1 Introduction

Present day mobile devices have access to multiple sources of location information. The possible sources include GPS sensors, WLAN location information, GSM localization, indoors positioning systems, dead reckoning. However, the sources of information are not always available. For example, GPS sensor is unstable indoors, it is battery consuming, and users usually turn it on only for driving from one place to another. In turn WLAN is often available at home and in office buildings, but it is rarely used outdoors. Therefore, for location awareness it is vital to fuse the location measurements from different sources. Location data fusion can remedy both with the unavailability of location information sources and the lack of precision of location information.

---

[*] For the color version of the figures, please refer to the online version of the paper.

S. Andreev et al. (Eds.): NEW2AN/ruSMART 2012, LNCS 7469, pp. 51–62, 2012.
© Springer-Verlag Berlin Heidelberg 2012

The area of mobile location awareness faces two major challenges that we are going to address in this paper:

- Detecting the places of interest out of location measurements that are fused from multiple sources.
- Finding the model of places of interest that allows feasible mobile real-time recognition using the measurements from multiple sources.

We used Nokia Mobile Data Challenge (MDC) dataset [10] as a benchmark, in order to evaluate location awareness algorithms, as well as to test the long-run performance of the application by imitating the sensor feed. We used ECSTRA toolkit [4] for context awareness and context information sharing.

The paper is structured in the following way. Section 2 proposes the approach to extract relevant places from location information. Section 3 discusses the architecture of ContReMAR – mobile context awareness application that we developed to evaluate the proposed approaches. ContReMAR has embedded capabilities for location awareness and location data fusion. Section 4 provides evaluation and demonstration of the proposed application. We use Nokia MDC data [10] in order to configure the parameters of clustering algorithms and increase the precision of place recognition. Section 5 discusses the related work. Section 6 summarizes the results, provides further work directions and concludes the paper.

2   Mobile Location Awareness

Location awareness is an important part of context awareness. The information about current or previously visited places can be used to provide timely assistance and recommendation for the user. The outcomes of location awareness can be also used as a baseline for activity recognition or context prediction. One of the main aspects of location awareness is detecting and labeling the places of interest.

In order to detect the relevant places, we need to identify the concept of a relevant place first. The place where the user has spent significant time is likely to be the place of interest for the user. The latter can be obtained by clustering the location measurements. We propose the following algorithm for extraction and identification of the relevant places. Evaluation of the proposed approach is provided in section 4.

Place Recognition Approach.
Step 1. Fusion of location measurements. We complemented the information obtained from GPS data with the location information obtained from WLAN. The GPS entries, which corresponded to high-speed movement, were removed from consideration.

Step 2. Cluster the location information. When logging is complete (e.g. at the end of the day), the application finds the clusters of location measurements to detect the places of interest. The nature of the task enforces following requirements for the clustering approach:

1. The user stays in the relevant place for significant time. So, the place of interest can be characterized by a relatively dense cluster of location measurements.

2. The number of places, that user has visited during a day, is not known in advance. Therefore, the number of clusters is initially unknown.

3. Some measurements do not correspond to any place of interest (e.g. user just walks on a street). Therefore, the clustering algorithms should not try to attribute all points to some cluster.

The approach, which satisfies all the constraints, is density-based clustering. DBSCAN [6] and OPTICS [3] are the most well-known algorithms of that approach. Those algorithms have the following parameters:

-        *MinPts* – minimum number of points in vicinity, in order for the point to be in the core of the cluster.

-        *Eps* – vicinity radius.

Step 3. Analysis of relevance. This step identifies, whether the detected clusters are the places of interest or not. For example, staying at the traffic light can result in multiple GPS measurements around the same spot, but it is not a place of interest. Irrelevant places can be filtered out by applying a threshold on the time spent at the place. More advanced aspects of relevance analysis are discussed in section 4.

Step 4. Auto-labeling the detected places. The detected place should be presented to the user in a meaningful way. Our approach combines two ways to auto-label the detected places.

Step 4.1. Obtain the list of possible relevant places. We use Google Places API [15] to detect the relevant places in the vicinity. The obtained list of places can be then filtered and ordered depending on at what time user was at that place.

Step 4.2. Analyze the time spent at a certain place of interest. In order to identify, what place of interest does the cluster correspond to, it is important to notice when the user was at that place. For example, if the user spent entire Wednesday at some place, it is very likely to be his/her work or study place, any night is very likely to be spent at home, and break in the middle of the working day is very likely to be lunch.

Step 5. Save the place description for later real-time recognition.

Section 3 discusses ContReMAR application, in which we implemented this location awareness approach. The evaluation of the proposed approach is presented in section 4.

3   ContReMAR Application

In order to prove, test and validate our approach, we designed and developed ContReMAR (CONText REasoning for Mobile Activity Recognition) application – a solution for mobile context awareness. The scope of ContReMAR is broader than just location awareness. However, in this paper we are going to focus on location awareness capabilities of ContReMAR, and its other aspects of mobile context awareness are out of scope of this paper.

The location awareness approach, proposed in section 2, was embedded into the ContReMAR application. The architectural and implementation solution are addressed in details in the subsections 3.1-3.3.

3.1   ContReMAR Architecture

The structure of ContReMAR application is depicted in figure 1. The blocks, which we designed and implemented specially for ContReMAR are depicted in green. The third-party solutions that we used are presented in yellow.

The application is divided into server part (which resides, for example, on a stationary computer or a laptop) and client part (which resides on the user device). Server side is responsible for computationally heavy and memory heavy parts of the work: logging the GPS data and identifying the clusters. The client side is responsible for real-time activity recognition. ECSTRA (Enhanced Context Spaces Theory Based Reasoning Architecture) framework [4], which is the basis of context reasoner, is available for Android platform. In order to evaluate the algorithms, we also used a simulation of mobile device, where the sensor feed was replaced with Nokia MDC [10] data flow.

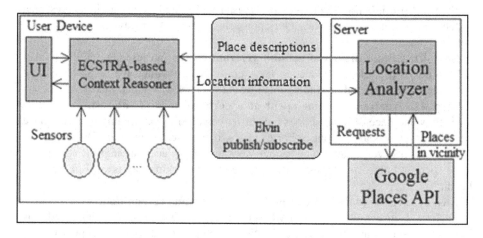

Fig. 1. ContReMAR Application Architecture

Figure 1 illustrates that ContReMAR application consists of the following components:

- **ECSTRA-based context reasoner.** We developed context reasoner for real-time location awareness, situation awareness and activity recognition. The detailed description of context reasoner component is provided in the section 3.2.
- **Location Analyzer.** We implemented location analyzer in order to infer new meaningful places from raw location data. Location analyzer is discussed in details in section 3.3.

- **User Interface.** It is a currently prototyped application component, which shows the results of context reasoning. The design of user-friendly interface is a subject for future work.
- **Elvin publish/subscribe environment.** In order to facilitate the communication between mobile device and the server side, we employed Elvin publish/subscribe protocol [16]. Our application used Avis [17] open source implementation of Elvin protocol. The proposed solution ensures seamless communication, even if both mobile device and server (which can be situated on a laptop) are moving between the coverage areas of different WLAN spots.
- **Sensors.** In non-simulated environment the sensors are merely the sensors situation on a mobile device. In order to evaluate our approach using Nokia MDC data [10] we also developed a simulated mobile device, which imitates sensor feed by substituting it with Nokia MDC data flow.

The application uses Google Places API [15] in order to detect the possible places of interest in the vicinity of the location.

3.2   Context Reasoner

We proposed and developed context reasoner for real-time inference of location, situations and activities. Context reasoner is based on ECSTRA framework [4], and has many of it components reused or extended from ECSTRA. ECSTRA is a general purpose context awareness and situation awareness framework, which acts a backbone for context reasoning in ContReMAR. The structure of context reasoner is depicted in figure 2. The components specially designed for ContReMAR are depicted in green. Components reused from ECSTRA framework are depicted in blue.
   Context reasoner consists of following components:

- **Context collector.** We redeveloped context collector component based on the similar component of ECSTRA framework. It is responsible for sensor fusion, proper reporting of the newly arrived sensor data and preliminary analysis.
- **Application space.** Application space was reused from ECSTRA framework. Application space is responsible for handling the reasoning requests, which come from UI component. Application space contains complete current context information and encompasses situation spaces – the possible interpretations of sensor data.
- **Situation spaces.** Situations are generalizations of sensor information. Situation spaces are ESCTRA components, which are responsible for reasoning about one situation each. The situations of interest for this paper are the relevant places. However, in general case situations can also represent activities, events, etc.
- **Location importer.** We developed location importer to ensure proper introduction of new places of interest (i.e. new situation spaces) into the application space.
- **Elvin connector.** Elvin connector is ECSTRA component (part of Elvin support in ECSTRA), designed to send and receive context information. It was extended comparing to ECSTRA in order to incorporate exporting and importing the place descriptions.

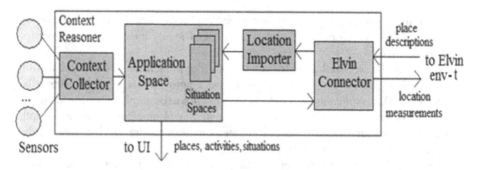

Fig. 2. Context Reasoner Architecture

More details on ECSTRA and its components can be found in [4].

3.3   Location Analyzer

We designed and implemented location analyzer for extracting relevant places out of location data. Location analyzer is also responsible for relevance analysis, place type analysis and interacting with Google Places to detect the places in proximity of the location measurements cluster. Figure 3 shows location analyzer architecture. ContReMAR-specific components are depicted in green. The components, reused from ECSTRA, are depicted in blue. Location analyzer contains the following components:

  - **Elvin connector.** The component originates in ECSTRA, but it was extended to handle new functionality of exchanging the situation descriptions. Elvin connector was already described in section 3.2.
  - **Clusterer.**   The clusterer component is one of the core components in ContReMAR. We designed the component to cluster location measurements in order to define the places of interest. Clustering was facilitated by the libraries of Weka toolkit [7]. Our application use density-based clustering, which was justified in section 2. The parameters of clustering algorithms are discussed in section 4. The clusterer is also responsible for analyzing the type of the place and communicating with Google Places service in order identify possible places, which the cluster can correspond to.
  - **Location exporter.** Location exporter is responsible for translating clusters of points into situation descriptions and providing the descriptions to Elvin connector for further sharing with user device.
  - **Logger.** We implemented logger component to provide detailed reports of how the application worked. It is mostly used for monitoring, debugging and evaluation purposes.

Next section discusses the advanced aspects of location analysis and provides the evaluation of ContReMAR application.

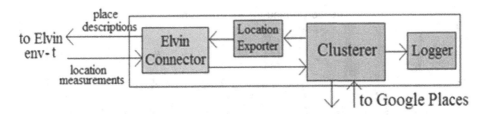

Fig. 3. Location Analyzer Architecture

4   Evaluation

The proposed location awareness approach contains multiple parameters, like the minimum number of points for the cluster core, radius of the location point proximity, minimum time for the place to be relevant. In order to determine the best values of the parameters, extensive user studies are needed. Still unlabeled data, provided by Nokia MDC [10], allow performing some analysis and establishing some error bounds and parameter recommendations.

4.1   Experiments

We performed a series of experiments to prove and evaluate our approach. We used ContReMAR application with imitated user device as a testbed. During the experiment we simulated the mobile device and used Nokia MDC [10] data as an imitated sensor feed. The settings of every experiment, as well as the results, are reviewed in the subsequent subsections.

Experiment 1. Shall the location information be fused from multiple sources? Or is there any dominant source of location information? Location measurements come from two sources: GPS measurements and WLAN access at known spots. Figure 4 shows proportion of GPS location data in the total number of location measurements. The users to be depicted on the plot in figure 4 were chosen randomly (uniformly among all users in the Nokia MDC database [10]).

The absence of obvious pattern in figure 4 allows deriving an important conclusion. Figure 4 proves that there is no dominant source of location information. Therefore, as it was expected, in practice we have to analyze both sources of location information and cannot simply use just one of those. This conclusion justifies the need for further experiments in order to determine, whether density-based clustering works well in presence of multiple-source location measurements.

Experiment 2. Can we use time thresholds to increase the precision of place recognition and minimize the number of false recognitions? What threshold value should it be? The relevance of the place is influenced by the amount of time

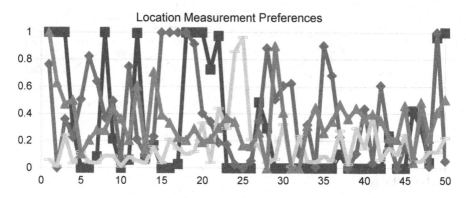

Fig. 4. Proportion of GPS data in location measurements

that the user spends at that place. If the user spends less than a minute at some place, most likely the user just passed it by. However, if the user spends around 15 minutes at some place, it is usually worth noticing. In the testbed application we implemented the following criterion – if the user spends less than certain amount of consecutive time at the location cluster, then the cluster is rejected and removed from consideration. The dependency between the number of detected clusters over time and the consecutive time threshold is presented in figure 5. The user was chosen randomly, but the trend holds for the vast majority of the users. We analyzed the possible time threshold values of 1 minute, 3 minutes, 5 minutes, 10 minutes and 15 minutes. The analysis of the trends in the number of recognized places allowed us adjusting the value of the time threshold.

We analyzed figure 5 according to the following criteria. The proper identification of relevant places should notice the most frequently visited places (like home, work, favorite lunch place, etc.) in first week or few weeks. After that the rate of detecting new places should slow down. If the rate stays high all the time, it can be a sign of numerous false recognitions. To summarize, if the number of recognized places grows with constant trend, it is a sign of possible massive false recognitions.

Figure 5 shows that the threshold of 1 minute leads to constant growth of recognized places. The threshold of 3 or 5 minutes also shows the same problem for most users. For the thresholds of 10 minutes of 15 minutes the number of recognized places almost stops growing in the first three weeks, and it matches the expected behavior of the correct location awareness algorithm. Therefore, the time threshold values of 10-15 minutes should be preferred.

Experiment 3. How can we configure clustering algorithm using unlabelled data? Can we do it by analyzing the fraction of revisited places? The efficiency of the place recognition algorithm can be evaluated by analyzing the number of revisited places. If the place is visited only once, it means that this place will just hamper recommendation and prediction algorithms. The possible unsupervised criterion for place recognition efficiency is the count of places, which were later revisited. Higher is that count, more significant places are detected.

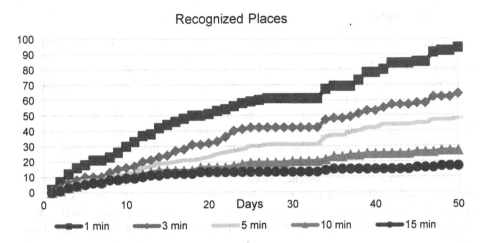

Fig. 5. Recognized places over time for random user, depending on the time threshold

In table 1 we show the results of location measurements analysis for four randomly chosen users. We took all the places recognized in the first 25 days and analyzed how many times were they re-encountered during the first 50 days (i.e. during the 25 of active place recognition and 25 days after that, when we only analyzed the visits to the already recognized places). For the experiment we used the clustering algorithm with parameters *Eps*=50m, *MinPts*=3 and without consecutive time restrictions. Due to the very relaxed constraints the algorithm tends to recognize as a place everything that even remotely resembles a place of interest. We are going to refer to that algorithm as the benchmark clustering algorithm.

Table 1. Proportion of revisited places

User (anonymized)	a	b	c	d
Revisited places (%)	49%	53%	38%	19%

As table 1 shows, the benchmark clustering algorithm will result in 50-60% of false recognitions (up to 80% in some cases). This information can be later used for auto-configuration purposes of location awareness algorithms. The parameters of the clustering algorithm can be preliminary evaluated by comparing it to the benchmark clustering approach. If the clustering approach under consideration leaves out in average 50-60% of clusters, recognized by the benchmark clustering algorithm, then it shows the expected behavior of a correct place clustering approach. Still, it should be noted that this criterion can be used only for preliminary estimation, and more precise parameter choice needs extensive user studies.

4.2   Demonstration and Evaluation Summary

To summarize, the experiments and demonstration lead to the following conclusions:

-   GPS and WLAN location measurements are equally important for location awareness on the mobile phone. Density-based clustering is a feasible approach for detecting relevant places.
-   In order to avoid false recognitions, the threshold can be put on the consecutive time that the user remained at some area. Using the threshold values of 1-5 minutes exhibits the signs of massive false recognitions, while the threshold values of 10-15 minutes show no visible problems.
-   The clustering algorithm parameters can be evaluated by comparing its results to the benchmark clustering algorithm. The benchmark clustering algorithms is likely to produce from 50-60% to 80% of false recognitions. If the clustering algorithm differs from benchmark by that amount, it can be a preliminary indication of appropriate performance.

The space requirements do not allow extensive demonstration. For the proof of concept we show the following example. Figure 6 shows the results of location measurements analysis for randomly chosen user for day one. The system was able to determine most likely home and work places. For example, the application detected that the user works in EPFL Lausanne. The system suggested 3 possible places, where the user is likely to work (all of them are subdivisions of EPFL) and filtered out the places which couldn't be user's workplace (like the nearby streets).

5   Related Work

There is a large body of work in various techniques for trajectory mining, e.g. pattern discovery, similarity measures, clustering and classification for GPS data streams [8,9,11].

Spaccapietra et. at [13] introduced "stops and moves" model for reasoning from GPS data. The application allowed detecting user visits to place of interest, but the user had to specify the relevant places manually. Palam et. al. [12] proposed spatio-temporal clustering method, based on speed in order to find interesting places automatically. SeMiTri system [14] focused on processing heterogeneous trajectories, integrating information from geographic objects and accommodating most existing geographic information sources. The GeoPKDD1 [1] address semantic behaviors of moving objects.

Andrienko et. al. [2] developed a generic procedure for analyzing mobility data in order to study place-related patterns of events and movements.

An important novelty of our approach is fusion of information from multiple sources to build a comprehensive picture of user's location. Our approach also scales to large amounts of data to find the model of places of interest that allows feasible mobile real-time recognition using the measurements from multiple sources. One more novel feature of our approach is auto-labeling the places of interest based on both location and time analysis.

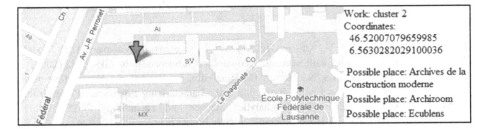

	Work: cluster 2
	Coordinates:
	46.52007079659985
	6.5630282029100036
	Possible place: Archives de la Construction moderne
	Possible place: Archizoom
	Possible place: Ecublens

Fig. 6. ContReMAR application detected the workplace of the user

Next section concludes the paper and provides the direction of future work.

6   Conclusion and Future Work

In this paper we proposed a novel technique for mobile location awareness. In order to prove the feasibility and efficiency of the proposed approach we developed an application called ContReMAR. The outcomes of ContReMAR can enhance mobile activity recognition, context prediction and mobile context-driven recommender systems.

We used Nokia MDC dataset [10] as a simulated sensor feed in order to prove the soundness of our approaches, configure the proposed algorithms and evaluate the application performance. The evaluation showed the feasibility of density-based clustering approach to place identification and place recognition. Subsequent analysis led us to designing a set of experiments in order to test and evaluate the algorithms in absence of labeled data. In turn, it allowed us to configure the parameters of the algorithms.

We identified the following major directions of the future work:

- Improving the quality of place recognitions by performing user studies.
- Location-driven mobile activity recognition.
- Designing friendly user interface.

Present day mobile phones are able to sense the large amount of location information, as well as other diverse context. The proper analysis and verification of that information will significantly improve the capabilities of mobile devices and enable full-scale mobile location and context awareness, as well as support for advanced services and applications as has been demonstrated in [5].

Acknowledgements. We'd like to thank Shonali Krishnaswami for her support and mentorship. We'd also like to thank Basel Kikhia for providing valuable testing and visualization tools.

References

1. Alvares, L.O., Bogorny, V., Kuijpers, B., Macedo, J., Moelans, B., Vaisman, A.: A Model for Enriching Trajectories with Semantic Geographical Information. In: GIS, p. 22 (2007)
2. Andrienko, G., Andrienko, N., Hurter, C., Rinzivillo, S., Wrobel, S.: From Movement Tracks through Events to Places: Extracting and Characterizing Significant Places from Mobility Data. In: IEEE Visual Analytics Science and Technology (VAST 2011) Proceedings, pp. 161–170. IEEE Computer Society Press (2011)
3. Ankerst, M., Breunig, M.M., Kriegel, H., Sander, J.: OPTICS: Ordering Points To Identify the Clustering Structure. In: Proceedings of the 1999 ACM SIGMOD International Conference on Management of Data (SIGMOD 1999), pp. 49–60. ACM, New York (1999)
4. Boytsov, A., Zaslavsky, A.: ECSTRA – Distributed Context Reasoning Framework for Pervasive Computing Systems. In: Balandin, S., Koucheryavy, Y., Hu, H. (eds.) NEW2AN/ruSMART 2011. LNCS, vol. 6869, pp. 1–13. Springer, Heidelberg (2011)
5. Boytsov, A., Zaslavsky, A.: Formal verification of context and situation models in pervasive computing. Pervasive and Mobile Computing (available online March 23, 2012), ISSN 1574-1192, doi:10.1016/j.pmcj.2012.03.001
6. Ester, M., Kriegel, H., Sander, J., Xu, X.: A density-based algorithm for discovering clusters in large spatial databases with noise. In: Proc. KDD, vol. 96, pp. 226–231. AAAI Press (1996) ISBN: 1577350049
7. Hall, M., Frank, E., Holmes, G., Pfahringer, B., Reutemann, P., Witten, I.H.: The WEKA data mining software: an update. SIGKDD Explor. Newsl. 11(1), 10–18 (2009)
8. Han, J., Lee, J.-G., Gonzalez, H., Li, X.: Mining Massive RFID, Trajectory, and Traffic Data Sets (Tutorial). In: KDD (2008)
9. Jeung, H., Yiu, M.L., Zhou, X., Jensen, C.S., Shen, H.T.: Discovery of Convoys in Trajectory Databases. In: VLDB, pp. 1068–1080 (2008)
10. Laurila, J.K., Gatica-Perez, D., Aad, I., Blom, J., Bornet, O., Do, T.-M.-T., Dousse, O., Eberle, J., Miettinen, M.: The Mobile Data Challenge: Big Data for Mobile Computing Research. In: Proc. Mobile Data Challenge by Nokia Workshop, in Conjunction with Int. Conf. on Pervasive Computing, Newcastle (June 2012)
11. Li, Z., Ji, M., Lee, J.-G., Tang, L.-A., Yu, Y., Han, J., Kays, R.: MoveMine: Mining Moving Object Databases. In: SIGMOD, pp. 1203–1206 (2010)
12. Palma, A.T., Bogorny, V., Kuijpers, B., Alvares, L.O.: A Clustering-based Approach for Discovering Interesting Places in Trajectories. In: 23rd Annual Symposium on Applied Computing (ACM-SAC 2008), Fortaleza, Ceara, Brazil, March 16-20, pp. 863–868 (2008)
13. Spaccapietra, S., Parent, C., Damiani, M.L., de Macedo, J.A., Porto, F., Vangenot, C.: A Conceptual View on Trajectories. Data and Knowledge Engineering 65, 126–146 (2008)
14. Yan, Z., Chakraborty, D., Parent, C., Spaccapietra, S., Aberer, K.: SeMiTri: A Framework for Semantic Annotation of Heterogeneous Trajectories. In: 14th International Conference on Extending Database Technology (EDBT 2011), pp. 259–270 (2011)
15. Google Places API documentation, https://developers.google.com/maps/documentation/places/ (accessed on April 14, 2012)
16. Elvin Protocol Specifications, http://www.elvin.org/specs/index.html (accessed on April 14, 2012)
17. Avis Event Router Documentation, http://avis.sourceforge.net/documentation.html (accessed on April 14, 2012)

Where Are They Now – Safe Location Sharing
A New Model for Location Sharing Services

Dmitry Namiot[1] and Manfred Sneps-Sneppe[2]

[1] Lomonosov Moscow State University
Faculty of Computational Mathematics and Cybernetics
Moscow, Russia
[2] Ventspils University College
Ventspils International Radioastronomy Centre
Ventspils, Latvia
{dnamiot,manfreds.sneps}@gmail.com

Abstract. This paper describes a new model for sharing location info without revealing identity to third party servers. We can describe it as a safe location sharing. Proposed approach splits location info and identity information. Service operates with a special kind of distributed data store, where identity info is always local. It eliminates one of the main concerns with location-based systems – privacy. This article describes a model itself as well as its implementation in the form of HTML5 mobile web application.

Keywords: location, lbs, mobile, HTML5, geo coding, social networks.

1   Introduction

It is a well-known fact that the question "where are you" is one of the most often asked during the communications. 600 billion text messages per year in the US ask, "where are you?" – as per Location Business Summit 2010 data. A huge amount of mobile services is actually being built around this question so their main feature is user's location exchange.

Location, while being only one of the sensor readings of a modern smart phone, is probably the first attribute (candidate) to share for mobile users.   The typical applications are well known and include for example geo-tagged context, friend-finder, recommendation systems, turn-by-turn navigation, etc.

In location-based service (LBS) scenarios we can describe the following actors [2]:

- Intended recipient, e.g., the service company, friends, parents, etc. This usually involves the use of a service provider that offers to forward your location to the intended recipient.

- Service provider, e.g., Google providing you with the Latitude application, or a restaurant recommendation system for near-by places. In contrast to the intended recipient, users usually do not have a primary goal of letting the service provider know their location – it is a by-product of getting a restaurant review or staying in touch with friends.

S. Andreev et al. (Eds.): NEW2AN/ruSMART 2012, LNCS 7469, pp. 63–74, 2012.

- Infrastructure provider, e.g., your mobile operator. While self-positioning systems such as GPS can work without an infrastructure provider, mobile phone users are often implicitly located in order to provide communication services (for example, route phone calls)

Some papers mentioned also so-called unintended recipients [2]. For example, we can mention accidental recipient, illegal recipient and law enforcement.

Interesting also, that in the most cases talking about LBS we assume that for a given system, the infrastructure provider needs to be trusted. In other words the need for sharing location data with infrastructure providers is non-discussable.

In the most cases location sharing is implemented as the ability for the mobile user (mobile phone owner) write down own location info in the some special place (special mobile application).

But it means of course, that user must be registered in this service (download some special application). And even more important – everyone who needs this information must use the same service too [1].

One of the biggest concerns for all location-based services is user's privacy. Despite the increased availability of these location-sharing applications, we have not yet seen wide adoption. It has been suggested that the reason for this lack of adoption may be users' privacy concerns regarding the sharing and use of their location information.

For example, the widely cited review of social networks practices [3] concluded that location information is preferably shared on a need to know basis, not broadcast. Participants were biased against sharing their location constantly, without explicit consent each time their location is requested. This suggests that people are cautious about sharing their location and need to be reassured that their private information is only being disclosed when necessary and is not readily available to everybody.

The key point for any existing service is some third party server that keeps identities and locations. We can vary the approaches for sharing (identity, locations) pairs but we could not remove the main part in privacy concerns – the third part server itself.

As mentioned in [4] peer opinion and technical achievements contribute most to whether or not participants thought they would continue to use a mobile location technology.

One possible solution is using peer-to-peer location sharing. The easiest way to apparently "solve" location privacy problems is to manually or automatically authorize (or not) the disclosure of location information to others. But we should see in the same time the other privacy issue that is not eliminated. Your location will be disclosed to (saved on) some third party server. For example, you can share location info in Google Latitude on "per friend" mode, but there is still some third party server (Google) that keeps your location and your identity

Typically we have now two models for location sharing in services. At the first hand is some form of passive location monitoring and future access to the accumulated data trough some API. It is Google Latitude for example. Possible problems are privacy - some third party tool is constantly monitoring my location and what is more important – saves it on the some external server as well as the shorted life for handset's batteries.

Another model for location sharing is check-in. It could be an active (e.g. Foursquare), when user directly sets his/her current location or passive (e.g. Twitter) when location info could be added to the current message. A check-in is a simple way to keep tabs on where you've been, broadcast to your friends where you are, and discover more about other people in your community. But here we can see not only privacy issue - all my friends/followers can see my location but also a noise related issue – my location info could be actually interested only for the physical friends. For the majority of followers my location info (e.g. Foursquare status in Twitter's time line) is just a noise [1]

Lets us describe some existing approaches in LBS development that targets the privacy.

One of the most popular methods for location privacy is obfuscation [5]. Obfuscating location information lowers its precision, e.g., showing only street or city level location instead of the actual coordinates, so that the visible (within our system) location does not correspond to the real one. For example, in Google Latitude we can allow some of the users get our own location info on the city level only. Sometimes even the random noise could be added to the real location data [6]. But once again – it is just a visible location. The central point (points) for such a system can have all the information.

Another popular approach in the area of location privacy is "k-anonymity" [7]. As per this approach the actual location is substituted by a region containing at least k − 1 other users, thus ensuring that a particular request can only be attributed to "1 out of k" people. Of course, this approach has the disadvantage that if the region contains too few people, it has to be enlarged until it contains the right number of people. But in general k-anonymity protects identity information in a location-oriented context [8]. The big question again is it core-level protection or just a view. In other words what kind of data do have inside of our system – anonymous location info right from the moment data being put into our system or it is just a view and data internally saved in raw formats.

Of course, the deployment of location privacy methods depends on the tasks our system is going to target. For example, obfuscating location information in case of emergency help system could not be a good idea. But from other side many geo-context aware applications (e.g. geo search) can use approximate location info.

Also we need to highlight the role of identity in LBS. It looks like combining identity with location info is just an attempt for delivering more targeted advertising rather then the need of the services themselves. It is obviously for example, that local search for some points of interests (e.g. café) should work for the anonymous users too.

Our idea of the signed geo messages service (geo mail, geo SMS) based on the adding user's location info to the standard messages like SMS or email. Just as a signature. So with this service for telling somebody 'where I am' it would be just enough to send him/her a message. And your partner does not need to use any additional service in order to get information about your location. All the needed information will be simply delivered to him as a part of the incoming message.

It is obviously peer-to-peer sharing and does not require any social network. And it does not require one central point for sharing location with by the way. Our location signature has got a form of the map with the marker at the shared location. And what is important here – the map itself has no information about the sender and recipient. That information exists only in the message itself. The map (marker) has no information about the creator for example. That is all about privacy [1]

In terms of patterns for LBS this approach targets at the first hand such tasks as 'Friend finder' and the similar. In other words it is anything that could be linked to location monitoring.

The biggest danger of such systems is the recording of location information by service providers. Because every time a location update is shared, the service provider gets an update and is thus able to create detailed behavioral profiles of its customers (Google Latitude). As it is mentioned in [2] an ideal privacy-aware location sharing system should be able to share location information even without a central service provider receiving a copy of the entire movement track. It is exactly what Geo Message does.

Geo Messages approach works and really eliminates identity revealing problems but it is pure peer-to-peer. What can we do if we need to monitor several participants simultaneously? It is simply not very convenient to jump from one message to another.

Here we bring a new peer-to-peer service that solves the privacy issues and lets you deal with several location feeds simultaneously.

2   The Model

What if we split the locations and identity? In other words rather than using one server that keeps all our data we will switch to some distributed architecture.

WATN (Where Are They Now) [9] requires no sign-in. It combines anonymous server-side data with local personalized records.

We can separate location info and identity data just in three steps:

a) assign to any participant some unique ID (just an ID, without any links to the personality)
b) save location data on the server with links to the above-mentioned IDs
c) keep the legend (descriptions for IDs, who is behind that ID) locally

In this case any participant may request location data for other participants from third party server (as per sharing rules, of course), get data with IDs and replace IDs (locally) with legend's data. With such replacement we can show location data in the "natural" form. For example: name (nick) plus location. And in the same time the server (third party server for our users) is not aware about names.

What does it mean technically?

Server keeps two things.

a) location info with meaningless IDs:

ID1 -> (latitude, longitude)
ID2 -> (latitude, longitude)
ID3 -> (latitude, longitude)
Etc.

Just a set of current coordinates for users (presented via own IDs)

and

b) social graph – who is sharing location to whom:

ID1-> (ID2, ID3)
ID3 -> (ID1)
Etc.

Just a set of records states (as in example above) that user marked as ID1 shares location data with users ID2 and ID3

In the same time any local client keeps the own legend:

ID1 -> (name or nick)
ID2 -> (name or nick)

Note, that in this approach each client keeps own legend info. And because our clients are not aware about each other and there are no third party servers that know all registered clients. It means, obviously, that in this model the same ID may have different legends. Each client technically can assign own name (nick) for the same ID. Our social graph saves information (links between participants) using our meaningless IDs only. And the human readable interpretation for that graph can vary of course from client to client.

But that is probably very close to the real life, where the same person could be known under different names (nicks) in different contexts (e.g. compare some work environment and family space).

In general it is like keeping social graph, location and identity info in distributed database. But it is distributed on the level server-client rather than server-server.

On practice the structure could be a bit more elaborated. For example, in the current implementation we are saving the history - historical set of (latitude, longitude) pairs, we can keep some text messages associated with the current position etc. But it is just a set of features that does not changed the main idea – server-side store for anonymous location data and distributed client-side store with personal data.

3   The Algorithm

WATN has been implemented as mobile web application. HTML5 is significant there. Application uses W3C geo location [10] and local storage specification [11]. As per W3C documents HTML5 web storage is local data storage, web pages can store data within the user's browser.

Earlier, this was done with cookies. However, Web Storage is more secure and faster and our data is not included with every server request, but used only when asked for. It is also possible to store large amounts of data, without affecting the website's performance. The data is stored in key/value pairs, and web pages can only access data stored by them.

Storage is defined by the WhatWG Storage Interface as this:

```
interface Storage {
    readonly attribute unsigned long length;
    [IndexGetter] DOMString key(in unsigned long index);
    [NameGetter] DOMString getItem(in DOMString key);
    [NameSetter] void setItem(in DOMString key, in DOMString data);
    [NameDeleter] void removeItem(in DOMString key);
    void clear();
};
```

The DOM Storage mechanism is a means through which string key/value pairs can be securely stored and later retrieved for use. The goal of this addition is to provide a comprehensive means through which interactive applications can be built (including advanced abilities, such as being able to work "offline" for extended periods of time).

Mozilla-based browsers, Internet Explorer 8+, Safari 4+ and Chrome all provide a working implementation of the DOM Storage specification.

We are using local storage for saving legends for IDs as well as for the saving own ID.

As soon as the client calls the application we can restore his own ID from local storage (or obtains a new one from the server).

After that client saves location data on the server (it is check-in) and obtains shared location data (by the social graph). Server side part returns social graph data with ID's as JSON array. It is some like this:

```
[ {"id":ID1, "lat":lat1, "lng":lng1},
  {"id":ID2, "lat":lat2, "lng":lng2},

  ...

]
```

For our server-side database it is just a plain select (no joins) where our own ID is a key. It is very important, because complex database queries in geo systems can seriously affect the performance.

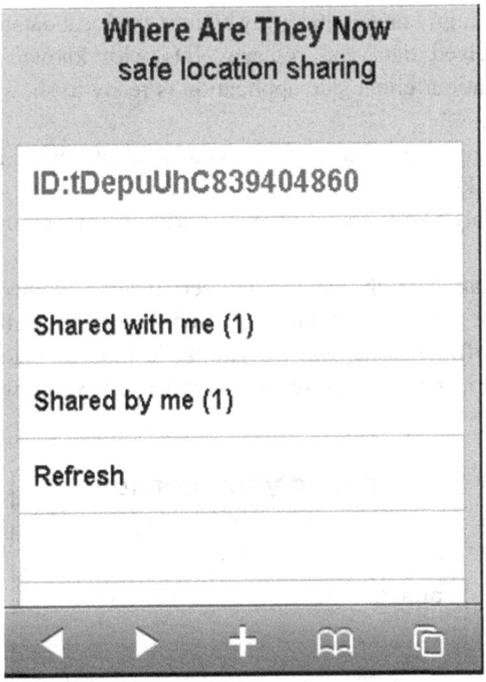

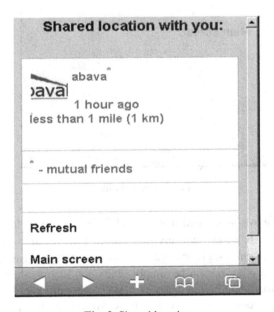

Fig. 1. Main screen

Fig. 2. Shared location

After that we can simply match that array against the local database with identities. Client modifies received data and replaces IDs with known names from local database. So, after that our client side application is ready to show location data with names instead of IDs.

If our system is unaware about some legend, than of course it shows "raw" ID instead of name or nickname.

We can see (control) who is sharing location with us, as well as who can read our location info.

Note, that using native JSON parsing and serialization methods provided by the browser, we can save the obtained data itself too. And technically it let us use the whole application in offline mode, playing with the last known data.

And by the similar manner we can see to whom we share our own location info:

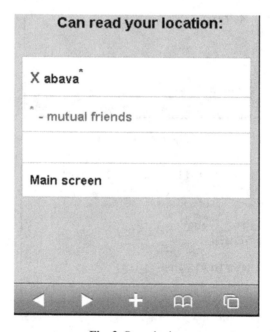

Fig. 3. Own sharing

Where are the above-mentioned names (nicks) for IDs come from? WATN uses peer-to-peer sharing. It means that any user shares own location to another person directly. There are no circles, groups, lists etc. As soon as some user is going to show own location info to other person he simply sends notification about this to another email address (phone number in case of SMS). Actually the location could be shared to any person with known email address. So, this notification plays a role of invitation too.

Such notification contains some text with explanation "what is it" and, what is obviously should be the main part of this process, a special link to WATN. This link contains an ID for the request originating party.

As soon as this link is fired, WATN application (client) becomes aware about two IDs: own ID1 for this client (it is restored from the local storage – see description above) and ID2 from the "shared with you" link (originated request ID). So, if notification is accepted, we can add social graph record (on the server) like

ID2 -> ID1

(client with ID2 shares own location info with the client with ID1. Or, what is technically equal, client with ID1 identity may read location info for client with identity ID2)

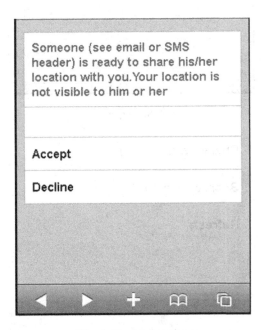

Fig. 4. Sharing location

But because the notification link comes from some message (email or SMS), the receiver is aware about the context. Simply, he knows either email header ('From') or phone number or name in address book SMS comes from. It means, that based on that info, our receiver may assign some nick (name) for ID in "shared with you" link. Actually it is a part of confirmation: confirm and set some name. And that name (nick) we can save locally. So, it is like "two phase commit" in databases – save a new social graph record on the server and create a new legend (record for identity) locally.

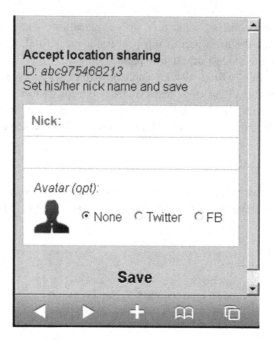

Fig. 5. Setting name

| Shared with me (1) |
| Shared by me (1) |
| Refresh |
| |
| Share |
| Leave a message |
| |
| Erase data |

Fig. 6. Data erasing

And as a source for profile images (remember – there is no registration and profiles) we can use social networks (e.g. Twitter and Facebook). If you set a nick for new share that corresponds to Twitter's (Facebook's) account, the system can attach public photo from the social network.

Of course, as per above described scheme, the mutual location sharing could not be set automatically. The message with location sharing link is email (SMS) delivered outside of this application. So the application itself is completely unaware who is sending sharing message to whom.

It is obviously also, that in this schema each client has got own legends. We can have different names for the same ID (each client can technically assign own name)

Additional options include messaging and data clearing. As soon as you share your location info, you can leave messages attached to your location. WATN users that can read your location data will see your messages too.

And any time you can delete ID (as well as erase all the associated data) from the system. Note, that in case of any reconnection in the future WANT will assign a new ID for the user. There is no way to reuse sometimes deployed ID. This feature also increases privacy.

4   Conclusion

This article describes a new model for sharing location info in mobile networks. The proposed approach creates a special form of distributed database that splits location info and identity information. In this distributed data store identity info is always saved locally where the social graph data store is still centralized. The proposed service eliminates one of the main problems with security in LBS applications: the need for third party server that keeps users identities and locations. This service could be described as a safety location sharing.

Service is implemented as HTML5 mobile web application and is compatible with all the modern HTML5 mobile web browsers (iPhone, Android, Bada, etc.).

We are not aware about any another LBS implementation with the similar features.

Acknowdgement. The paper is financed from EDRF's project SATTEH (No. 2010/0189/2DP/2.1.1.2.0./10/APIA/VIAA/019) being implemented in Engineering Research Institute «Ventspils International Radio Astronomy Centre» of Ventspils University College (VIRAC).

References

1. Namiot, D.: Geo messages. In: 2010 International Congress on Ultra Modern Telecommunications and Control Systems and Workshops (ICUMT), pp. 14–19 (2010), doi:10.1109/ICUMT.2010.5676665
2. Scipioni, M.P., Langheinrich, M.: I'm Here! Privacy Challenges in Mobile Location Sharing. In: Second International Workshop on Security and Privacy in Spontaneous Interaction and Mobile Phone Use (IWSSI/SPMU 2010), Helsinki, Finland (May 2010)

3. Wagner, D., Lopez, M., Doria, A., Pavlyshak, I., Kostakos, V., Oakley, I., Spiliotopoulos, T.: Hide and seek: location sharing practices with social media. In: Proceedings of the 12th International Conference on Human Computer Interaction with Mobile Devices and Services, Lisbon, Portugal, September 07-10 (2010)

4. Tsai, J.Y., Kelley, P., Drielsma, P., Cranor, L.F., Hong, J., Sadeh, N.: Who's viewed you?: the impact of feedback in a mobile location-sharing application. In: Proceedings of the 27th International Conference on Human Factors in Computing Systems, Boston, MA, USA, April 04-09 (2009)

5. de Paiva, A.C., Monteiro, E.F., Leal Rocha, J.J., de Souza Baptista, C., Silva, A.C.: Location Information Management in LBS Applications. In: Encyclopedia of Information Science and Technology, 2nd edn., pp. 2450–2455 (2009)

6. Duckham, M., Kulik, L.: A Formal Model of Obfuscation and Negotiation for Location Privacy. In: Gellersen, H.-W., Want, R., Schmidt, A. (eds.) PERVASIVE 2005. LNCS, vol. 3468, pp. 152–170. Springer, Heidelberg (2005)

7. Krumm, J.: Inference Attacks on Location Tracks. In: LaMarca, A., Langheinrich, M., Truong, K.N. (eds.) Pervasive 2007. LNCS, vol. 4480, pp. 127–143. Springer, Heidelberg (2007)

8. Gruteser, M., Grunwald, D.: Anonymous usage of location-based services through spatial and temporal cloaking. In: MobiSys 2003: Proceedings of the 1st International Conference on Mobile Systems, Applications and Services, pp. 31–42. ACM, New York (2003)

9. Langheinrich, M.: Privacy in ubiquitous computing. In: Krumm, J. (ed.) Ubiquitous Computing, pp. 95–160. CRC Press (September 2009)

10. WATN, http://watn.linkstore.ru (revised: March 2012)

11. Geolocation API Specification (September 7, 2010), http://www.w3.org/TR/geolocation-API/

12. Casario, M., Elst, P., Brown, C., Wormser, N., Hanquez, C.: HTML5 Solutions: Essential Techniques for HTML5 Developers 2011, pp. 281–303 (2011), doi:10.1007/978-1-4302-3387-9_11

Survey on Congestion Control Mechanisms
for Wireless Sensor Networks

Ekaterina Dashkova and Andrei Gurtov

University of Oulu, Centre for Wireless Communication
Oulu, Finland
{edashkov,gurtov}@ee.oulu.fi

Abstract. Congestion is very common in 6LoWPAN networks, but classical congestion control techniques do not suit well to the resource constrained environments. The main goal of this paper is to make an overview of the existing congestion control techniques for constrained environment and propose a need for the development of new flexible technique. It shall address all restrictions set by the environment and at the same time be generic enough. If the congestion control mechanism will be implemented at the application layer, i.e. as an improved extension of Constrained Application Protocol, then such solution will not be generic, as only this protocol will have modification. It would be more beneficial if the research will result in a new general solution suitable for the Constrained Application Protocol congestion control mechanism and other protocols.

In this paper an overview of congestion control techniques for constrained environment is made, scenarios of network performance, when congestion can appear and become a catastrophic are underlined. An idea of improving congestion control mechanism of Constrained Application Protocol is proposed.

Keywords: Congestion control, M2M, COAP extension, Scalability, Sensors, Ubiquitous, 6LOWPAN.

1 Introduction

Rapid development of wireless sensor networks (WSNs) pushes an idea of connecting WSNs to the Internet. It requires the new REST architecture that would satisfy restrictions of the resource constrained nodes (e.g., 8-bit microcontrollers with only a small RAM and ROM) with weak network connection (e.g., 6LoWPAN with the speed of 250 Kbit/s). To address this need the Constrained Application Protocol (CoAP) [1] has been proposed. It is a generic web protocol that satisfies special requirements of constrained environment, especially considering energy, building automation and other M2M applications. The CoAP protocol can be seen as an implementation of REST architecture for the specific environmental conditions, which is delivered in a number of ways, e.g. by compressing HTTP. The CoAP protocol is based on unreliable UDP transport layer, which does not provide internal congestion control (CC) mechanisms so the congestion control has to be provided by the upper layers.

S. Andreev et al. (Eds.): NEW2AN/ruSMART 2012, LNCS 7469, pp. 75–85, 2012.

Wireless sensor networks often experience significant congestion [2]. To study congestion control in large-scale networks that consist of small devices with tiny processors and small amount of RAM and ROM we have to make some assumptions concerning the observing system, and construct an effective mechanism to predict and avoid congestion.

We believe that a combination of Active Queue Management mechanism (AQM) - Beacon Order-Based Random Early Detection (BOB-RED) [3] and Explicit Congestion Notification (ECN) bits usage can be very effective combination for congestion control mechanism. The most of this solutions can be deployed by the routers or some other intermediate devices that behave as sub-network coordinators. We assume that these devices are not energy constrained and can handle most part of the congestion control mechanism calculations.

The rest of the paper is organized as follows: first we study limitations of the system in Section 2, define the scenarios in Section 3 then make survey on the existing methods in Section 4, and finally make our proposal in Section 5. Conclusion finalizes our work.

2   Background of the Study

On the first step we will describe the stack of protocols which is implemented to perform in the network (Fig. 1 is taken from the source [4]).

Layer	Protocol
Application	CoAP
Transport	UDP
Network	IPv6
Adaptation	6LoWPAN
MAC	IEEE 802.15.4
Physical	IEEE 802.15.4

Fig. 1. Stack of protocols

Physical and MAC layers are IEEE 802.15.4 [5] (it means that bandwidth is roughly 250Kbit/sec and maximum packet size is 127 bytes on the physical layer). Due to specification of IEEE 802.15.4 MAC layer uses CSMA/CA with optional TDMA mechanism [6]. On the network layer is IPv6 and transport layer is provided by unreliable UDP, both of them are compressed by Adaptation layer of 6LowPAN and at the top of the stack is CoAP application protocol. Below more detailed description of each layer is presented.

A. Physical and Data Link Layers

Implementation of the IEEE 802.15.4 standard of MAC layer [5] specifies two types of modes: non-beacon mode and beacon enable mode. In non-beacon mode, 802.15.4 uses CSMA/CA with optional TDMA mechanism; CCA (Clear Channel Assessment) is carried out before sending on the radio channel; if the channel is occupied, a node forced to wait for a random period of time, before trying to retransmit data one more time. In a beacon-enabled mode, a super-frame structure is introduced; time is divided into different transmission periods (Beacon, CAP (Contention Access Period), CFP (Contention-free Period) and inactive).

Survey on Congestion Control Mechanisms for Wireless Sensor Networks

B. Network Layer

On the network layer is IPv6 protocol. IPv6 header is 40 bytes long and this is too much for the IEEE 802.15.4 standard as after all deductions there is just 41 bytes left for the transport and application layers. Due to this problem an adaptation layer is used to compress IPv6 header to just 2 bytes.

Usage of ECN bits of the IPv6 header (Fig. 2) for congestion detection is described in details in [7] where one can find detailed description of ECN bits usage only in cooperation with TCP transport entity. One solution for the considering model is to provide capability of ECN approach to work in cooperation with UDP protocol. The first limitation is to make mandatory capability of ECN bits processing to all devices, in the network. One more direction research can follow is dropping old non-relevant information even during multicast to save network resources.

indentation in octets	indentation in bits	0 1 2 3 4 5 6 7	8 9 10 11 12 13 14 15	16 17 18 19 20 21 22 23	24 25 26 27 28 29 30 31
0	0	Version	Traffic Class	Flow Label	
4	32	Payload Length		Next Header	Hop Limit
8	64	Source Address			
C	96				
10	128				
14	160				
18	192	Destination Address			
1C	224				
20	256				
24	288				

Fig. 2. IPv6 header format

C. Transport Layer

Why is there UDP on the transport layer? UDP has several benefits that from the viewpoint of the constrained devices are quite important [6]:

- has a low overhead, its header is just 8 bytes;
- is well suitable for applications for which memory footprint is prioritizing;
- provides multicast delivery;
- UDP has several drawbacks as well:
- doesn't have any recovery mechanism from the packet loss, that occurs in the network quite regularly;
- doesn't have any congestion control mechanism leaving this function to the upper layers;
- doesn't have any mechanism to regulate the size of packets (it should be very small for 6LoWPAN network) – there is no segmentation and re-assembly mechanism.

WSNs are usually deployed by applications for which freshness of information is more crucial then its completeness. It is expected that application will periodically receive fresh data and it is more important to process just coming packets, then to wait for previous ones. And UDP protocol is more suitable in this case.

IPv6 and UDP headers are compressed by 6LoWPAN adaptation layer to be suitable for the limitations of the environment. More detail specification of 6LoWPAN compression can be found in [8].

D. Application Layer

On the application layer of the stack is CoAP protocol, which is easily translated to HTTP and satisfies all limitations of the constrained environment [1]. CoAP protocol already has some CC mechanism, proposed in [9]. One of the goals of the current research is to find a new way to improve its performance and do research in the field of congestion control mechanisms.

CoAP has some primitive congestion control technique described in [9]. It deploys primitive stop and wait mechanism as a retransmission technique, constant values of the retransmission counter and retransmission timer. This mechanism can be improved through using non-constant values of the retransmission timer and the retransmission counter maximum thresholds as well as minimum and maximum thresholds of the intermediate devices buffers. So, new mechanisms of calculating threshold can be implemented. This mechanism may focus on different parameters and techniques, such as Round Trip Time (RTT), analyzing Congestion Experienced bits from the IPv6 header, error rate, and active queue management algorithms.

E. Network Topology and General Assumptions

Assumption that network topology is star will help to study limits of intermediate devices and servers and amount of nodes they can support. The bottlenecks of our system are the channel that connects a group of sensors with the intermediate device and intermediate devices themselves (Fig.3).

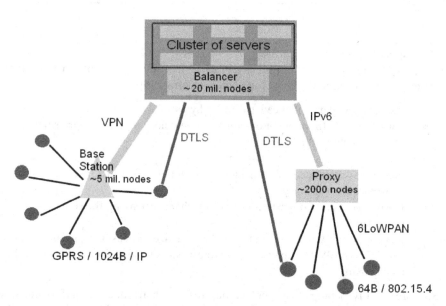

Fig. 3. The network architecture

It is important to experiment with different values of timeouts, amount of retransmissions and intermediate devices' buffer sizes. The main point is discovering an opportunity to connect as many sensors as possible with the intermediate device by reliable, congestion-free channel and use this channel effectively. Each device in the network has a unique 64-bit extended address or allocated 16-bit short address [5]. We assume that some nodes are battery equipped and some of them are not energy limited (sub-network coordinators). One additional consideration is mobility, and we assume that sensor nodes are static.

The congestion control mechanism in wireless sensor networks should be light and efficient at the same time. Congestion can appear in two main cases: the routers or intermediate device buffers are overflowed or channel collisions took place. Transport layer of the proposed model is unreliable UDP, so congestion control mechanism is going to the application layer and may be supported by information which is gathered and processed on the lower layers (a so called cross-layer solution).

To create a general solution for detecting and avoiding congestion in WSNs it is not enough to propose a new algorithm only on the application layer because in this circumstance innovation can be used only for one particular application case. We propose an idea of developing new mechanisms in congestion control that could later be interpreted as a general solution for the wireless sensor network stack (6LoWPAN networks). Extension of research in the area of congestion control for CoAP by cross-layer congestion control development gives better perspectives and matches the main goals of the research.

3   Scenarios of Congestion in Sensor Networks

First of all it is important to divide traffic on downstream (from the sink/server to the sensors) and upstream (from the sensors to the sink/server). Obviously, that the downstream traffic has one-to-many nature while upstream many-to-one. The upstream traffic can be classified into four categories: event-based, continues, query-based and hybrid [10].

It is easy to imagine a sensor network for example spread through a national park (the purpose can be just controlling temperature to avoid fire dissemination). Sensors are spread randomly through the territory and below we describe key drawbacks of this network. Devices are energy constrained, but it is hard to maintain them. Because of randomness it is hard to predict how many sensors will be connected with one router or some intermediate device which plays a role of the sub-network coordinator. There can be several scenarios of network performance, effluent from the hybrid data delivery including continuous, event-based, and query-based.

1)   Sensors regularly exchange service information or data and sleep for most periods of time. If our network is idle most of the time, we can consider that one intermediate device (proxy) can support 10000 of nodes and process all information from them correctly.

2)   Some event took place and a lot of nodes (if not all) in one particular area have to send information to the server. In this situation a congestion collapse can take place. All information will be important and should be delivered in time. "Emergency" data can give additional information as where the fire started and how it is spread. This information will be helpful not only for the decreasing extent of damage but also for investigation of the incidence. This scenario influences greatly on amount of intermediate devices in the network as if all sensors at once will become active. Then limits of the intermediate devices buffer size and processing capability will become critical parameters of the whole system.

3)   Server can query information at specific time, nodes can start to send
      information, interrupt each other, and collision in the channel will occur.
      Even if it will be enough intermediate devices and there won't be any buffer
      queues, this will definitely lead to congestion.

All sensors are divided between routers, proxies or some other intermediate devices
which play the role of sub networks coordinators [5]. At each moment sensor can
connect some sub network or disconnect from it (for example, one device can broke
up, while a new one can be sent to the territory).

4   Congestion Control Mechanisms

New research directions and resent solutions solving the congestion problem in
wireless sensor networks were studied [11] - [16]. The first fact that is underlined in
most part of discovered sources is that classical TCP-based congestion detection and
avoidance technique is not suitable for the wireless sensor networks, as it consumes a
lot of resources and is very aggressive from the view point of constrained devices and
unstable environment.

There are two types of congestion in wireless sensor networks: buffer congestion
and channel collision. Channel collision can be overcome using mechanisms
employed by the data link layer: Carrier Sense Multiple Access (CSMA), Frequency
Division Multiple Access (FDMA), Time Division Multiple Access (TDMA). These
mechanisms help to share medium through frequency division FDMA, time division
TDMA and sampling medium on the existence of the transmission of some other node
CSMA. Below these techniques will be described in more details.

A large number of techniques exist which were invented especially for the wireless
sensor networks. These methods are deployed by different layers of the OSI stack.

A. Data Link Layer Techniques
TDMA-based techniques as **Self-organizing Medium Access Control (SMACS)**
[17]; TDMA techniques should be included to the data link layer congestion control
mechanism as nodes have to switch-off for some time, to avoid idle listening and
through this avoid energy starvation of the device. This is an important case because
listening and transmitting are both very energy-expensive operations in a low-power
radio;

On-demand TDMA extension of IEEE802.15.4 MAC layer with priority-based
communication scheduling mechanism in nearby routing devices is described in [18].
This approach proposes an idea of extending existing active period of work, by using
additional communication period (ACP), in the inactive period of the standard
IEEE802.15.4 MAC superframe;

Hybrid TDMA/FDMA-based medium access [19] - this scheme balances
between optimal number of channels and gives the minimal power consumption;

CSMA. There are two most popular modifications of CSMA protocols: Carrier
Sense Multiple Access with Collision Avoidance (CSMA/CA) and Carrier Sense
Multiple Access with Collision Detection (CSMA/CD). Ethernet networks use
CSMA/CD mechanism as their links are full duplex, but wireless networks use

CSMA/CA, for half-duplex channels. One more difference between them is that CSMA/CA sampling channel by the jam signal and by this way preventing any data loss. When CSMA/CD use strategy of sending real data packets and only after collision occurs (two devices start transmission in one time) this algorithm conducts device to wait for random period of time and send a jam signal.

B. Network Layer Techniques

Beacon Order Based RED (BOB-RED) - Active queue management techniques such as BOB-RED are effective in networks with dozens of sensors connected to few intermediate devices (routers). BOB-RED in comparison with classical RED has strong advantage as it divides traffic on real-time and non-real time. With the help of such virtual queues it becomes easier to calculate priority of each particular piece of data and mark or drop packet when buffer overflows. Through marking packets because of the buffer overflow it becomes easier to inform sensors about congestion that router or some other intermediate device experiences. That can influence on the retransmission counter and retransmission timer values and slow down amount of upcoming packets to the congested intermediate node and filter emergency information. This approach consists of a virtual threshold function, a dynamic adjusted per-flow drop probability, a dynamic modification of beacon order (BO) and super-frame order (SO) strategy that decrease end-to-end delay, energy consumption, and increase throughput when there are different traffic type flows through the intermediate node [3]. Technique allows adapting BO and SO of each intermediate device individually to satisfy the requirements of each intermediate device in the WSN. BOB-RED assigns BOi, SOi for each neighbor node i, of the particular intermediate device. It is highly efficient in large-scale wireless sensor networks as deployed by the intermediate devices (which have more resources and calculating capabilities). Authors of [3] propose a relationship between qualities of service with the parameters (minimum and maximum thresholds, queue types and etc.) They divide all the factors influencing end-to-end delay into five levels depending on their values.

Several performance metrics were measured in [3]:

- average end-to-end can be decreased by decreasing the packet retransmissions;
- packet delivery ratio (PDR) which equals the ratio of received packets to the send packets is used for denoting the network performance;
- energy consumption which is measured only on the intermediate device.

With the help of numerous of simulation results and performance metrics analysis in [3] it was proven that BOB-RED can lessen the congestion by early detecting the queue status to drop packet to decrease the retransmission of arriving packets.

C. Transport Layer Techniques

Datagram Congestion Control Protocol (DCCP) developed by the IETF, the standard was accepted in year 2006 [20]. Recently it supports TCP-like congestion control mechanism and TCP-Friendly Rate Control (TFRC).

Pump Slowly Fetch Quickly (PSFQ) [21] - transport protocol, suitable for constrained devices. It includes three main functions: message relaying, relay-initiated error recovery and selective reporting. Main drawbacks of this approach are: it is not compatible with IP (minimal requirements on the routing infrastructure) and needs precise time synchronization between sensor nodes. PSFQ designed with the assumption that sensors application generates light traffic. This proposal opposite to the main use case we consider (preventing congestion collapse when network is under heavy load).

Sensor Transmission Control Protocol (STCP) – protocol focuses on the sensor transmission requirements – [22] it is a general protocol of the transport layer which satisfies requirements of the constrained devices. The most part of the functionality is realized on base stations or intermediate devices. Functionality includes mechanisms for early congestion detection and avoidance, variable reliability and support of several applications in one network.

Light UDP - [23] transport layer protocol, the main feature of which is that damaged packets are not dropped but delivered for the application layer for further analysis. This approach can be effectively deployed by applications for which delivery of all data has more priority than its integrity (multimedia protocols, stream video, voice IP). The main issue of this approach is that CheckSum field doesn't cover the whole packet but the current part of the header which is important for the future transmissions.

Reliable UDP - [24] transport layer protocol, the main feature of which is that it is working on the UDP/IP stack and provides reliable in order delivery. This protocol doesn't support classical congestion control technique or "slow start" mechanisms.

D. Techniques with Cross-Layer Nature

Fusion technique [25] combines tree mechanisms which cover different layers of the classical stack: hop-by-hop flow control (transport layer), rate limiting source traffic and prioritized data link layer that gives backlogged node priority over non-backlogged nodes for access to the shared medium.

Congestion Detection and Avoidance (CODA) [16] technique combines☐ three mechanisms: receiver-based congestion detection; open-loop hop-by-hop backpressure; and closed-loop multi-source regulation. As it is proved by ns-2 simulation results this mechanism can be very effectively deployed by event driven networks, which perform under the light load most of the time, but after some critical event become heavy loaded.

5   Outline of Our Approach

A lot of techniques exist, but because of the protocol stack that we have and specific scenario that we consider there can be incompatibility with some of these techniques. We propose an idea of using ECN bits in the IPv6 header together with CoAP acknowledgement and active queue management technique BOB-RED on the side of the intermediate devices such as routers.

There are several types of data [26]:

- regular data that is sent by the sensors. In this case, it will be more crucial to keep data up-to-date than to gather it all but with delays because of the retransmissions;

- sometimes data that is sent by the sensor contains information about event that occurs in the network. Delivery of this particular piece of information should have priority in the channel as due to the situation, even delay of such data can be dangerous;
- requests for some service information from sensors to the proxies/routers and servers have to be fully delivered as it is crucial for the further correct work of the whole network. This data can be delayed, as it is non-urgent. This data is important because sensors are going into sleep mode to avoid idle listening and spending energy resources;
- at the same time the flow of service information from the proxies, routers, other intermediate devices and servers has to be fully delivered to all sensors for the same reason listed above;

Due to the cases described above it is crucial to deliver as much data as possible, but at the same time remember about the relevance of the transmitted data and follow its priority.

We can consider several steps of congestion detection by examining the packet coming to the queue of the intermediate device:

1) If the queue is almost empty, then packet is accepted and waits for its turn to be processed and sent to the end point. After receiving such packet an end point replies with the CoAP acknowledgement.

2) If the queue is filled with some packets, but its length doesn't exceed the maximum threshold defined in the device, then the packet is marked with the Congestion Experienced bit and sent to the end point. The end point sends an acknowledgement but with the marked field (or just bit). This mark says to the sender the amount of retransmissions should be decreased by 1 and value of the retransmission timeout should be increased till the random value from the interval [1,5*recent_value; 1,9*recent_value]

3) If queue is filled with some packets, and its length exceeds the maximum threshold defined in the device, then the upcoming packet is dropped by the intermediate device. In this case end point doesn't receive anything and no acknowledgement will be sent. If on the sender side after the retransmission timer expires and there will be no acknowledgement received, the retransmission timer and retransmission counter should be halved.

This idea is quite simple and is based on the already existing technology (BOB-RED), but at the same it defines a small modification to the proposed technique. All static parameters became dynamic now, and their values change depending on the type of congestion.

Considering the security aspect of the system one more parameter was taken into account. Adversary can try to flood the network by "urgent traffic" and prevent network further correct work. In this scenario all resources of the network will be governed to process only "urgent traffic". This will lead to the situation when data from the "honest nodes" will be lost or delayed while adversary data will flood the network and can finally lead to the network collapse. To avoid this scenario one more limit should be taken into account, it is amount of the "urgent traffic" (real-time data) that can be transmitted by each sensor node. Due to this aspect there are two options that authors are considering:

- calculating ratio of the "urgent traffic" to the whole generated traffic of the node to prevent sending more "real-time data" than some fixed threshold;
- implementing specific timer after expiration of which a packet with "urgent" data can be sent.

Decision of which of two ways to choose is left for the future work and further discussions.

6   Conclusion

Wireless sensor networks often experience congestion, so an advanced congestion control solution is required. The CC mechanism should differ from its sibling deployed in the Internet. A lot of research and solutions were published targeted to solve the congestion problem in resource restricted communications. For our purpose most of them are not suitable because of the fixed protocol stack and assumptions concerning network topology and mobility.

We made a survey on congestion control mechanisms for wireless sensor networks and formulate an effective algorithm satisfying all underlined assumptions. The key approach is using BOB-RED active queue management mechanism to predict the overflow of the intermediate devices buffers. Modifications to the application layer CC mechanism were proposed. At the moment model for testing proposed technique is being built.

Future work is to develop and prototype the congestion control solution for CoAP, measure and evaluate its performance by simulations and in a large-scale sensor testbed.

Acknowledgments. Authors would like to thank the Massive Scale Machine-to-Machine Service (MAMMOTH) project and Tekes for financial support. We would like to thank Jussi Haapola for conducting the organization process of the research.

References

1. Shelby, Z., Hartke, K., Bormann, C.: Constrained Application Protocol (CoAP), Internet-Draft draft-ietf-core-coap-07, Expires: January 9 (2012)
2. Moon, S.-H., Lee, S.-M., Cha, H.-J.: A Congestion Control Technique for the Near-Sink Nodes in Wireless Sensor Networks. In: Ma, J., Jin, H., Yang, L.T., Tsai, J.J.-P. (eds.) UIC 2006. LNCS, vol. 4159, pp. 488–497. Springer, Heidelberg (2006)
3. Lin, M.-S., Leu, J.-S., Yu, W.-C., Yu, M.-C., Wu, J.-L.C.: BOB-RED queue management for IEEE 802.15.4 wireless sensor networks. EURASIP Journal on Wireless Communications and Networking (2011)
4. Kovatsch, M., Duquennoy, S., Dunkels, A.: A Low-Power CoAP for Contiki. In: Proc. of the IEEE Workshop on Internet of Things Technology and Architectures (October 2011)
5. IEEE Std 802.15.4-2011 IEEE Standard for Local and metropolitan area networks— Part 15.4: Low-Rate Wireless Personal Area Networks (LR-WPANs)
6. Vasseur, J.-P., Dunkels, A.: Interconnecting Smart Objects with IP. Morgan Kaufmann Publishers (2010) ISBN: 978-0-12-375165-2
7. Ramakrishnan, K., Floyd, S., Black, D.: The Addition of Explicit Congestion Notification (ECN) to IP, RFC 3168 (September 2001)

8. Montenegro, G., Kushalnagar, N., Culler, J.: Transmission of IPv6 Packets over IEEE 802.15.4 Networks, RFC 4944 (September 2007)
9. Eggert, L.: Congestion Control for the Application Protocol (CoAP). Internet Draft draft-eggert-core-congestion-control-01, Expires July 31 (January 27, 2011)
10. Wang, C., Li, B., Sohraby, K., Daneshmand, M.: Upstream Congestion Control in Wireless Sensor Networks Through Cross-Layer Optimization. IEEE Journal on Selected Areas in Communications 25(4), 786–795 (2007)
11. Michopolous, V., Guan, L., Oikonomous, G., Phillips, I.: A comparative study of congestion control algorithm in IPv6 wireless sensor networks. In: Proc. of the 3rd International Workshop on Performance Control in Wireless Sensor Networks (June 2011)
12. Chakravarthi, R., Gomathy, C., Sebastian, S.K., Pushparaj, K., Mon, V.B.: A survey on congestion control in wireless sensor networks. International Journal of Computer Science and Communication 1(1), 161–164 (2010)
13. Lunden, M., Dunkels, A.: The Politecast Communicative Primitive for Low-Power Wireless. In: ACM SIGCOMM, vol. 41(2) (April 2011)
14. Heikalabad, S.R., Ghaffari, A., Hadian, M.A., Rasouli, H.: DPCC: Dynamic Predictive Congestion Control in wireless sensor networks. IJCSI International Journal of Computer Science Issues 8(1), 472–477 (2011)
15. Sheu, J.-P., Chang, L.-J., Hu, W.-K.: Hybrid Congestion Control Protocol in Wireless Sensor Networks. Journal of Information Science and Engineering 25, 1103–1119 (2008)
16. Wan, C.-Y., Eisenman, S.B., Campbell, A.T.: CODA: Congestion Detection and Avoidance in Sensor Networks. In: Proc. of SenSys 2003 (November 2003)
17. Sohrabi, K., Gao, J., Ailawadhi, V., Pottie, G.J.: Protocols for Self-Organization of a Wireless Sensor Network. In: Proc. of the 37th Allerton Conference on Communication, Computing and Control (September 1999)
18. Zhong, T., Zhan, M., Hong, W.: Congestion Control for Industrial Wireless Communication Gateway. In: Proc. of International Conference of Intelligent Computation Technology and Automation 2010, pp. 1019–1022 (2010)
19. Shih, E., Cho, S.H., Ickes, N., Min, R., Sinha, A., Wang, A., Chandrakasan, A.: Physical Layer Driven Protocol and Algorithm Design for Energy-Efficient Wireless Sensor Networks. In: Proc. of ACM Mobicom 2001, pp. 272–286 (July 2001)
20. Kohler, E., Handley, M., Floyd, S.: Datagram Congestion Control Protocol (DCCP), RFC 4340 (March 2006)
21. Wan, C.-Y., Campbell, A.T., Krishnamurthy, L.: PSFQ: a Reliable Transport Protocol for Wireless Sensor Networks. In: Proc. of First ACM International Workshop on Wireless Sensor Networks and Applications (WSNA 2002), USA, Atlanta, pp. 1–11 (September 2002)
22. Iyer, Y.G., Gandham, S., Venkatesan, S.: STCP: A Generic Transport Layer Protocol for Wireless Sensor Networks (2005)
23. Larzon, L-A., Degermark, M., Pink, S., Fairhurst, G.: The Lightweight User Datagram Protocol (UDP-Lite), RFC 3828 (July 2004)
24. Bova, T., Krivoruchka, T.: Reliable User Datagram Protocol. Internet Draft draft-ietf-sigtran-reliable-udp-00, Expires August 1999 (February 1999)
25. Hull, B., Jamieson, K., Balakrishnan, H.: Mitigating Congestion in Wireless Sensor Networks. In: Proc. of SenSys 2004 (November 2004)
26. Ilyas, M., Mahgoub, I.: Handbook of Sensor Networks: Compact Wireless and Wired Sensing Systems. CRC Press (2005); Su, W., Cayirci, E., Akan, Ö.B.: Overview of Communication Protocols for Sensor Networks
27. Gurtov, A., Ludwig, R.: Lifetime packet discard for efficient real-time transport over cellular links. ACM Mobile Computing & Communications Review 7(4), 32–45 (2003)

On IEEE 802.16m Overload Control
for Smart Grid Deployments

Vitaly Petrov[1], Sergey Andreev[1], Andrey Turlikov[2], and Yevgeni Koucheryavy[1]

[1] Tampere University of Technology
Korkeakoulunkatu 1, FI-33720 Tampere, Finland
{vitaly.petrov,sergey.andreev}@tut.fi, yk@cs.tut.fi
[2] State University of Aerospace Instrumentation
Bolshaya Morskaya 67, 190000 St. Petersburg, Russia
turlikov@vu.spb.ru

Abstract. The evolved vision of the Internet-of-Things predicts tremendous growth in the number of connected human-unattended objects over the following years. However, existing wireless network protocols were not originally designed for such excessive numbers of nodes. Therefore, currently deployed networking solutions are not optimized for Machine-to-Machine (M2M) communications. To bridge in this gap, IEEE 802.16m cellular technology targets extensive improvements to support M2M communications. As such, performance assessment of an 802.16m-based network with respect to the expected M2M scenarios is topical. In particular, the contention-based procedure for network admission control is very likely to become a bottleneck for the entire system. Therefore, we focus performance assessment of the standardized collision resolution algorithm, used in the random access channel, with respect to M2M. In particular, this paper addresses a Smart Grid use case, where a large number of devices access the network within a relatively short period of time. We discuss specific performance requirements that apply to the considered scenario. Then, an evaluation methodology to investigate the adaptability of the 802.16m system is presented. Finally, we conclude on the capability of the 802.16m technology to support an excessive number of human-unattended devices.

1 Introduction and Background

1.1 General Background

Generally, the Internet of Things (IoT) is a technology trend, that refers to a future Internet, where diverse objects and locations are equipped with sensors, actuators, and processors, made discoverable though the network and able to communicate both with humans and each other [1]. In fact, such communications between human-unattended devices or *Machine-to-Machine (M2M)* communications are the crucial novelty of the trend. Implementation of this technology helps humans deal with the routine and, therefore, looks very promising and

S. Andreev et al. (Eds.): NEW2AN/ruSMART 2012, LNCS 7469, pp. 86–94, 2012.

attractive. Responding to the consumer needs, and due to the huge market potential of IoT, many companies develop their own solutions, both on hardware and software sides.

So far, the amount of M2M capable devices is predicted to reach about 7 trillion by 2020 [2], that refers to a *thousand* devices per human being. With respect to this explosion, leading standardization committees (IEEE [3] and 3GPP [4]) are now focusing the development of the protocol enhancements to support M2M communications. In particular, IEEE develops the 802.11ah [5] standard, that primary targets M2M communications. Simultaneously, several Work and Study Items (including [6], [7] and [8]) are being open in 3GPP to investigate possible enhancements for M2M communications in future releases of HSPA+ [9] and LTE-Advanced [10].

1.2 Motivating Scenarios

Being a more mature technology, IEEE 802.16m [11], is however expected to better face the challenges of M2M communications. In particular, emerging IEEE 802.16p proposals address enhancements for the core IEEE 802.16m standard to extensively support M2M applications. These enhancements primary embrace the following motivating scenarios [12]:

- **Secured Access and Surveillance.** M2M applications of this category help prevent the unauthorized access into buildings, vehicles, and locations. Smart areas are equipped with intrusion detection sensors, that automatically send an alert signal to the M2M server in case of emergency.
- **Tracking, Tracing, and Recovery.** The respective use cases are mainly related to services that rely on location-tracking information. In particular, vehicles are equipped with M2M devices that send status data (e.g. location, velocity, local traffic, etc.) periodically or on-demand to the M2M server. The collected information is then analyzed by the M2M application server. Analysis results can be further provided to the end user via wired or cellular network. The examples of the emerging vehicular tracking services are navigation, traffic information, road tolling, automatic emergency call, pay as you drive, etc. The distinct feature of these scenarios is that the M2M application server needs to monitor the status and/or position of an individual vehicle or group of vehicles.
- **Payment.** The huge growth in M2M communication increases the level of flexibility in deploying point-of-sale/ATM terminals, including vending and ticketing machines, parking meters, etc. Whenever it is possible, such systems may perform without any human interaction. In addition, wireless M2M also allows payment facilities to overcome the lack of wired infrastructure.
- **Healthcare.** One of the most promising M2M applications unites different mechanisms now used for patient monitoring and tracking, doctor responsiveness, and hospital maintenance. Moreover, healthcare M2M devices might help patients with advanced age, complicated physical conditions, and chronic diseases to overcome limitations and live the full life. Information

from the body medical sensors can be transmitted in real-time to the M2M-capable healthcare management system and analyzed there, providing remote patient monitoring.

- **Smart Grid.** This category includes applications primarily used in energetics to inform the maintainer of its equipment status. By implementing the M2M-aware infrastructure the efficiency of remote maintenance and control services might be significantly improved. Also the proper meter readings could be sent to the customer or even used by the automated system in order to optimize the resource utilization and minimize total expenses.

1.3   Network Entry Issue

Among the mentioned scenarios, smart grid is one of the most important, providing huge benefits to the consumer as well as to the utilities. Such smart grid applications as network automation, distributed energy resources and storage, and control and support for electric vehicles, are very likely to be deployed in the near future. These applications require multiple connected devices, including meters, sensors, and terminals. Given that such applications are generally characterized by infrequent and delay-tolerant traffic, the appropriate use of existing network design can support the aggregate traffic from up to 35000 devices per sector [14], which is sufficient in typical situations.

However, there are certain cases, when near simultaneous network entry by the large number of devices occurs, such as power outage event or alarm reporting [15]. Then the highly correlated network entry attempts by excessive number of devices may easily overwhelm the dedicated network capacity.

In this paper, we indicate the bottlenecks of IEEE 802.16m protocol assuming the highly correlated network entry attempts by large number of devices. Our results are obtained by protocol-level simulation with respect to the methodology described in [12].

2   Considered Scenario

2.1   Network Topology

In this paper, a single cell of IEEE 802.16m network is considered. The M2M devices are placed randomly within the cell and communicate with the base station directly, when needed, without using any kind of relays to forward the traffic (see Figure 1).

2.2   Revision of Network Entry Procedure

Random-based network access procedure in IEEE 802.16p [13] inherits the respective scheme by IEEE 802.16m and is performed as follows. Firstly, the client waits for two service messages from the base station: *Downlink Channel Descriptor (DCD)* and *Uplink Channel Descriptor (UCD)*. Then if it would like to establish a connection with the network, it initiates the Backoff procedure

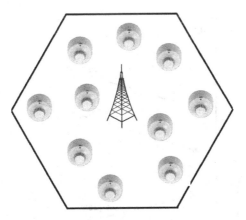

Fig. 1. Considered network structure

and randomly select a slot to transmit its connection request. The base station attempts to decode this request and if neither collision nor error occurs then it answers with an ACK. If the base station is not able to decode a particular request,ACK is not to be sent, and after a timeout the client starts retransmission. More detailed illustration of the client connection process is given in Figure 2).

2.3 Metrics

In our analysis, the following performance metrics of the network entry process are considered:

1. **Conditional collision probability** — probability of connection request collision given that a client transmits it.
2. **Access success probability** — probability of successful network entry within the allowed number of transmission attempts.
3. **Mean entry delay** — mean network entry delay regardless of entry success or failure.
4. **Conditional mean entry delay** — mean network entry delay for a client if it accesses the network successfully.
5. **Mean number of retries** — mean number of request retransmissions to access the network successfully.

3 Numerical Results

3.1 Simulation Methodology

To obtain the numerical results, a time-driven protocol level simulator was developed. It was designed in a scalable and configurable manner, in order to allow flexible changes to signaling, timings, and other controlled parameters.

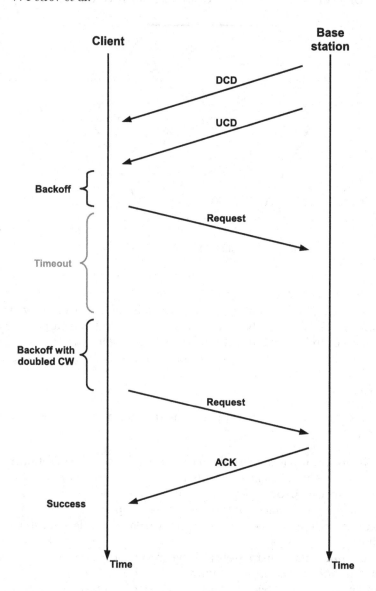

Fig. 2. Network entry signalling in 802.16p

During the simulations, highly correlated network entry attempts by a very large number of clients (up to 30K) were studied. Also the system was explicitly configured to support the maximum number of access opportunities per second. With respect to [12], the request arrival process at the clients was assumed to follow the Bernoulli distribution with the following arrival rates: 1/60s, 1/10s, and 1/0.5s. More system parameters could be obtained from Table 1.

When the simulation starts, all the clients are reset to the idle state and are waiting for a request. Then, after receiving DCD and UCD messages from the base station a client with a request to send starts a backoff procedure and randomly chooses a slot for the initial transmission. Depending on the transmission outcome: *success*, *collision*, or *error*, the client either increases the backoff window and starts retransmission, or returns to the idle state waiting for a new request (see Figure 3).

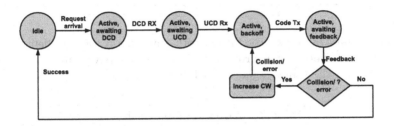

Fig. 3. Client network entry state machine

Table 1. Simulation parameters

Parameter	Value
Bandwidth	5MHz
CP	1/4
Number of ranging channel opportunities per second	6400
Decoding error probability	0.01
Nodes number	10K, 30K
Mean arrival rates	1/60s, 1/10s, 1/0.5s
Arrival distribution	Bernoulli
Initial backoff window	1
Final backoff window	16
Maximum number of retries	16
Waiting time before retransmission	1 frame
BS processing latency	1 frame

3.2 Obtained Results

Random-based network entry mechanism in IEEE 802.16m was not originally designed to support a surge in access attempts by a large number of devices. Therefore, even with the maximum number of access opportunities, the conditional collision probability increases dramatically, when the arrival rate is high enough (see Table 2).

Table 2. Conditional collision probability

Arrival rate per M2M client per second	10K devices	30K devices
1/60	0.17	0.35
1/10	0.5	1
1/0.5	1	1

The probability of the successful network entry for a tagged client is presented in Table 3. Note that it drops sharply when the arrival rate grows and is extremely low for the values higher than 1/0.5s.

Table 3. Access success probability

Arrival rate per M2M client per second	10K devices	30K devices
1/60	1	1
1/10	1	0.05
1/0.5	0	0

Such low values of access success probability prohibit network entry for most clients. Therefore, the mean network entry delay and number of retries are high, especially for 30K of M2M clients (see Tables 4 and 5, respectively).

Table 4. Mean delay, ms

Arrival rate per M2M client per second	10K devices	30K devices
1/60	30	30
1/10	50	650
1/0.5	680	680

Table 5. Mean number of retries

Arrival rate per M2M client per second	10K devices	30K devices
1/60	1.4	1.8
1/10	2	2
1/0.5	16	16

With respect to the results presented above, the arrival rate of 1/10s per a single M2M device can be considered as a turning point in the system behavior.

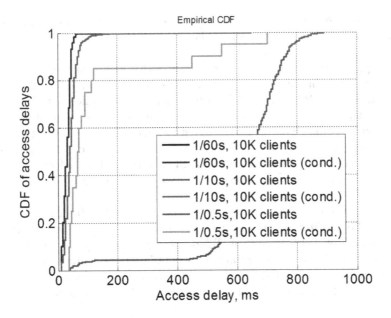

Fig. 4. Empirical CDF for access delays

The CDF of the access delays for such arrival rate is presented below (see Figure 4). Conditional delays are plotted for the users, which are successful in network entry.

3.3   Results Analysis

Our results summarized in the previous section demonstrate that the current system capacity can be overwhelmed with the arrival rates higher than 1/10s. It practically means that with the traffic load of greater than 6 packets per minute from a single M2M client, the communication through the network becomes un-reliable. According to the scenarios presented in Section 1, such traffic patterns can easily occur during a correlated network entry in typical smart grid deployments. Therefore, some further work within IEEE 802.16p is needed to mitigate the effect of the indicated bottleneck.

4   Conclusions

Generally, the concept of the Internet of Things and M2M communications present a new perspective view on the Internet of the future, where non-human oriented applications will play a more significant role. In particular, the smart grid use case promises huge benefits to both consumer and industry. However, the support for M2M communications in the contemporary wireless networks is a challenging task primarily in light of the changed traffic patterns. Whereas some recent releases of wireless networking standards are being adopted for M2M communications, there are still several bottlenecks that prevent the ubiquitous deployment of M2M networks.

In this paper, the limitations of the IEEE 802.16m network entry protocol were studied while handling a surge in near-simultaneous access attempts from a large number of M2M devices. Our findings demonstrate differences in network behaviour, depending on the arrival rate, and also highlights the requests intensity network is stable. The results reported in this paper have been important in supporting our contributions to the development of IEEE 802.16p technology [14], [15] and emphasize the need for further research on effective ways to de-correlate extensive access requests from many M2M devices.

Acknowledgments. This research was conducted in the Internet of Things program of Tivit (Finnish Strategic Centre for Science, Technology and Innovation in the field of ICT), funded by Tekes. The authors are also grateful to Dr. Nageen Himayat (Wireless Commun. Lab./Commun. Tech. Lab., Corp. Tech. Group, Intel Corp., Santa Clara, CA, USA) for initializing their interest in this topic, as well as for helpful discussions and insightful comments in the course of this work.

References

1. Finnish Strategic Centre for Science, Technology, and Innovation, Internet of Things Strategic Research Agenda (IoT–SRA) (2011)
2. Vodafone, Vodafone M2M usaceses prediction,
   http://m2m.vodafone.com/images/2011/videos/m2m_city/index.jsp
3. IEEE, http://www.ieee.org/index.html
4. 3GPP, http://www.3gpp.org/
5. Status of Project IEEE 802.11ah. Sub 1GHz,
   http://www.ieee802.org/11/Reports/tgah_update.html
6. 3GPP SI, Study on RAN improvements for MTC
7. 3GPP WI, Study on enhancements for MTC
8. 3GPP WI, System Improvements to MTC
9. Evolved High-Speed Packet Access, http://www.3gpp.org/HSPA
10. LTE-Advanced, http://www.3gpp.org/lte-advanced
11. IEEE 802.16m-2011, Advanced Air Interface (2011)
12. Cho, H.G.: M2M Communications Technical Report, IEEE 802.16p-10/0005
13. IEEE 802.16's Machine-to-Machine (M2M) Task Group
14. Himayat, N., Talwar, S., Johnsson, K., Andreev, S., Galinina, O., Turlikov, A.: Proposed 802.16p Performance Requirements for Network Entry by Large Number of Devices. IEEE 802.16 C80216p-10/0006 (2010)
15. Himayat, N., Mohanty, S., Srivasan, R., Talwar, S., Johnsson, K., Andreev, S., Galinina, O., Turlikov, A.: Proposed Performance Requirements for Network access with Large Number of Devices. IEEE 802.16 C80216p-10/0036 (2011)

An Overview of Information Extraction from Mobile Wireless Sensor Networks

Abdelrahman Abuarqoub, Mohammad Hammoudeh, and Tariq Alsboui

School of Computing, Mathematics & Digital Technology
Manchester Metropolitan University
Manchester, UK
{a.abuarqoub,m.hammoudeh,tariq.al-sboui}@mmu.ac.uk

Abstract. Information Extraction (IE) is a key research area within the field of Wireless Sensor Networks (WSNs). It has been characterised in a variety of ways, ranging from the description of its purposes, to reasonably abstract models of its processes and components. There has been only a handful of papers addressing IE over mobile WSNs directly, these dealt with individual mobility related problems as the need arises. This paper is presented as a tutorial that takes the reader from the point of identifying data about a dynamic (mobile) real world problem, relating the data back to the world from which it was collected, and finally discovering what is in the data. It covers the entire process with special emphasis on how to exploit mobility in maximising information return from a mobile WSN. We present some challenges introduced by mobility on the IE process as well as its effects on the quality of the extracted information. Finally, we identify future research directions facing the development of efficient IE approaches for WSNs in the presence of mobility.

Keywords: Mobile, Wireless Sensor Networks, Information Extraction, Information Attributes, Mobile Nodes, Data Collection.

1   Introduction

The main goal of Wireless Sensor Networks (WSNs) is to collect data from the environment and send it to end user's applications, where it is analysed to extract information about the monitored conditions. The success of such applications is dependent on knowing that the information is available, the type of information, its quality, its scope of application, limits to use, duration of applicability, likely return, cost to obtain, and a host of other essential details. Information Extraction (IE) is a practical multistage process at which end user's applications operate using a structured methodology to discover and extract the information content enfolded in data [1]. Every stage has a particular purpose and function. This paper gives the reader a feel for the process: what goes in, what goes on, and what comes out of collected data. While much of this discussion is at a conceptual level, it offers some practical advice on how to measure the quality of information and covers the main concerns and interrelationships between the stages.

S. Andreev et al. (Eds.): NEW2AN/ruSMART 2012, LNCS 7469, pp. 95–106, 2012.

To reduce communication costs, many concepts have been developed in the context of distributed WSNs such as data aggregation and fusion, e.g. [2]. The in-network processing of sense data has led to clear distinction between data and information. For this reason there is no set and commonly agreed definitions of information and information attributes. The variety of definitions identified in the literature stem from the highly context specific nature of databases and data mining fields and the complexity of their operationalisation and conceptualisation. Differences in the definitions can lead to a variance in research focus and performance evaluation comparison approaches. Out of this need, we propose standard, unified definitions of information and information attributes; the particular definitions adopted by this study will depend on the mobile WSNs (mWSNs) discipline and level of investigation.

To address the need for common definitions of some of the issues surrounding the concept of IE from mWSNs we define data as a collection of unstructured chunks of facts and statistics measured by sensor nodes. Information is the processed and interpreted data that provides conceptual explanations of data, i.e. it converts information encapsulated in data into a form amenable to human cognition [1]. Data comes in a variety of forms and carries different amounts of information. For instance, scalar sensor measurements from a proximity sensor about a monitored object can be limited to just 'present' or 'not-present'. In this case the user will see less information than the 'very close', 'close', 'far', 'very far', and 'not-present'.

IE is considered a costly task as it involves the collection and processing of often large amounts of unstructured or semi-structured sensor data. There is a wide body of literature about IE from static WSNs; we refer interested readers to [3], a recent survey that provides a comprehensive review to IE approaches. However, there has been few attempts to address the problem of IE from mWSNs, e.g. [4, 5]. These approaches dealt with mobility as the need arises and do not attempt to deal with the fundamental challenges and variations introduced by mobility on the WSNs. This paper focuses on the unique characteristics of mWSNs and sheds light on their effect on the quality of the extracted information. We believe that better understanding the above issues will help researchers in developing more efficient approaches to IE from mWSNs.

The rest of this paper is organised as follows: Section 2 describes the processes involved in converting data to information. Section 3 presents and defines the attributes of mWSNs information. Section 4 identifies the benefits of introducing mobility to WSNs. Section 5 presents the challenges inherent with introducing mobility to WSNs. Section 6 discusses some future research directions and concludes the paper.

2   The Process of Converting Data to Information

WSNs have large numbers of sensor nodes to provide full coverage of the monitored area. Consequently, the WSN returns large volumes of data that is imperfect in nature and contains considerable redundancy [6]. This data needs to be processed in order to extract information relevant to a user query. Converting data to information goes into variant and nested stages of processing, including: data retrieval, filtering, collection, and processing. In the following, we explain these processes in detail.

Data Retrieval (DR): The DR process begins by specifying the information needed by a user/application. Often, this takes the form of a query or event trap. DR selects a relevant subset of data or nodes carrying data that is relevant to some required information from a larger set. Specifically, it identifies the nodes that carry data that is significant to the needed information. Nodes can be classified based on their soft-state (sensor readings), e.g. [7], or on their physical location when the information is spatial in nature, e.g. [8]. For example, consider a situation where the end user is interested in locations where the temperature is above 50 °C. Identifying nodes carrying data that satisfy this condition has many benefits. First, it becomes easier to choose the best path over which to transmit data - in terms of energy consumption, link reliability, and end-to-end delay. Second, if relevant data can be identified, then the others can be abandoned. This reduces the load on the rest of the system and improves information accuracy.

Data Filtering (DF): The DR process returns data that is erroneous and redundant; if left untreated, it could affect the accuracy of the final IE process results. The DF is mainly used to remove redundant or unwanted data from a data stream. Data aggregation, fusion and compression are clearly candidates for this task, but other techniques from active networking could also be used. Continuous and cumulative sensory readings can be filtered instantly in the network at different levels, e.g. at cluster heads or at a sink, to avoid the expensive transmission of inappropriate data. Essentially, filtering attempts to trade off communication for computation to reduce energy consumption since communication is the most energy-expensive task. Moreover, a filter may attempt to remove data, which are artifacts of the DR process.

Data fusion has been placed forward as a technique to improve bandwidth utilisation, energy consumption, and information accuracy [9]. It combines and integrates multi-sensor and multi-sourced data to produce higher accuracy and comprehensive information [10]. The fusion technique achieves high information accuracy by fusing redundant observations. To produce comprehensive information, the fusion technique fuses readings from different sensors that are related to the same event. For example, in road monitoring applications, to discover whether there is a frost or not, a combination of temperature, humidity and solar radiation data is needed. However, since WSNs are resource limited, data fusion techniques has to be resource efficient to be deployed in such networks. Furthermore, WSNs are application dependent networks, hence, it is infeasible to devote one data fusion technique to work with all situations. There has been persistent research efforts to develop data fusion algorithms that suites several applications [2, 11].

Data fusion is usually coupled with another process called data aggregation. Data aggregation is the process of combining data coming from different sources in order to eliminate redundancy and minimise the number of transmissions [12]; thereby, conserving the scarce energy resources. Data aggregation is affected by the way that data is gathered at the sensor nodes and the way that packets are routed through the network [13]. Aggregation techniques can be broadly classified into two classes, lossless aggregation and lossy aggregation [14]. The former is when multiple data segments are merged into a single packet in order to reduce bandwidth utilisation. In the later, data is combined and compressed by applying statistical processes, such as average, minimum, and maximum, before transmission in order to keep the number of transmissions as low as possible.

Data Collection (DC): DC is the most energy-expensive process of IE. There are two abstract models of DC, central at a sink, or hierarchical at cluster heads. In mWSNs, the data collection model is tightly coupled with how data is routed through the network, patterns of node mobility, and the underpinning communication paradigm. DC from mWSNs can be classified based on the nature of mobility into three classes: flat-tier; two-tier; and three-tier [15].

A flat-tier model, consists of a set of homogeneous sensor nodes, which can be static or mobile. Sensory data is routed from the originator sensors to a central sink in a multi-hop ad-hoc fashion. Centralised data collection is not desired because it increases the delivery delay of the extracted information, it causes communication bottlenecks around the sink, and it is not suitable for large networks.

In the two-tier hierarchical model, static sensor nodes occupy the bottom tier and mobile nodes occupy the top tier. Mobile nodes act as data collectors that move towards the sink to deliver their data. However, moving data collectors all the way to the sink stops the collection process for a while causing coverage holes. This results in high information delivery delay, which could negatively impact the validity of the extracted information. Moreover, coverage holes could result in reduction of information availability.

In the three-tier hierarchical model, static nodes occupy the bottom and the top tiers. Whereas the middle tier hosts the mobile nodes, which act as access points. The mobile nodes collect sensory data from the bottom tier and forward it to the top tier. The top tier delivers the data to the sink. This model could greatly improve the overall system performance as it reduces the data delivery delay, and hence, increases the information validity. Furthermore, since the distances covered by the mobile nodes become shorter, the number of coverage holes will be reduced, which results in higher information availability.

Data Processing: Data processing is the stage before passing information to end users or application. It refers to a class of programs to organise, sort, format, transform, summarise, and manipulate large amounts of data to convert them into real-world information. These programs define operations on data such as algorithmic derivations, statistical calculations, and logical deductions that exists in the user application. This stage can be followed by other processes like information visualisation and display. Data processing includes a second phase of data filtering and could include a second iteration of data fusion or integration. Data processing can be performed centrally at the data sink, hierarchically on cluster heads or in a distributed manner [16].

In centralised processing models, data is first gathered at a sink, where data processing is applied. This approach produces high quality information as the entire network data is used for extracting the information. However, a fully centralised data collection and processing is not always feasible. This is because centralised data processing incurs a significant data transfer cost and introduces delay to information delivery. However, some information is of a spatio-temporal nature, which requires centralised data collection and processing; extracting information for characteristics of such nature where local processing is not enough should be feasible.

To solve the apparent problems posed by the centralised model, hierarchical data processing was proposed. It exploits local processing resources to effectively reduce the amount of data transmitted across the network. In this approach, sensor nodes are divided into multiple clusters. Measurements are transferred and processed on sparsely distributed cluster heads. These cluster heads either send processed data to a fusion centre for decision making or collaborate with each other to make decisions. The hierarchical approach helps in reducing information delivery delay as data processing is performed on multi processors simultaneously. Moreover, since the amount of data transferred across the network is reduced, the energy usage will be reduced leading to improved information affordability.

In some scenarios, it is desirable or necessary to process data on site and, as a result, distributed processing provides a critical solution to in-field data analysis. Distributed data processing was proposed to exploit sensor nodes computational power. Sensor nodes process the data and collaborate to transform the data into information. Therefore, processing takes place in the network and only the results are returned. This approach improves the extracted information timeliness by distributing the computational work over all the nodes in the network. However, the information accuracy could be degraded due to lack of computational resources.

Fig. 1 shows an illustration of the described processes. It shows the cooperation between the major entities and processes involved in the IE process.

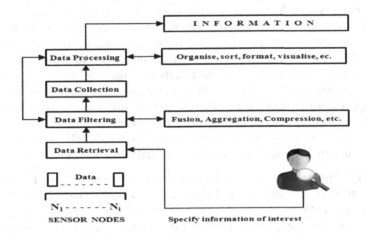

Fig. 1. A general model of IE process

3   Attributes of mWSNs Information

In this section, different attributes for mWSNs information are identified and defined. Most of these attributes have been used constantly in the literature. However, inconsistencies in these definitions have led to problems in measuring the quality of extracted information. Hence, standard definitions are needed in to facilitate the comparison of different approaches. In this paper, we only choose the attributes that are relevant and useful in evaluating the quality of extracted information from mWSNs.

Accuracy: Accurate information allows the end user to take correct actions by providing a realistic reflection of the actual sensed environment. The term accuracy, also known as correctness, has been widely used in the Quality of Information field [17]. The authors in [18] define accuracy as the level of detail (precision) in the sensed data. Similarly, in [17], accuracy is the degree of correctness, which provides the level of detail in the network. However, the above definitions do not differentiate between data accuracy and information accuracy. This problem has evolved from the confusion of the terms data and information. Moreover, the level of details is controlled by the user and missing a detail does not necessarily affect the reported information accuracy. The accuracy of information is not only achieved by the accuracy of data, data processing models can significantly affect the information accuracy. We adopt the definition in [19] as it covers exactly what we mean by information accuracy in this work. Information accuracy is the degree of deviation of the extracted information from the actual current state of the monitored environment.

Completeness: In the literature, the definition of information completeness is linked to data integrity, which is the absence of accidental/malicious changes or errors in data. In [17], information completeness is defined as the characteristic of information that provides all needed facts for the user/application during the process of information construction. The authors in [20] define information completeness as a measure of the fraction of all generated reports that arrive to the end user. Each of these definitions refer to the information as complete information when the delivered information represents all the sensed data without any diminution. In other words, they define completeness as the ratio of the received reports over the sent reports. However, if part of the environment is not covered by sensor nodes, then this part is not represented in the extracted information. Therefore, it is important to incorporate the sensing converge in the definition of information completeness. To include sensing coverage, we re-define completeness as the degree of obtaining all the desired information that represent the actual current state of the full monitored environment.

Affordability: Affordability refers to the cost of collecting sensed data [18], i.e., it is the expensiveness of information. In [17], affordability is the characteristic of information associated to the cost of measuring, collecting and transporting of data/information. We define affordability as the ability to afford the cost of information in terms of resource utilisation from the stage of sensing the environment to the stage of extracting the required information.

Timeliness: Information timeliness is a crucial and decisive criteria in time critical applications. In [18], timeliness describes how timely the data is provided to be useful to the end users or applications. To incorporate the scale of a multi-hop network, timeliness is measured as the time normalised against the average time for a single-hop along the shortest path from a sensor to the sink [21]. In the above definitions, timeliness accommodates different types of delay including: loading, propagation, queuing, and processing delay. However, in mobile approaches, the mobile sink has to travel to a specific point to collect information. This introduces considerable delays on data delivery; therefore, the time the sink spends travelling toward sensor nodes should be also considered. We extend the definition in [17], timeliness is an indicator for the total time required from when the first data sample is generated in the network until the information reaches the user for decision making. This includes the time that the mobile node spends travelling towards the target nodes.

Availability: The term availability has been widely used in computer networks as a primary QoS measure. Network availability refers to the overall up-time of the network, or the probability that the network is available to use [22]. In [23], availability is defined as the fraction of time that a network is able to provide communications services. However, we are not only concerned about the availability of the nodes communication links or the network in total; but also in the availability of information. The network generated data could contain the desired information but the inability of the user or the lack of the powerful IE tools could lead to absence of some information. Moreover, nodes mobility is an important factor that impacts availability of information. If the node that carries the desired information is not in the vicinity of the mobile sink, information from that node will be unavailable. Other factors that has an impact on information availability are: sensor nodes; communication links; sensors generated data; and IE techniques. An inefficient factor from this list could lead to unavailable information.   We define information availability as the fraction of time the network is able to acquire and deliver the end user's desired information.

Validity: Information validity refers to whether the information is useful to the end user or not. There are many factors that could result in invalid information.   For instance, information based on un-calibrated sensor readings, corrupted packets, or noisy data is unbeneficial and even confusing to the end user. Furthermore, in time critical applications, delaying the information invalidates it. For instance, in target tracking application, information could be received indicating that the target is in location x, but when the information was received the target has moved to location y. The extracted information is valid if its content is entailed by its data.

4   Mobility Benefits

Although mobility requires a lot of management, it has advantages over static WSN such as: better energy efficiency [24], improved coverage [25], enhanced target tracking [8], greater channel capacity [26], and enhanced information fidelity [27].

In many WSN applications, the node location is important as it is useful for coverage planning, data routing, location services, and target tracking [28]. An appropriate node deployment strategy can effectively reduce the network topology management complexity and the communication cost. Sensor nodes can be placed on a grid, randomly, or surrounding an object of interest [29].   In applications where nodes need to be deployed in harsh or remote environments,   nodes deployment can not be performed manually or accurately. Therefore, if a node runs out of battery, the data from the dead nodes would be lost, which negatively affects the accuracy and completeness of the extracted information. Some approaches tried to solve node's energy depletion problem by exploiting node redundancy. This class of approaches requires dense node deployment, which increases the system cost and management complexity. Node mobility presents effective solution to the above problem at low cost. Mobile nodes can redeploy the network by connecting disjointed areas created by dead nodes without the need of very dense deployment. Some approaches, e.g. [25], move nodes to provide better coverage by filling in holes in sensing coverage. They relocate redundant nodes to areas where node density is low. A complete coverage results in high information accuracy and completeness as every point in the environment has data to represent it. Furthermore, relocating nodes to substitute dead nodes helps in tolerating node failure. That maintains high information availability and completeness.

Unfair coverage caused by random nodes distribution results with high traffic load in some parts of the network. In traditional static networks, the nodes located around the sink become bottlenecks due to the many-to-one multi-hop communication. Bottlenecks introduce information delivery delay and causes energy depletion in some parts of the network or could even lead to the network partitioning problem [30]. This decreases the level of information completeness and availability. Furthermore, the probability of error increases with the number of hops that a packet travels over [15], which lowers the information accuracy. mWSNs are believed to provide more balanced energy consumption than static networks [15]. Node mobility offers a solution by moving nodes as needed to optimise the network performance.

Moving the sink to data sources or moving the sensor nodes towards the sink is one way to avoid the communication bottlenecks. Approaches such as [4, 31] suggest moving the sink close to data sources to perform data collection and analysis. This has been shown to be an effective way of reducing network congestion levels and relaying information in partitioned networks. Keeping the network connected leads to better sensing coverage, and hence maintains the higher information completeness level. Furthermore, moving the sink closer to sensor nodes helps conserve power by reducing the bridging distance between the node and the sink [32]. This also increases the performance of the network by saving retransmission bandwidth [29]. Moreover, information accuracy also increases due to the fact that the probability of error in the data decreases when decreasing the number of hops [15]. Other approaches, e.g. [33], suggest using mobile nodes to collect data from the monitored field and deliver it to a fixed sink. In these approaches, mobile nodes send data over short range communication, which involves less transmission power. This leads to reduced energy consumption and communication overhead. Since the cost of transporting data is reduced due to using single hop communication, the total information cost is reduced, resulting in more affordable information. Moreover, introducing mobility adds load balancing capabilities to the data transmitted towards the sink, which helps in buffer overflow prevention [33]. However, the above mentioned approaches have some drawbacks: First, some nodes could have data to send but the mobile sink or data collector is not around, this negatively impacts the timeliness and validity of information. Second, moving nodes usually consumes more energy than sensing, computation, and communication. The mobile sink or data collector could move towards some nodes which have no data to send, this would be a waste of energy and time. Therefore, if the movements of nodes are not planned in an efficient way, they could deplete the limited node's energy; which can diminish the gains in quality of information.

5   Mobility Challenges

Localisation: Many WSNs applications rely heavily on the node's ability to establish position information. The process of obtaining the position of a sensor node is referred to as localisation. Localisation has been identified as an important research issue in the field of WSNs. Localisation algorithms use various available information from the network in order to calculate the correct position of each sensor node. The location information is a key enabler for many WSN applications, e.g. target tracking, and it is useful for managing deployment, coverage, and routing [8]. Location information enables binding between extracted information and physical world entities. If the positions of sensor nodes can be determined more accurately, it will leverage the achievement of meaningful use of extracted information. The location of an event can be determined by knowing the location of nodes that report it. Thus, the locations of nodes that carry information of spatio-temporal nature need to be considered. Obtaining the nodes' locations helps in identifying nodes that carry data relevant to a certain piece of information.

In mobile environments, locations of nodes keep changing over the time. This introduces additional challenges that need to be addressed. (1) Localisation latency: the localisation algorithm should take minimal time to cope with mobility speed. For instance, if a node is moving at speed of 10 meters per second and the localisation algorithm needs 3 seconds to complete execution, the node will be 30 meters away from the calculated location. In this example, if the radio range needed to keep the node connected is less than the distance that the node has travelled, the node would be lost; the information from that node will be inaccurate or even might be unavailable. (2) Increased control messaging: managing node location information requires communication and transmission of control packets. When a node location changes frequently, the control packet overhead will be increased leading to higher energy consumption. This negatively affects the affordability of extracted information.

Trajectory Calculation: In mWSNs, the trajectories of nodes can be random, fixed, or dynamic. Some approaches, e.g. [34], assume that mobile nodes are mounted on objects moving chaotically around the network. Due to the fact that nodes cannot communicate unless they are in the radio range of each other, all nodes in the network need to keep sending periodic discovery messages to keep their routing tables updated. Transmitting a large amount of discovery and control messages consumes more energy. Furthermore, as nodes needs to be aware of all changes in the network topology, they can not switch their transceivers to sleep mode to conserve energy.

Approaches such as [4] propose mobility models to move the sink or data collector in a fixed trajectory. Data or information is conveyed to rendezvous nodes that are closer to the data collector trajectory, where it is cached until the mobile data collector passes by and picks it up. Sensor nodes can turn their transceivers off when the mobile data collector is away. However, in these approaches the fixed trajectory need to be defined. This needs a complex algorithm to calculate the most appropriate route that the node should follow.

When the trajectory is dynamic as in [31], nodes can move according to pre-computed schedule, or based on occurrence of an event of interest. However, calculating a dynamic trajectory is a complex problem, since it should satisfy the spatial and temporal constraints of the monitored phenomena. Knowing the trajectory of mobile nodes is very important as it helps to predict the nodes' locations. Therefore, this helps to plan for more efficient data collection leading to energy savings and maximising the network lifetime. Sensor nodes could be pre-configured with a sleep-wake cycle that is based on the location of the mobile node; a node goes to sleep when the mobile node is out of its radio range.

Velocity Control: Commonly, in mobile WSNs, nodes move in constant speed [35]. Velocity of the mobile node effects the information delivery time. However, some data collection approaches, e.g. [5, 36], assume that the speed of the mobile nodes is variable and also has different accelerations in order to optimise the movement of mobile nodes to reduce the information delivery time. The velocity is controlled by the task that the mobile node performs. If a mobile node performs data collection task, its velocity should be low compared to a mobile node that performs fire sensing task. Controlling velocity helps in utilising the available resources and results in more efficient WSN system. For instance, consider that there is a sensor node that generates a reading every one minute, and a data collector visits that node every 15 seconds; in this case, a lot of the data collector's energy is being wasted. However, by optimising the speed of the mobile node to best match the data generation rate, the data collector visits that node every one minute; hence, the data transmission of the network will be more efficient. Moreover, determining the velocity of mobile nodes is crucial in many of mWSNs applications. For instance, in tracking moving targets, the mobile sensor node should stay close to the target in order to maintain constant coverage.

6   Conclusion and Research Directions

In [3], we gave an overview of existing, state-of-the-art IE approaches for both static and mobile networks. That study formed the motivation for this framework. We identified that there is no clear common definition for IE. Also, there is ambiguity about how to measure the goodness of extracted information. This prevents consistent evaluation and comparison of various IE approaches. We believe that a solid framework for IE over WSNs in the presence of node mobility is missing. Such a framework should have the ability to process dynamic sensor data streams rapidly in an energy efficient manner against a set of outstanding and continuous queries. It is desirable to be able to optimise and adapt IE approaches based on problem domain requirements in conjunction with knowledge of the spatio-temporal relationships of sensed information Another research direction is to develop new approaches to perform IE in an interactive mode to control the data collection directions (e.g., on clustering) and even the accuracy (e.g., on classification) and efficiency. This includes the definition of new spatio-temporal primitive operations along with distributed algorithms to adapt query execution plans to changing characteristics of the data itself due to nodes mobility. Such approaches and algorithms have to work in a distributed setting and be space, time, and energy efficient.

References

1. Pyle, D.: Data Preparation for Data Mining. Morgan Kaufmann Publishers, Inc. (1999)
2. Olfati-Saber, R.: Distributed Kalman filtering for sensor networks. In: 2007 46th IEEE Conference on Decision and Control, pp. 5492–5498 (2007)
3. Alsboui, T., Abuarqoub, A., Hammoudeh, M., Bandar, Z., Nisbet, A.: Information Extraction from Wireless Sensor Networks: System and Approaches. Sensors & Transducers 14-2, 1–17 (2012)
4. Pantziou, G., Mpitziopoulos, A., Gavalas, D., Konstantopoulos, C., Mamalis, B.: Mobile Sinks for Information Retrieval from Cluster-Based WSN Islands. In: Ruiz, P.M., Garcia-Luna-Aceves, J.J. (eds.) ADHOC-NOW 2009. LNCS, vol. 5793, pp. 213–226. Springer, Heidelberg (2009)
5. Sugihara, R., Gupta, R.K.: Optimal Speed Control of Mobile Node for Data Collection in Sensor Networks. IEEE Trans. on Mobile Computing 9, 127–139 (2010)
6. Hammoudeh, M., Newman, R., Mount, S.: An Approach to Data Extraction and Visualisation for Wireless Sensor Networks. In: Proceedings of the 2009 Eighth International Conference on Networks, pp. 156–161 (2009)
7. Hammoudeh, M., Alsbou'i, T.A.A.: Building Programming Abstractions for Wireless Sensor Networks Using Watershed Segmentation. In: Balandin, S., Koucheryavy, Y., Hu, H. (eds.) NEW2AN/ruSMART 2011. LNCS, vol. 6869, pp. 587–597. Springer, Heidelberg (2011)
8. Sivaramakrishnan, S., Al-Anbuky, A.: Analysis of network connectivity: Wildlife and Sensor Network. In: 2009 Australasian Telecommunication Networks and Applications Conference (ATNAC), pp. 1–6 (2009)
9. Gupta, V., Pandey, R.: Data fusion and topology control in wireless sensor networks. In: Proceedings of the 5th Conference on Applied Electromagnetics, Wireless and Optical Communications, pp. 135–140 (2007)
10. Llinas, J., Hall, D.L.: An introduction to multi-sensor data fusion. In: Proceedings of the IEEE Int. Symposium on Circuits and Systems, ISCAS 1998, vol. 6, pp. 537–540 (1998)
11. Yue, J., Zhang, W., Xiao, W., Tang, D., Tang, J.: A Novel Cluster-Based Data Fusion Algorithm for Wireless Sensor Networks. In: 2011 7th International Conference on Wireless Communications, Networking and Mobile Computing (WiCOM), pp. 1–5 (2011)
12. Krishnamachari, L., Estrin, D., Wicker, S.: The impact of data aggregation in wireless sensor networks. In: Proceedings of the 22nd International Conference on Distributed Computing Systems Workshops 2002, pp. 575–578 (2002)
13. Fasolo, E., Rossi, M., Widmer, J., Zorzi, M.: In-network aggregation techniques for wireless sensor networks: a survey. IEEE Wireless Comm. 14, 70–87 (2007)
14. Padmanabh, K., Vuppala, S.K.: An Adaptive Data Aggregation Algorithm in Wireless Sensor Network with Bursty Source. In: Wireless Sensor Network, pp. 222–232 (2009)
15. Munir, S., Dongliang, X., Canfeng, C., Ma, J.: Mobile Wireless Sensor Networks: Architects for Pervasive Computing. InTech (2011)
16. Gaber, M., Roehm, U., Herink, K.: An analytical study of central and in-network data processing for wireless sensor networks. Inf. Process. Lett. 110, 62–70 (2009)
17. Sachidananda, V., Khelil, A., Suri, N.: Quality of Information in Wireless Sensor Networks: A Survey. In: ICIQ (2010)
18. Bisdikian, C.: On Sensor Sampling and Quality of Information: A Starting Point. In: Fifth Annual IEEE International Conference on Pervasive Computing and Communications Workshops, PerCom Workshops 2007, pp. 279–284 (2007)

19. Ismat, M., Uthman, B., Naseer, A.R.: Cautious rating for trust-enabled routing in wireless sensor networks. EURASIP J. Wirel. Commun. Netw. 2009, 1–16 (2009)
20. Hoes, R., Basten, T., Tham, C.-K., Geilen, M., Corporaal, H.: Analysing qos trade-offs in wireless sensor networks. In: Proceedings of the 10th ACM Symposium on Modeling, Analysis, and Simulation of Wireless and Mobile Systems, pp. 60–69 (2007)
21. Luo, J., Panchard, J., Sun, P., Seah, W.K.G., Lee, P.W.Q.: Efficient Data Delivery with Packet Cloning for Underwater Sensor Networks. In: Underwater Tech. and Workshop on Scientific Use of Submarine Cables and Related Technologies, pp. 34–41 (2007)
22. Mihajlović, B.: Compression and Security Platform for the Testing of Wireless Sensor Network Nodes. Dept. of Electrical & Computer Eng. McGill University (2008)
23. Green, H., Hant, J., Lanzinger, D.: Calculating network availability. In: 2009 IEEE Aerospace Conference, pp. 1–11 (2009)
24. Munir, S.A., Biao, R., Weiwei, J., Bin, W., Dongliang, X., Man, M.: Mobile Wireless Sensor Network: Architecture and Enabling Technologies for Ubiquitous Computing. In: 21st International Conference on Advanced Information Networking and Applications Workshops, AINAW 2007, pp. 113–120 (2007)
25. Coskun, V.: Relocating Sensor Nodes to Maximize Cumulative Connected Coverage in Wireless Sensor Networks. Sensors 8(4), 2792–2817 (2008)
26. Grossglauser, M., Tse, D.N.C.: Mobility increases the capacity of ad hoc wireless networks. IEEE/ACM Transactions on Networking 10(4), 477–486 (2002)
27. Zhu, C., Shu, L., Hara, T., Wang, L., Nishio, S.: Research issues on mobile sensor networks. In: 2010 5th International ICST Conference on Communications and Networking in China (CHINACOM), pp. 1–6 (2010)
28. Hightower, J., Borriello, G.: Location systems for ubiquitous computing. Computer 34(8), 57–66 (2001)
29. Kansal, A., Rahimi, M., Estrin, D., Kaiser, W.J., Pottie, G.J., Srivastava, M.B.: Controlled mobility for sustainable wireless sensor networks. In: 2004 First Annual IEEE Communications Society Conference on Sensor and Ad Hoc Communications and Networks, IEEE SECON 2004, pp. 1–6 (2004)
30. Dini, G., Pelagatti, M., Savino, I.M.: An Algorithm for Reconnecting Wireless Sensor Network Partitions. In: Verdone, R. (ed.) EWSN 2008. LNCS, vol. 4913, pp. 253–267. Springer, Heidelberg (2008)
31. Sabbineni, H., Chakrabarty, K.: Datacollection in Event-Driven Wireless Sensor Networks with Mobile Sinks. Int. Journal of Distributed Sensor Networks 2010
32. Kinalis, A., Nikoletseas, S., Patroumpa, D., Rolim, J.: Biased Sink Mobility with Adaptive Stop Times for Low Latency Data Collection in Sensor Networks. In: Global Telecommunications Conference on GLOBECOM 2009, pp. 1–6. IEEE (2009)
33. Gu, Y., Bozdag, D., Ekici, E., Ozguner, F., Lee, C.-G.: Partitioning based mobile element scheduling in wireless sensor networks. In: 2005 Second Annual IEEE Communications Society Conference on Sensor and Ad Hoc Communications and Networks, IEEE SECON 2005, pp. 386–395 (2005)
34. Tzevelekas, L., Stavrakakis, I.: Sink mobility schemes for data extraction in large scale WSNs under single or zero hop data forwarding. In: 2010 European Wireless Conference (EW), pp. 896–902 (2010)
35. Yang, Y., Fonoage, M.I., Cardei, M.: Improving network lifetime with mobile wireless sensor networks. Comput. Commun. 33(4), 409–419 (2010)
36. Sugihara, R., Gupta, R.K.: Speed control and scheduling of data mules in sensor networks. ACM Trans. Sen. Netw. 7(1), 4:1–4:29 (2010)

VR-Smart Home
Prototyping of a User Centered Design System

Mohammadali Heidari Jozam[1,*], Erfaneh Allameh[1], Bauke De Vries[1],
Harry Timmermans[1], and Mohammad Masoud[2]

[1] Department of Built Environment, Eindhoven University of Technology,
Eindhoven, The Netherlands
{m.heidari.jozam,e.allameh,b.d.vries,h.j.p.timmermans}@tue.nl
[2] Department of Architecture and Urban Design, Art University of Isfahan, Isfahan, Iran
m.masoud@aui.ac.ir

Abstract. In this paper, we propose a prototype of a user centered design system for Smart Homes which lets users: (1) configure different interactive tasks, and (2) express activity specifications and preferences during the design process. The main objective of this paper is how to create and to implement VR Smart Home prototype as a platform to increase user contribution in the earliest phases of design. The presented prototype has the capability of visualizing smart technologies (Smart BIM), performing real-time interactions and tasks (Smart Design System), and revealing users' preferences (Activity Preference plug-in). The long-term goal of developing this prototype is to bridge the gap between the designers and the clients in a Smart Home design process. Eventually, it is expected that the research will lead to match Smart Home functionalities with users' demands and therefore an improved user acceptance of Smart Homes.

Keywords: Smart Home, User Centered Design, Smart BIM, Smart Design System, User Preferences, Activity Data Record.

1   Introduction

Smart Homes confront many challenges moving from a vision to a reality. Current researches on user acceptance demonstrate that even if innovative functions are accessible to people, there is no inherent guarantee that they will actually be accepted and used (Punie, 2003). The same situation is happening for Smart Homes with the lack of success in being accepted by people. Poor understanding of Smart Home by both designers and end users cause that many people resist accepting Smart Home as their new housing, although they could benefit from it. Filling this gap, the user participation in Smart Home design process is demonstrated.

This paper attempts to challenge the established practice of design and engineering of Smart Homes by offering a new design system tool which is based on users' experiments, attitude and preferences. According to Bruce Mau (1998), designers should develop their own tools in order to build unique things. Even simple tools can yield

---

* Corresponding author.

S. Andreev et al. (Eds.): NEW2AN/ruSMART 2012, LNCS 7469, pp. 107–118, 2012.

entirely new avenues of exploration. He believes that tools amplify designers' capacities, so even a small tool can make a big difference. Hence, we propose a prototype of the user centered design system for Smart Homes, while most of the current Smart Home design tools are concentrating on technical issues. Applying this prototype in design process can improve users' understanding of Smart Homes as well as designer' understanding of users' preferences.

Hence, the outline of this paper is as follows: we describe what Smart Homes and their technological features are and how they can be related to existing BIM (Building Information Model) components. Following, we discuss the three stages of developing the user centered design system prototype for Smart Home. First of all, an interactive virtual space named Smart BIM is developed to simulate the functions of smart technologies. Smart BIM is different from conventional 3D space in that the created virtual space embodies smart objects. These objects are capable of doing some functions and reacting toward users' interaction according to their available property sets. Secondly, a Smart Design System is proposed in which users can perform real-time interactions in a task-based process. Smart Design System can be used to simulate not only how smart spaces will look like but also how users interact with them. When users can directly utilize a task in the virtual model, they can deliver a better comprehension in how smart spaces will be designed, constructed and utilized. Thirdly, we are going to develop our system in order to find out users' preferences. Hence, we add an Activity Preferences plug-in to our system. This plug-in inserts several activity lists corresponded to different zones. It allows users to specify their Activity-Arrangement and Activity-Schedule in different given time period context. The resulted activity data record is helpful for further design stages. Finally, we draw conclusion on the consequence of this system prototype for the design process and the building industry of Smart Homes.

2   Smart Home

A Smart Home contains several highly advanced smart technologies and interconnected devices. Hence, the environment of a Smart Home has the abilities of perception, cognition, analysis, reasoning and anticipation about a user's activities and can accordingly take proper reactions (Ma et al. 2005). All of the interactions and responses will support users' needs and preferences to increase their quality of life. In 2007, the Smart Home Association in the Netherlands defined Smart Home as the integration of technology and services through home environments for higher comfort and quality of living at home (Bierhoff et al. 2007).

But what are technological changes involved in a Smart Home? As a potential answer to this question, we determine a Smart Home as a home environment which consists of both Ambient Intelligent Space (AmI-S) and Virtual Space (VR-S) combined with Physical Space (PS) (Allameh et al. 2011). Then consider the technologies involved in these spaces:

The Virtual Space consists of ICT appliances such as smart walls and smart furniture that are connected to an information network. It supports information-related activities, such as social networking, tele-shopping, tele-working and tele-learning.

The Ambient Intelligent Space refers to environments that are equipped with computers and sensors, in such a way that they can adapt to user activities through an automated form of awareness. An example of this space is the context around smart kitchen table. This kind of space will assist daily activities such as cooking and personal activities like caretaking of elderly and child caretaking.

The Physical Space is the traditional space where people actually are with their bodies.

3   Prototyping of a User Centered Design System

As any Smart Home will be eventually used by end users, providing a method to consider users' activity preferences is indispensable; especially for addressing several key problems in *User Acceptance of Smart Home*. As it is depicted in Figure1, there are many challenges in a way moving from *Smart Home Vision* to *Smart Home Reality*. Venkatesh studies (2008) show that in many cases consumers are unaware of the benefits of new smart technologies. He believed that by growing this awareness, the demand for Smart Homes' products will rise. Hence, the *Smart Home Design* requires collaborative efforts in integration of design process with *Users' Contribution*. Understanding of users' living in a Smart Home is a key to ensure proper acceptance. Since the general orientation of end users will depend on their lifestyles and the way in which they organize their activities in time and space. While the common justification for much of the researches in the *Smart Home Design* are the technological facilitation of devices, regardless of User Feedbacks. Applying a *User Centered Design System tool* instead of technical tools will result a better home design. Figure1 shows the role of our design system prototype in the design process of Smart Homes from a vision to reality.

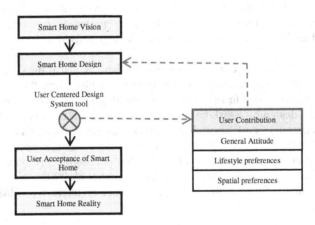

Fig. 1. The role of User Centered Design System tool in design process of Smart Homes

Using VR (Virtual Reality) prototypes in the domain of Smart Home Design is not new and System prototyping has since become a principal research approach in this emerging area. There are other examples such as:

— ViSi for SM4All is a 3D environment with interactive and pro active devices. It is able to adapt the behavior of devices to the needs of the home inhabitants. For example, a movie may be automatically paused when the subject leaves the room, and then launched again when he/she is back; or the windows are automatically opened to regulate the air condition. (Lazovik et al. 2009)
— ISS is an Interactive Smart Home Simulator. A 2D application tool focuses on controlling and simulating the behavior of an intelligent house. It determines the optimal sensor network and device placement. (Van Nguyen et al. 2009)
— CASS is Context-Aware Simulation System for Smart Home. A 2D application tool which generates the context information associated with virtual sensors and virtual devices in a Smart Home domain. By using CASS, the developer can determine optimal sensors and devices in a smart home. (Park et al. 2007)

These simulators propose to reduce the testing costs by replacing actual home services with virtual objects to visualize the behavior of the Smart Home. Relatively most of the developed VR Smart Homes prototypes are used in functionality tests of Smart Homes with less attention to the users' point of view. On the other hand, there are some researches which aim to simulate and predict occupants' activities in the given building and evaluate the building performance including evacuation, circulation, building control system (Shen et al. 2011). While many of them focus on office environments like Tabak's research (2010) and non-smart environments, the efforts toward developing user centered prototypes of VR Smart Homes are rare. V-PlaceSims is one of the rare examples of VR Smart Homes prototype which pays intensive attentions toward users (Lertlakkhanakul et al. 2008).This model enables virtual users as agents to perform specific behaviors autonomously for each spatial building entity. The interaction level between space and users takes place through their avatars. This prototype explores how to create and implement virtual space as a platform to simulate Smart Home configuration. Applying this prototype in design process improves users' understanding on the Smart Home and their involvement in the communication with designers. But it still does not consider users' preferences toward the activities they preferred to do. It does not let users specify their activities in the given Smart Home setting.

The added value of our system presented here is that it does not only support the design of Smart Home spaces and functions but also elicit the activities that a user wants to perform within the contextual conditions. The system collects users' feedback of the design by a task-based interaction between user and building. It improves users' understanding of Smart Home functionalities by performing real time interactions and helps them specify their activities in the given new home setting through an Activity Preference plug-in. This leads to the "paradigm shift in user role from a passive listener to an active actor" (Lertlakkhanakul et al. 2008). As a start point for developing our VR Smart Home prototype, we need an interactive BIM.

3.1   Smart BIM

Most BIM systems serve designers well up until now but will have to evolve toward a more user-centered design, focused on interactive spaces rather than focusing on digital representation. There is still lack of information needed in order to create a virtual environment which can interact with users. To create a Virtual Smart Home environment, we need to develop an advanced BIM system called Smart BIM which consists of several interactive smart objects. In today's design process, BIM systems support spatial design that is accommodate by smart technology. Usually this smart technology is added after the spatial design in the final design stage by the installations expert. In our research, we want to turn this process around; the smart technologies are accommodated by spatial design. Therefore, we develop a design system with a library of smart components such as smart wall, smart floor, smart kitchen table and smart furniture. The difference between smart technologies and standard building components is that smart technologies interact with the building users. Digital representations of interactive systems are not entirely new. On the internet we can find many examples (often implemented in Flash) of commercial products that one can view and virtually operate by clicking on buttons or hot spots. But in our view, Smart-BIM presents a virtual space with a wide range of smart technologies. While performing tasks in the virtual model, users express how these certain technologies will fit within their scheme of daily activities and give their requirements and feedbacks.

Developing BIM is not possible without standardization of building components. Many building component libraries have been developed for different aspects of the building design, such as spatial design, structural design, installations, etc. Often these libraries are included in Architecture, Engineering and Construction (AEC) tools, or they are provided by product suppliers. To support the quest for data exchange models between different AEC tools, standardization efforts have focused on building components. The most widely spread ISO certified building component library is the Industry Foundation Classes (IFC) standard. Building component libraries like IFC have developed from traditional catalogues of building products. Building components libraries are intensively used in the AEC industry today. However, since they are based on traditional building components, they prohibit fast adoption of new building components.

All the objects and technologies inside the Smart Home support people carrying out their everyday activities, tasks and rituals in an easy, intelligent and interactive way [1]. Accordingly, the difference between a 'normal' library component and a smart library component is the interaction between the component and the building users. Smart objects contain their own functions in their property set to interact with users and other objects. If we take as an example the IFC -Wall Standard Case and its property set, this wall type will contains a geometry description and a list of properties such as: Acoustic Rating, Fire Rating, Combustible, Surface Spread of Flame, etc. The wall geometry and properties are determined in a long standardization process by analyzing the most common wall types that are found in today's building sector. In case of a smart wall, there are hardly any precedents. The definition of a smart wall should not only describe the geometry and material, but also the interaction with its intended users. We think that designing a smart building requires interactive building components which respond to touch, remote control, motion detection, or whatever method is used to interact. Interactive building components are often integrated components consisting of constructive parts and embedded technologies. These embedded

---

[1] More information about smart technologies has been argued in our previous paper: (Allameh et al. 2011).

technologies can range from LCD screens to micro sensors. For more realistic evaluation of a virtual building model, the smart building components should be able to
receive input from the users' interaction and to act accordingly.

Technically, digital building component libraries need to be extended with interactive 2D or 3D models to turn these into smart components. In current digital libraries,
we can also find multiple representations for the same product. Multiple representations have proven to be useful for different levels of detail and for hiding irrelevant
information. An interactive model can be seen as yet another view on the same product. An example of interactive digital model of a smart wall is presented in Figure 2.

A smart object consists of several smart components with a property set that specifies its capabilities to respond to user activities. Figure 2 shows a smart wall with
three interactive surfaces as smart components. A smart wall 'senses' the activities
that are executed and it will act accordingly for instance by switching on screens for
different purposes such as tele-communication, tele-shopping, entertainment, etc.

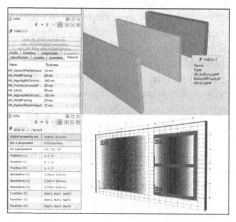

Fig. 2. Interactive digital model of a smart wall: A smart object contains several components
with property set

3.2   Smart Design System

Using VR technology, the platform is capable of visualizing smart technologies and
performs real-time interactions with the home. The Smart Design System proposed in
this paper is based on a task-based model in which users interact with the system and
experience how smart objects respond to typical domestic activities. According to Oxman et al (2004), there are three design paradigms to induce interaction process with
virtual environment: task-based design, scenario-based design and performance-based
design. Implementing each of these paradigms enhances users experience in virtual
environment and improves the human sense of "being there" (Oxman, 2011). While
scenario-based design and performance-based design are based on specific predefined
situations, task-based results could be closer to users' real reactions. Task-based
process enables users to imagine the scheme of their daily activities in the task context
and react accordingly. It also enables context reaction toward users' interaction and
functions as found in physical smart space. Finally, an improved understanding of

smart technology usage is expected from both end users and designers through this task-based interaction.

Applying this Smart Design System needs Smart BIM, that is, a model that includes smart objects and specific functions. Smart objects are part of a building model. A *smart object* consists of one or more smart components that have specific capabilities. Examples of *HasCapabilities* are: Displaying, Heating, Lighting, etc. A *user* will execute many tasks in the home environment. Each *task* is determined by the combination of a *zone* in the *building model* and an *activity*. The *activity* refers to the main activity executed in that *zone* (e.g. cook, work, and relax). An *activity* has one or more *NeedCapabilities*. Examples of *NeedCapabilities* for cooking activity are: Baking, Recipe displaying, Air cleaning, etc. The structure of this Smart Design System is presented in Figure 3. The interaction between user and its Smart Home is established through matching the *NeedCapabilities* with the *HasCapabilities*. In our prototype design systems, this matching procedure is a simple rule-based system, because it aims to experience smart objects, but avoid too much complexity for the user. In reality, however this matching process is performed by an intelligent home system that lets smart technologies communicate with each other.

In a smart design session, the designer will first create the home interior like with any CAD program but with use of smart objects. After the user is satisfied with the spatial design, the designer will create specific zones in the home and enter some general information about the user. In discussion with the user, the designer determines the tasks. Then the user is requested to navigate through the digital building model. At any spot in the home, he/she can execute a task. Therefore, he/she can use a virtual mobile phone to select a predefined task. After task selection, the smart objects will respond. The type of response is determined by the rule-based system that is called upon by the Smart Design system.

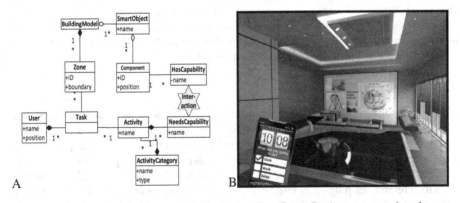

Fig. 3. A= Smart Design data model, B= Screen shot: Smart Design prototype interface

As an example, the subject will navigate to the smart kitchen table (Figure 3). At the spot zone, the virtual mobile phone will pop up. After selection of cooking as the task to be executed, the kitchen table will present a flexible cooking area. The area position and temperature can be adjusted by the subject interactively. At the same time, one of the Smart Walls nearby makes a connection to the social network and another one shows a web site with cooking recipes for users' diet. According to the subject target group, the system is able to active more capabilities. As an example for an elderly subject, the floor surface will be sensible for falling and the lighting is set to a high level for a better vision.

Thereby, users can experience a real time interaction with the Smart Home and gain an overview toward the functionalities of smart technologies. In the case of explained experiment, an improved understanding of Smart Kitchen Table capabilities is expected when the subject see the VR environmental reaction toward the cooking task executed by him/her. After doing the task, some positive improvement for understanding of the technology functions is expected. For instance:

— Experiencing the flexible cooking areas with wireless power and touch screen capabilities improves the understanding of safety functionality,
— Experiencing seating adjustment capability or wheelchair turning around the kitchen table improves the understanding of comfort functionality,
— Experiencing tele communication capability during the task improves the understanding of sociability functionality especially for aged people living alone,
— Experiencing online diet recipe presentation, suggested by doctor, improves the understanding of health functionality,
— Experiencing the supportive capabilities such as camera network, alarm facilities, lighting adjustment and smart floor facilities improves the understanding of protection functionality. It will result in encouragement of aged people to do their tasks independently,

At this stage, it is possible to measure users' general attitude toward the smart technologies and spaces by a simple questionnaire. It will increase users' participation in further design stages. The presented Smart Design System let users experience different tasks but it cannot still record the activities doing by users. Tasks indicate the activities users are doing not the activities they are willing to do. Only by adding a plug-in of Activity Preferences to our Smart Design System, users are able to select the activities they preferred to do. As a result, designers can analyze how people really use the smart technologies before constructions steps. It also leads to determine the relations among peoples' attitudes and their actual use.

3.3   Activity Preferences Plug-in

Further development of Smart Homes needs users' contribution in the design process (see figure 1). The proposed Smart Design System supports users to have a better understanding of Smart Homes but still cannot support designers to have better overview toward users actual use (see figure 3). Having a better understanding of how users really act in a Smart Home, designers need a supportive instrument. Accordingly, we add an Activity Preferences plug-in to our Smart Design System in order to support designers with an activity data record. The most popular assessment instruments in use today for studying the activities of people in natural settings are self report, recall surveys, time diaries, direct field observation, and experience sampling (Intille et al, 2003). These methods are not applicable for Smart Homes' experiments because users rarely have natural experiences of this domain. Developing a system to measure Activity Preferences virtually, is helpful method to have more realistic results.

Our developed system let users to experience Smart Homes virtually and specify their daily activities. Hence, the user do not perceive the architectural space as an image, but as a composition of various elements in which he/she can select his/her preferred activity. This experiment consists of two assignments: Activity Arrangement and Activity Schedule. Figure 4 shows the structure of the added plug-in and its interface. This plug-in inserts several activity lists in which the user is able to select his/her preferred activities. The designer can link the activity lists to the different zones which have already been defined in Smart BIM. This connection creates an awareness possibility for recognizing the position of user. Then, the Activity List related to that zone will pop up and ask the subject to select the most preferred activities he/she will do in this zone. The activity list contains two types of activities: Main activity: its sub activities (e.g. work: tele meeting, personal work activity and short-term work activity) and Secondary activity: its sub activities (e.g. E-activity: internet surfing, Personal caring: going to toilet, Family caring: child caring, Entertainment: do hoppy). Secondary activities are the activities that people usually do during their main activities (Figure 4).

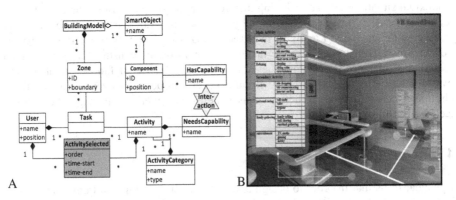

Fig. 4. A= The Activity Preference plug-in data model. B= Screen shot: Activity lists of zones.

Repeating it for each zone, the subject can have a complete Activity Arrangement. Selecting activity from activity list results in an Activity Arrangement data record. Such a data record can be used for behavioral research of users' future lifestyle and technology use researches which are essential for developments of Smart Homes. Through it, the following analysis can be resulted:

— The level of multi-functionality for each zone can be measured in such a way that more types of activity categories offer more levels of multi-functionality.
— The levels of flexibility for each type of activity can be measured in such a way that the more locations selected for one activity offer the more levels of flexibility for that activity.
— The demands of different target groups can be measured. By knowing the differences among the Activity Arrangement of each target group, their real demands can be revealed.

But there is still some missing information in this Activity Arrangement data base such as:

— The amount of time spending in each zone.
— The level of multitasking which is related to the main activities and secondary activities executed at the same time period.
— The types of conflicts among the executed activities in one zone. Understanding the incompatibilities in the zones can lead to the practical solutions in the design process.

To provide these analyses, we need to let users create their Activity Schedule as the second assignment. In Activity Schedule, users can set their activity duration and orders. Entering the data in the schedule is done outside the VR environment, in a textual form. This activity specification assignment is set up in a given time period and lifestyle context. For instance, the user can be said that it is a "typical day evening (7pm to 11 pm) coming back home to have relaxing, short work and a fast cooking" or a "week-day (8am to 11 pm) staying at home to work", a "relaxing day in the weekend (8am to 11 pm)". Hence, the resulted activity data record covers different contextual situations. By contributing different target groups of users, the activity data record covers the lifestyle and technology use patterns available among the end users more comprehensively. Finally, it is expected that some other non-functional properties in Smart Homes can also be taken in to consideration using the activity data record. For instance, by having the Activity Schedules of all family members and overlapping them, the concurrency circumstances can be recognized. Knowing these circumstances before construction phases helps designers to manage the spatial-technical solutions accordingly.

4   Conclusion

It is clear to us that in order to design and engineer successful smart environments, it is necessary to have a system which addresses not only technological, but also user activity layout. While most of the current design systems and Smart Home simulators are focusing on technical issues with little attention to users' contribution, we propose a User Centered Design System which let users to explore smart objects and real-time responses in a virtual model. In this design system, users are able to express appreciation, misunderstanding or disapproval considering their lifestyle and makes suggestions for improvement in design process. Accordingly, the building process will change because in a smart design process, it is impossible to design without the involvement of the client/user. Participatory design has not been very successful so far, but for smart design it is a precondition. At various stages of the design process, user's acceptance needs to be evaluated. User involvement through shared virtual models is possible today and may provide designer and engineer with much new information upfront.

We believe that the proposed system is also a powerful and economical assessment toolset for Smart Home designers in order to account users' preferences and their emerging lifestyles. By means of it, designers are able to adjust the spatial/technical layout of the Smart Home such that it accommodates the user's lifestyle optimally and properly. It will surely influence on the design construction process and give the opportunity to experiment greater flexibility and choices in creating the optimal spatial/technical layout: the location of smart technologies, the position of sensors, cameras and lights, the adjustments of technology functions. The final result is having a more realistic Smart Home design offering higher compatibility with users' lifestyles. We see this as a process of domestication of Smart Home technologies and a process of users' acceptance improvement.

We imagine that the Smart Design System can also be used for other purposes such as measuring users' spatial preferences, space utilization analyses, user behavior research and flexibility issues (time invariant, preferences changes, added smart objects). In our future research, we will investigate these possibilities and the resulted analyses.

References

1. Allameh, E., Heidari Jozam, M., de Vries, B., Timmermans, H., Beetz, J.: Smart Home as a smart real estate: a state of the art review. In: 18th Annual European Real Estate Society Conference, Eindhoven, The Netherlands (2011)
2. Bierhoff, I., Van Berlo, A., Abascal, J., Allen, B., Civit, A., Fellbaum, K., Kemppainen, E., Bitterman, N., Freitas, D., Kristiansson, K.: Smart Home environment. In: Roe, P.R.W. (ed.) Towards an Inclusive Future, Impact and Wider Potential of Information and Communication Technologies, COST, Brussels, pp. 110–156 (2007)
3. Intille, S., Rondoni, J., Kukla, C., Iacono, I., Bao, L.: A Context-Aware Experience Sampling Tool. In: CHI 2003 Extended Abstracts on Human Factors in Computing Systems (2003)
4. Lazovik, A., Kaldeli, E., Lazovik, E., Aiello, M.: Planning in a Smart Home: Visualization and Simulation. In: Application Showcase Proceedings of the 19th (ICAPS) Int. Conf. Automated Planning and Scheduling (2009)
5. Lertlakkhanakul, J., Won Choi, J., Yun Kim, M.: Building data model and simulation platform for spatial interaction management in Smart Home. Automation in Construction 17, 947–957 (2008)
6. Ma, J., Yang, L.T., Apduhan, B.O., Huang, R., Barolli, L., Takizawa, M.: Towards a Smart World and Ubiquitous Intelligence: A Walkthrough from Smart Things to Smart Hyperspaces and UbicKids. International Journal of Pervasive Computing and Communication 1, 53–68 (2005)
7. Mau, B.: An Incomplete Manifesto for Growth. In: BMD's Design Process (1998)
8. Oxman, R.: Design Paradigms for the Enhancement of Presence in Virtual Environments. In: Wang, X., Tsai, J.J.-H. (eds.) Collaborative Design in Virtual Environments. ISCA, vol. 48, pp. 41–49. Springer, Heidelberg (2011)
9. Oxman, R., Palmon, O., Shahar, M., Weiss, P.L.: Beyond the reality syndrome: designing presented in virtual environments. In: Orbak, H. (ed.) Proc. ECAADE, Copenhagen, pp. 15–18 (2004)
10. Park, J., Moon, M., Hwang, S., Yeom, K.: CASS: A Context-Aware Simulation System for Smart Home. In: Fifth International Conference on Software Engineering Research, Management and Applications, pp. 461–467 (2007)
11. Punie, Y.: A social and technological view on Ambient Intelligence in Everyday Life: What bends the trend? In: European Media, Technology and Everyday Life Research Network (EMTEL2) Key Deliverable Work Package 2, EC DG-JRC IPTS, Sevilla (2003)
12. Shen, W., Shen, Q., Sun, Q.: Building Information Modeling-based user activity simulation and evaluation method for improving designer–user communications. Automation in Construction 21, 148–160 (2011)

13. Tabak, V., de Vries, B.: Methods for the prediction of intermediate activities by office occupants. Building and Environment, 45 (2010)
14. Venkatesh, A.: Digital home technologies and transformation of households. Information Systems Frontiers Journal 10(4) (2008)
15. Van Nguyen, T., Kim, J.G., Choi, D.: ISS: The Interactive Smart home Simulator. In: Advanced Communication Technology, 11th ICACT (2009)

Appendix

Home auto: http://www.interinter.com/haihouse.swf
Hettich   House: http://www.hettich.com/discoverhettich/de/LivingRoom.html
LGHomeNet: http://www.lghomnet.com/homnet/exper/exper.html

Smart Space Governing through Service Mashups

Oscar Rodríguez Rocha[1], Luis Javier Suarez-Meza[2], and Boris Moltchanov[3]

[1] Politecnico di Torino, Corso Duca degli Abruzzi 24, 10129. Turin, Italy
oscar.rodriguezrocha@polito.it
[2] Universidad del Cauca, Popayán, Cauca, Colombia
ljsuarez@unicauca.edu.co
[3] Telecom Italia, via G. Reiss Romoli, 274, 10148. Turin, Italy
boris.moltchanov@telecomitalia.it

Abstract. The rapid and constant evolution of Smart Spaces has introduced an increase in the complexity and quantity of its components, making the monitoring and management processes more challenging and complex. On the other hand, Service Mashups enable the design and development novel and modern Web applications based on easy-to-accomplish end-user service compositions.

We introduce the concept of Smart Space "governing": managing an Smart Space with a set of monitoring processes and specific actions performed by the system when some events or behaviors occur. Our approach to this concept allows the end-user to personalize those processes and actions by creating Service Mashups that can published into the platform's marketplace in order to be reused.

Keywords: Smart Spaces, Service Mashups, Service Creation, End-user driven Service Creation.

1 Introduction

Smart Spaces are defined as a multi-user, multi-device, dynamic interaction environment that enhances a physical space by virtual services [1]. Their rapid and constant evolution has introduced an increase in the complexity and quantity of its own components, making the monitoring and management processes more challenging and complex. There's an emerging need to find new ways to provide end-users the ability of managing in a easy and personalized way their own Smart Spaces. On the other hand, Service Mashups (an extension of common Web mashups), enable the design and development novel and modern Web applications based on easy-to-accomplish end-user service compositions [2].

Given this context, we introduce the concept of Smart Space "governing": an end-user customized monitoring process for an Smart Space, with the possibility of defining specific actions to be performed by the system when some events or behaviors occur. Our approach to this concept is achieved by allowing end-users to create Service Mashups which define rules and actions to be executed by the

S. Andreev et al. (Eds.): NEW2AN/ruSMART 2012, LNCS 7469, pp. 119–127, 2012.

system. Once a composition is created, it can be published on the platform's marketplace in order to be reused.

This work is an extension of the European project 4CaaSt [3], since for this paper, we have introduced a variation to the original mashup model.

We present some related works on section 2, while in section 3 we provide a more detailed description of our governing concept. The details on how Service Mashups are implemented are described on section 4. An overview of the system architecture is presented below and finally the use-case of a Telecommunications operator is provided.

2   Related Work

Authors in [4] provide a very interesting work illustrating uDesign, an architectural style for enabling end-users to quickly design and deploy software systems. We found many interesting ideas an concepts from this work. However, our vision of composition, is different as we treat of the elements of an Smart Space as set and not as single elements. Additionally our work proposes a simple visual editor (based on drag and drop functionalities) instead o a complex one based on pipes and boxes.

On [5], an approach for practical Smart Space deployment using Web technology is presented. We also agree with authors about the technical advantages of representing an space over the web. In fact, our platform was designed having in mind a Web architecture. However, with the advances of mobile devices and many other technologies, we don't want to limit the system to be available only on the web. We'll consider in a near future also to deploy on mobile devices.

A reference model for a context-aware smart space is presented by A. Smirnov et al. in [6]. We consider this is a very inspiring work specifically from the point of view of the concepts expressed to handle the context of each node in a smart space. The approach presented in this paper differs in the way of managing the context information: we govern it in a centralized way performing the end user conditions defined through mashups.

3   Smart Space Governing

In our vision of Smart Spaces' management, it is predicted that the end-user can define itself the rules, it means, the conditions under which the space components must operate. Likewise, should be able to define behaviors or actions that the system will have to execute in the case where conditions may not be fulfilled. We call this "governing": making decisions or actions based on the results from the monitoring process. For this, we've created a mechanism in which the end-user can define step by step, how to govern a specific Smart Space. The latter can be also defined dynamically by our platform by exploiting the context data of the devices. We call Mashup Service to the composition of the set of instructions. Basically, end-users can access the system's Visual Editor in order to graphically define the govern conditions (a detailed description is provided on section 5).

One of the special features of our proposal, is that, in order to formulate specific governing conditions, we take into account the complexity, diversity and number of components that make up an space. In a Smart Space with many components, it is easier to establish rules to identify only those elements that do not meet certain criteria, than to analyze individually the operating parameters of each one. Additionally, to further simplify the process, the management criteria is presented to the end-user through layers, each one representing a management category of the components' parameters (device management, economy, errors and malfunctions, usage, etc).

It is important to emphasize that our system not only expects end-users to create their own compositions to govern an Smart Space, but also enables them to use the existing Service Mashups (available on the platform's marketplace [7]) that suit their requirements. Since in this paper we have introduced changes to the 4CaaSt's mashup model, storing and recovery mechanisms were adapted. A description is provided in the next sections.

We believe this approach can be applied not only on common Smart Spaces (e.g. home, office and city), but also in new scenarios. In this paper, the use case of a Telecommunications operator is presented.

4 Service Mashup Model

The fundamental characteristic of our model is that it captures not only the semantics of inputs/outputs (and its functional dependency), and operators (*entities* in the Smart Space), but also the semantics of control operator structures (i.e., composition structure patterns). Our mashup model can be expressed as a tuple $m = \{userID, name, O, C, M, reputation\}$, where:

- $userID$ is the identifier of the user that performs the request
- $name$ is the unique name (identifier) of the mashup
- O is the set of operators used in the mashup
- C is the set of data flow connectors ruling the propagation of data among operators
- M is the set of data mappings of output attributes to input parameters of connected operators
- $reputation$ counts how many times the mashup m has been used (e.g., to compute rankings).

Formally we have the following:

Definition 1. Operators (O): *At a logical level, operators O_l are defined as a set $O_l = \{O_{li}|O_{li} = (name_i, T_i)\}$ with $name_i$ being the unique name of the operator o_{li} and T_i represents a description based on tags of the o_{li} (from the some user). However, at an executable level, i.e., of composition patterns, which include sequence operations, parallel operations, etc. $O_p = \{O_{pi}|O_{pi} = (In_i, Out_i, Op_i)\}$ is a non-empty set of operators, where $In_i = \{in_{i0}, ..., in_{ij}\}$, $Out_i = \{out_{i0}, ..., out_{ik}\}$ and $Op_i = \{Op_{i0}, ..., Op_{il}\}$ are respectively the sets of input, output, and operations of an operator op_i. Thus, the set of Operators O is defined as: $O = O_l \cup O_p$. We distinguish three kinds of operators:*

- **Source operators**, *which fetch data from the web or the local machine. They don't have inputs, i.e., $In_i = \emptyset$.*
- **Typical operators**, *which consume data in input and produce processed data in output. Therefore, $In_i, Out_i \neq \emptyset$.*
- **Control operators**, *which are composition structure patterns: Sequential, $AND - Split$ (Fork), $XOR - Split$ (Conditional), $AND - Join$ (Merge) and $XOR - Join$ (Trigger)* [8], *as follows:*

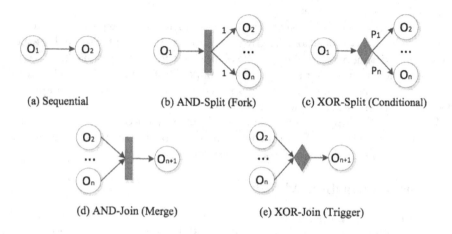

(a) Sequential   (b) AND-Split (Fork)   (c) XOR-Split (Conditional)

(d) AND-Join (Merge)   (e) XOR-Join (Trigger)

Fig. 1. Control Operator Structures

Definition 2. Data flow Connectors (C): *Let $C = \{c_m | c_m \in O \times O : C \cap O = \varnothing\}$ the data flow connectors that assign to each operator o_j its predecessor o_k (where: $j \neq k$) in the data flow.*

Definition 3. Data Mapping (M): *Let M the data mapping represents the set of data mapping of the data flow from output parameters of an operator o_j to input parameters of the predecessor operator o_k (where: $j \neq k$), as follows: $M = \{m_n | m_n \in In \times Out : In \cap Out = \varnothing, In = \cup_{i,j} in_{i,j}, Out = \cup_{i,j} out_{ik}\}$*

In order to better understand the formalisms defined above, the Figure 2 shows our proposal for a Mashups' meta-model, which is indeed very simple: only requires 13 concepts suffice to model its composition features at an executable level (abstractness).

Given the described model of Mashup M we create the mashup m meets the user's request from the large number and variety of resources on the *Smart Environment* in two ways: first, we generate a *Logical Mashup Model* (LMM) by analyzing the user's request by Drag and Drop and then, we generate an *Executable Mashup Model* (EMM abstractness) by AI Planner, which uses the modules *Situations Reasoner* and *Knowledge Extraction*. All this in order to

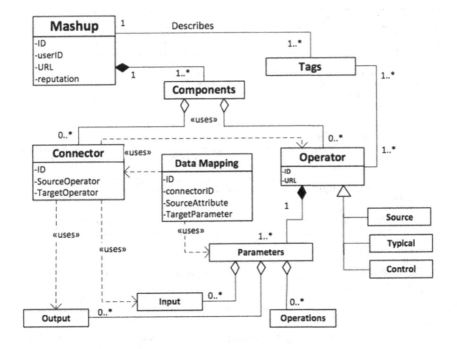

Fig. 2. Mashup metamodel

simulate the composition made by the user before running it in a real way. In case such that the simulation does not satisfy the wishes of the composer, the system will allow her to refine the *LMM* to achieve her goal.

The Algorithm on Figure 3, details this strategy and summarizes the logic implemented by the generation of *EMM*. In line 4, the *GetLogicalMashupModel()* function gets the *LMM* from the *Q*, which represents the user behavior during the writing process (drag and drop a component onto the canvas, or select a parameter to fill it with a value, connect a data flow connector with an existing target component, etc.). In line 5, the *GetEmmFromRepository()* function gets an *EMM* abstractness, which has been previously generated by the same user or other users of our platform. If the algorithm finds an exact or similar *EMM*, it is recommended to the user, avoiding the whole process of composition. In the absence of an *EMM* that satisfies the user's request, the *EMM* is composed based on retrieved operators and the *LMM* obtained (between lines 8 and 22). Finally, the *EMM* generated is stored in a *Repository* of abstractness *EMM*.

5   System Architecture

Figure 4 illustrates the modules that constitute our system architecture for governing Smart Spaces through mashups.

Algorithm 1. General EMM

---

1. **INPUTS**: Q, *User u*   /*Q informal query, which represents the desired results from *User u*/
2. **OUTPUT**: *EMM*   /*it is the executable mashup model EMM (abstractness)*/
3. **BEGIN**
4. *Let LMM* ← *GetLogicalMashupModel*(Q)   /* get LMM from Q */
5. *Let EMM* ← *GetEmmFromRepository*(Q)   /* get EMM previously generated by the *user* or another*/
6. **if** *EMM* != **null then**
7.   **return** *EMM*
8. **else**
9.   *Let Q(f)* ← *getFunctionalTags*(Q)   /* $Q(f) = \{q_{fi}, i=1,...,n\}$ */
10.   *Let Q(c)* ← *getControlTags*(Q)   /* $Q(c) = \{q_{ci}, i=1,...,m\}$ */
11.   *LMM* ← *getLogicalStructure*(Q, $Q(f) \times Q(c)$)   /* $Q(f) \cap Q(c) = \emptyset$, generate a logical mashup model(LMM)*/
12.   *Let O* = $\{o\}$ /* Operator Set*/;
13.   **for** each q_{fi} **in** $Q(f)$ **do**
14. $Q_f(O)$ ← *GetRankedOperatorsFromComponentRepository*(q_{fi})   /* $Q_f(O) \in \{Source, Typical\}$*/
15.   *Let* RQ_f ← $RQ_f \cup [Qf(O), q_{fi}]$ /*add $Q_f(O)$ to set RQ_f, ordered by q_{fi} */
16.   **end for**
17.   **for** each q_{ci} **in** $Q(c)$ **do**
18. $Q_c(O)$ ← *GetRankedOperatorsFromComponentRepository*(q_{ci})   /* $Q_f(O) \in \{Control\}$*/
19.   *Let* RQ_c ← $RQ_c \cup [Q_c(O), q_{ci}]$ /*add $Q_c(O)$ to set RQ_c, ordered by q_{ci} */
20.   **end for**
21.   *EMM* ← *genExecutableStructure*(LMM, RQ_f, RQ_c)
22. **end if**
23. *setEmmToRepository*(*EMM*, u)   /* stores the generated *EMM* in the repository*/
24. **return** *EMM*
25. **END**

Fig. 3. General EMM algorithm

5.1   Visual Editor

A visual editor, that features drag and drop functionalities, provides the user with a simple mechanism to integrate the different components that represent the entities embedded in the Smart Space. Such entities, according to the Mashup model defined previously, correspond to:

- the elements embedded in a Smart Space (sensors, devices, actuators, etc..).
- control structures (XOR-Split, AND-Split, AND-Join, OR-Split, etc..)
- monitoring elements (e.g., alarms).

The main goal of this module is to provide the user an intuitive interface for setting the rules of management, control and monitoring of Smart Spaces through the composition of the elements described based on Mashups.

5.2   Component Discovery

Given the characteristics defined in the mashups model described, it is possible to apply traditional recovery techniques for Web services in order to facilitate the selection of a wide variety of components that may exist in our Smart Spaces. This component is sensitive to all the information that the *KnowledgeExtraction* module can obtain from Smart Spaces.

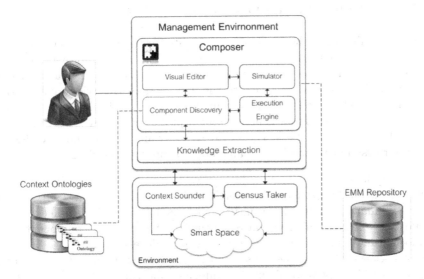

Fig. 4. System Architecture

5.3   Knowledge Extraction

It has the ability to infer rules that can be useful to users when composing the mashup. This module processes all the relevant information from the environment (Smart Space) to assist the user in the process of creating mashups from information collected from the environment and the context ontologies.

5.4   Context Sounder and Census Taker

This component is responsible of monitoring an Smart Space.

5.5   Simulator

Provides a simulation tool to ensure the proper creation of an LMM (Logical Mashup Model), so future errors can be corrected at the execution time of a created Mashup.

5.6   Execution Engine

Once the LMM has been validated by the user, this module is responsible for creating an EMM (Executable Mashup Model) from the mapping of real components (obtained through the Component Discovery).

5.7   EMM Repository

Stores all the created EMMs. This module was created to enable the recovery of existing EMMs, in this way, the creation process is improved through the reuse of previously composite structures. However, this aspect is not described as it is out of scope of this paper.

6   Telecommunications Operator Use Case

This use case illustrates the particular needs of a Telecommunications operator and how they may be filled through our approach. The operator provides mobile phone services (calls, sms, mms, mobile internet access etc.). For a year, has noticed remarkable low profits. Due to the wide variety of services offered, the large number of clients and mobile terminals, it is very difficult to identify the appropriate market areas that generate higher profits (it is impossible to make a customer by customer analysis, as they are so many and work would be endless). Instead our system can provide the answers that operator needs, how?:

- Let's consider the operator's mobile network as an Smart Space. The devices that make up the space are the users' mobile phones. Through them, we can retrieve many available information (such as active services, usage, device info ecc.). Additionally, we have another very important component, the information of mobile customer (age, gender, address, ecc).
- In this case the system's end-user will be the telecommunications operator. It will provide operational rules, parameters and conditions (a selection of users must send and average of 20 sms messages a week) to create a Service Mashup. Also, it will provide a set of actions that the system will perform when the latter are not completed (send to the marketing area, the info and the details of the users under the needed average).
- Govern it!. Maintaining in execution this compositions will help the profiling of users that generate low revenues.
- Decisions. The information extracted previously can be used for decision making.
- All the created compositions remain in available in the system even if they're not on execution. Thus they can be reused in another moment or if they're public, by another users of the system.

7   Conclusions and Future Work

We presented an approach to govern Smart Spaces. Our approaches considers the increasing number of components of each single space, making easier to identify only the ones that are not fulfilling certain conditions defined by the end-user. However, further tests of this proposal need to be done considering also more use-case scenarios. Evaluation trials to evaluate the prototype of the described system are currently running. The use of Semantics was not considered at this stage, when the system will be more mature, we plan to introduce Semantic descriptions to the generated compositions in order to facilitate and improve composition retrieval. Additionally it is planned to apply Semantics to describe data communicated by each component of the space.

References

1. Prehofer, C., van Gurp, J., di Flora, C.: Towards the web as a platform for ubiquitous applications in smart spaces. In: Second Workshop on Requirements and Solutions for Pervasive Software Infrastructures (RSPSI), at UBICOMB 2007, Innsbruck, September 16-19 (2007)
2. Benslimane, D., Dustdar, S., Sheth, A.: Services Mashups: The New Generation of Web Applications. IEEE Internet Computing 12(5), 13–15 (2008)
3. 4CaaSt European Project, http://4caast.morfeo-project.org/
4. Sousa, J., Schmerl, B., Poladian, V., Brodsky, A.: uDesign: End-User Design Applied to Monitoring and Control Applications for Smart Spaces. In: Seventh Working IEEE/IFIP Conference on Software Architecture, WICSA 2008, pp. 71–80 (2008)
5. Prehofer, C., van Gurp, J., Stirbu, V., Satish, S., Tarkoma, S., di Flora, C., Liimatainen, P.: Practical Web-Based Smart Spaces. IEEE Pervasive Computing 9(3), 72–80 (2010)
6. Smirnov, A., Kashevnik, A., Shilov, N., Boldyrev, S., Oliver, I., Balandin, S.: Context-Aware SmartSpace: Reference Model. In: The 5th International Symposium on Web and Mobile Information Services (2009)
7. 4CaaSt project's Marketplace, http://4caast.morfeo-project.org/wp-content/uploads/2011/11/4CaaSt_Marketplace_Whitepaper_v1.0.pdf
8. Yu, T., Zhang, Y., Lin, K.J.: Efficient algorithms for web services selection with end-to-end qos constraints. ACM Trans. Web 1(1) (May 2007)

Smart Space Applications Integration: A Mediation Formalism and Design for Smart-M3

Yury Korolev[1], Dmitry Korzun[2,3], and Ivan Galov[2]

[1] St.-Petersburg Electrotechnical University (SPbETU)
5, Professor Popov St., St.-Petersburg, 197376, Russia
yury.king@gmail.com
[2] Department of Computer Science, Petrozavodsk State University (PetrSU)
33, Lenin Ave., Petrozavodsk, 185910, Russia
{dkorzun,galov}@cs.karelia.ru
[3] Helsinki Institute for Information Technology (HIIT),
P.O. Box 19800, 00076 Aalto, Finland

Abstract. The Smart-M3 platform implements smart spaces environments with emphasis on the multi-device, multi-domain, and multi-vendor concept. One of the barriers against effective realization of the M3 concept is the lack of interoperability mechanisms between applications when they operate in different smart spaces. In this paper, we present an approach for applications integration; it provides interoperability such that one application uses services of another application. We extend the Smart-M3 space computing model with a mediation formalism of applications integration where ontology-driven knowledge exchange is performed between application spaces. We propose a generic architecture and design of application-specific mediator. Our approach makes a further step towards specification-based automated development in Smart-M3.

Keywords: Smart spaces, Smart-M3, interoperability, ontology-driven knowledge exchange, mediation.

1   Introduction

Smart spaces are environments where many devices participate to provide end-users with personalized and context-aware services [1, 14, 16]. Smart-M3 is an open-source interoperability platform for information sharing [4]. It implements smart spaces environments for applications that follow the M3 concept: Multi-device, Multi-domain, and Multi-vendor [8]. Application consists of agents that share and access local knowledge in the application smart space. Access to global knowledge is possible via gateway agents to the external world. The M3 concept supports provision end-users with personalized and context-aware services. The knowledge representation and reasoning mechanisms come from semantic web [5]: knowledge instances are described in RDF, usually according to ontologies expressed in OWL.

Within the same application its smart space can be effectively structured with a common ontology (though it can be modular, multi-domain, and composed from

S. Andreev et al. (Eds.): NEW2AN/ruSMART 2012, LNCS 7469, pp. 128–139, 2012.

multiple ontologies). Each agent applies its own "sub-ontology" to interpret its part of the shared content, e.g., see [10, 18]. The RDF representation allows easy linking and cross-agent interoperability. The case becomes more complicated when cross-application interoperability is required. In this paper, we consider a particular instance of the cross-application interoperability—applications integration when one application uses services provided by another application.

Smart-M3 has an architectural primitive for this type of integration. An agent-mediator is responsible for knowledge exchange between two application spaces. A reference solution we presented in [7]: the blogging service was integrated into the Smart Conference System for online discussion among conference participants. Nevertheless, the solution is customized and its transfer to other integration instances requires certain technical efforts for code development of the mediator. In this paper, based on our reference solution we continue the study and contribute a mediation formalism and corresponding mediator design for the applications integration problem in Smart-M3. We expect that the approach will result in service development of the next quality level required in such hot application areas of ubiquitous computing as smart airports, shopping malls, and healthcare environments.

Applications integration follows the known mediation principle for information systems [21], which is now also demanded in web environments [19]. A mediator performs knowledge exchange between spaces. The exchange problem is challenging for applications with heterogeneous data sources; in general it does not allow a full-automatic solution, e.g., see surveys [6, 17]. We apply ontology-driven exchange methods and develop a mediation formalism based on the Smart-M3 space computing model and ontology matching. The formalism embeds a notification mechanism to specify activation points of integration events. We describe basic steps a mediator must implement and introduce a mediator design pattern. It supports semi-automatic development when a domain-aware specification replaces lower-level coding.

The rest of the paper is organized as follows. Section 2 introduces the Smart-M3 space computing model. Section 3 states the applications integration problem and presents our mediation formalism based on knowledge exchange between smart spaces. Section 4 discusses the use of ontology methods applicable in this formalism. Section 5 contributes architecture and design of application-specific mediator. Section 6 makes comparison with the related work. Section 7 summarizes our findings.

2   The M3 Concept and Smart Space Applications

Smart-M3 space is dynamically shared by multiple users, devices, and services. Each user interacts with the surrounding digital environment and services continuously adapt to her/his current needs [1, 14, 16]. The core component is semantic information broker (SIB), an access point to the space. It maintains the smart space content using the RDF model. Smart-M3 employs term "knowledge": a smart space keeps habitual data, relations between them and even such information as computations. Participants access the space using software agents called knowledge processors (KP). Each connects to SIB by the Smart Space Access Protocol (SSAP) and performs insert, remove and update queries, and (un)subscribe operations.

Application is constructed as an ad-hoc assembly of KPs implementing collaboratively a service scenario to meet users' goal. Scenario steps emerge from actions taken by the KPs and observable in the smart space. For example consider two applications, adopted from [7]. Smart conference system (SCS, Fig. 1 (a)) maintains the visual content for participants: a current presentation slide on the conference projector (KP projector) and up-to-date session program on the conference whiteboard (KP whiteboard). Presenter changes the slides from her/his mobile device (user KP). Other participants browse the presentations, participants' info and conference program via user KPs. SCS automates time management and online control.

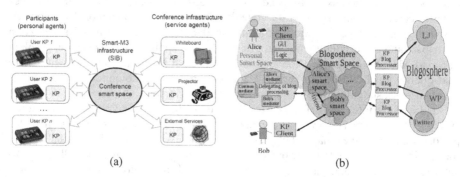

(a)   (b)

Fig. 1. Example Smart-M3 applications: (a) Smart Conference System assists and automates conference processes, (b) SmartScribo system allows a user or a group of them to benefit from semantic multiblogging

The second example is SmartScribo system that provides access to the blogosphere through its partial representation in the blogosphere smart space (Fig. 1 (b)). Blog content from multiple blogs on many blog services (via KP blog processors) as well as users' info, interests and context (via KP clients) are shared. SmartScribo user benefits from multi-blog data sharing and from semantic functions as searching personally relevant blogs and their proactive recommendation (via KP blog mediators).

Consider the Smart-M3 ontology-driven computing formalism. The generic smart space model was introduced in [14]. A space is $S = (n, I, \rho)$, where n is its name, I is information content and ρ is a rule set to deduce knowledge. In Smart-M3, I is represented using the RDF model (a set of RDF triples, or an RDF graph). The base case is when ρ is ontology O represented with the OWL model. Then I is structured with classes and properties from O. Deduction in S is based on the ontology instance graph. The latter is formed by individuals (nodes) that are interlinked with object properties (links) and have data properties (attributes).

In our formalism, an application smart space is $S = (I, O)$, where I is a set of RDF triples and O is an OWL ontology. A portion of knowledge $x \in S$ is an ontology instance graph that is a part of the deductive closure calculated from I according to O. Without confusion we also refer to S as to the space unique name. Notation $name(S)$ is used if technical attributes like SIB IP address and port are needed.

When a KP accesses S, it forms a query $q(S)$ that returns an ontology instance graph $x \in S$. The KP interprets it locally and then inserts new knowledge to S or updates some previous instances. A local sub-ontology $O' \subset O$ may be used when the KP is not interested in S as the whole. Rules of logic programming can be used for local reasoning with additional semantics to accompany ontology [12].

3   Applications Integration Problem

Smart-M3 allows composing scenarios from multiple applications. The global view on is illustrated in Fig. 2. The key point is the loose coupling between the participating KPs. The impact of each KP to others is limited by the knowledge the KP provides into the space. For instance, SmartScribo allows KP blog processors to join or leave the space making available or not the blog services. When an application needs a service from another Smart-M3 application, a KP mediator can be used to connect these spaces. This process is called *the applications integration*. Basically, an application that integrates a service from another application needs (i) to be aware what services are provided and available and (ii) to exchange knowledge between the spaces (export knowledge as input for the service and then import the result).

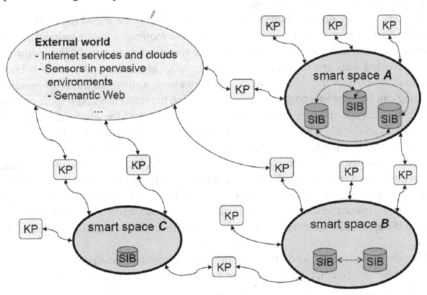

Fig. 2. Each space (e.g., A, B, C) is maintained by own set of SIBs to host an own set of applications. Each application consists of several KPs that publish and query content depending on observing actions of each other. Some KPs are wrappers for external services; some are used for knowledge exchange between spaces.

The notion of a KP mediator is crucial for applications integration. The mediator should understand appropriate knowledge in the source and target spaces, constructs a mapping between them, and makes the exchange. For a reference case, consider integration of SmartScribo with SCS [7]. In the integration scenario, a conference blog is created at a blog service. The blog consists of posts (one per a talk), derived from the conference program. During the conference, the participants read/write comments about the talks using SmartScribo clients. The conference blog mediator tracks the program in the conference smart space and transforms it to the blog posts in the blogosphere smart space. Whenever a new blog message appears in the blogosphere smart space or in the blog service then an appropriate blog processor performs mapping between the space and the blog service. The implementation in [7] assigns the latter mapping also to the conference blog mediator.

Such integration assumes that some knowledge from one space is imported by another space. It is a case of data exchange known in database applications, see [11] for recent results and references in the area. Data exchange problem [22]: given source and target schemas, a data instance over the source schema, and a set of dependencies describing source-to-target mappings and target constraints, find an instance over the target schema such that together with the source instance it satisfies the set of dependencies. In the ontology world, the term "ontology" is similar to "data schema" [6, 17]; both define structure of kept instances providing a vocabulary of terms and the meaning of terms used in the vocabulary. Ontology accompanied with logic programming rule sets can define constraints to which instances satisfy as well as specify mappings between the spaces.

Contrary to pure data exchange, which assumes the existence of data before running the exchange process, applications integration requires a computational process when some instances from the source space are transformed to new instances (or previously existed ones are updated) in the target space. The target space imports knowledge inferred from the source space. Nevertheless, known methods of ontology mapping can be used as a component for structuring the inference and import process.

The loose coupling property is another important issue of the applications integration. The mediator must know when to activate the exchange; it depends on observations of available knowledge. The subscription operation can be applied such that the mediator subscribes to certain instances in the spaces. A change notifies the mediator to activate the exchange. Application provides a notification set of instances to be subscribed by other applications. It forms an application integration interface (AII). For example, SmartScribo provides notifications of read/write events for blog messages. One application selects by its own to which instances in the other space to subscribe.

The above discussion leads to the following mediation formalism, which extends the Smart-M3 space computing model for the needs of applications integration.

Knowledge Exchange $S_1 \rightarrow S_2$ **by Mediator:** Given a source and target spaces $S_1 = (I_1, O_1)$ and $S_2 = (I_2, O_2)$. Mediator μ knows notifications $N = \{n_i\}$ and source queries $P = \{p_i\}$, source-to-target dependencies $D = \{d_i\}$, and target queries $Q = \{q_i\}$. For every n_i, mediator μ implements a transformation process t_i such that after any execution of t_i the spaces S_1 and S_1 satisfy

$$x = p_i(S_1), \quad y = q_i(S_2), \quad d_i(x, y). \tag{1}$$

We define a mediator for S_1 and S_2 as a specification

$$\mu = (O_1, O_2, N, P, Q, D, T),\tag{2}$$

where $T = \{t_i\}$ is a set of transformation processes, one per notification.

In our reference example, S_1 is the conference smart space and S_2 is the blogosphere smart space. The mediator subscribes to the conference program with notification $n \in I_1$. The query $x = p(S_1)$ returns all talks x_k of the recent program. The mediator performs t to transform x_k to post instances y_k for S_2; after publishing they conform the query $q(S_2)$. The dependency $\delta(x, y)$ states the one-to-one mapping between the conference talks and blog posts.

Note that (1) and (2) support the directionality: if applications integration requires bidirectional exchange then knowledge exchange is defined for each direction. Iterative exchange is also possible when y in the exchange $S_1 \to S_2$ affects notifications of the reverse knowledge exchange $S_2 \to S_1$. Generalization to the multi-target exchange $S_0 \to \{S_j\}_{j=1}^m$ is obvious.

4   Ontology Methods for Applications Integration

Knowledge exchange specified in (2) is ontology-driven. Let us consider ontology methods appropriate to the applications integration. Some RDF-based mechanisms are already supported by Smart-M3 on the SIB side. Higher-level OWL-based mechanisms are built into SmartSlog SDK [9]. Accompanied rule-based logic program reasoning is available in the ssls tool for Smart-M3 [12].

The notification model was presented in [10] for smart blogging. We generalize it for use in (2). A notification is an RDF triple or chain of them corresponding to an event observable in the source space. The multi-triple case is for notifications with additional attributes (to be used by its recipients). Agents subscribe to a notification if they need to react on the event. In the OWL model a notification is represented as an individual with several data properties.

If the application explicitly provides a set of its notifications then they form AII for other applications. It supports two notification types: service notifications that provide some service operation execution and event notifications that inform about events important when the service runs. The former type encourages a service running for integrated application while the latter type provides additional control for service execution (e.g., the context changes). AII simplifies the mediator since it and its developer do not need to know in detail the ontology of source space; a subontology that describes notifications and service-related knowledge can be used. The notification mechanism can be extended with the advanced SmartSlog subscription, e.g., for detecting the appearance in the space of new individuals of a given class.

Queries $P = \{p_i\}$ in (1)-(2) utilize matching algorithms and can be constructed using semantic query languages as SPARQL, see [3]. Additionally, SmartSlog knowledge patterns are applicable [9], being a high-level OWL analogue of RDF patterns in SPARQL. A knowledge pattern is a graph where nodes represent virtual individuals of the ontology. They are interpreted as masks or variables for actual individuals. The developer specifies only a part of the all properties to describe required instances in the space.

Source-to-target dependencies $D = \{d_i\}$ in (1)-(2) use ontology matching methods, see [6, 17]. For instance, ontology mapping indicates how elements from different ontologies are semantically related. It should be noted that matching procedures for ontologies used by smart space applications differ from traditional ontology matching operations. The point is that traditional ontology matching methods allow to match equivalent classes of two ontologies. However, the integration of applications in the smart spaces often requires logical relationship to be established between semantically different classes, basing on the concept of new use case of integrated systems. In our reference example, talks in conference program are one-to-one mapped to posts in conference blog. The mediator should implement the invariant mapping rules before running the exchange process. In general, development of these rules requires expert assistance, although there is some progress in automation so far [17]. In Smart-M3, automatic ontology matching is possible to a certain extent based on multi-model approach [18] to translate on-the-fly between ontologies.

Rule-based mapping formalisms [19] can be implemented in $T = \{t_i\}$ to automate runtime mapping. It is worth mentioning that the entity mapping rules in the smart spaces differ from ontology mapping rules. The target entity generation may often need some computations on the source entities. In particular, it requires mathematical operations to be carried out, identifiers to be generated, and string and date manipulations to be performed. Thus, when application of transformation rules gives rise to a new entity in the smart space, there is often a need to indicate expressly that a unique identifier must be generated for such an entity. For instance, Skolem functions allow creating new individuals IDs of classes in O_2 from properties in O_1. In addition, the transformation of text information often requires the source text parameters to be inserted in the string pattern. For example, in the case of integration the SmartScribo blogging service into SCS, the value of "text" attribute from "Post" class of SmartScribo ontology is generated on the basis of 6 source attributes of Smart Conference ontology by summarizing string values with a certain template.

Target queries Q reflect the correctness of execution of T and q_i need not to be represented directly in the mediator. Algorithmic of T requires (i) mediation state and (ii) local reasoning. Mediation state represents knowledge about the executed exchange. That is, the mapping between IDs of exchanged instances makes easier further updates between the spaces, e.g., keeping pairs $\left(\text{ID}_{\text{talk}}, \text{ID}_{\text{post}}\right)$ reduces the number of search queries to S_2 for our reference example. Moreover, even if some talks disappear form the conference program (e.g., due to cancellation or temporal uncertainty about presenter's presence), the conference blog can continue their discussion. Note that IDs mapping is one of the most frequent processing the mediator implements in T. Mediation state can be thought as an overlay space that allows continuing the mediation, though the space can be kept locally at the mediator.

Local reasoning is made over x queried from S_1, mediation state, and knowledge in S_2. For example, there is a new talk in the program that has no post at the conference blog. In general, reasoning is iterative when some knowledge is extracted from x and this portion drives the current iteration. The extraction can be based on SmartSlog knowledge patterns for local filtering: selecting the individuals from x which satisfy the pattern. The final goal is to construct y such that (1) becomes true. Importantly that $y = (y_{ins}, y_{upd})$ where y_{ins} is new instances to S_2 and y_{upd} is updates for the instances already kept in S_2. Note that the Smart-M3 update operation requires two arguments, for old and new values, respectively.

5   Mediator Architecture and Algorithmic Design

Applications integration development requires two phases. At the first one, an expert matches applications defining source-to-target dependencies. At the second phase, a mediator is developed to runtime knowledge exchange between smart spaces.

Consider the expert-based phase. First, the expert must identify a notification which will initiate a certain *integration scenario*. Such a notification is a sort of broadcast message intended to inform potential information users of important events in the application life cycle. Secondly, the expert needs to identify which information should be imported from the smart space after the notification is received. A graph query for information retrieval from the smart space is specified for this purpose, e.g., in SPARQL. Using the group graph pattern, the expert can specify a single query for data loading scenarios of any complexity. Thirdly, the expert sets the transformation rules of mapping source smart space instances to the target ones. Transformation mapping rules are "pattern => result" expressions that may be written in various logical calculus languages [23] (assertion-level logic, first-order logic and descriptive logic). Automatic logical inference [24] is the major requirement for such languages, so Prolog, SWRL and CLIPS are preferred. Of course, an expert may define several integration scenarios that will be performed on one mediator at the same time.

It should also be noted that dynamic transformation of knowledge from one smart space to another is not always necessary. Sometimes it is required to determine a static set of data to be inserted to the certain smart space when the relevant notification comes. In this case an expert may specify an empty graph query and a set of triples in the transformation rules that should be inserted in the smart space.

Consider the mediator architecture to implement the second phase. There are three principal modules; they are shown in Fig. 3. The data source interaction module interchanges data with semantic information brokers using SSAP protocol. When third-party RDF repositories are used, a wrapper is needed to interact with specific data source API. The interaction module is initiated by a set of notifications, with a graph query and related smart space name assigned. The number of such 3-tuples ‹smart space name – notification – query› is not limited but should not be less than one for each source smart space. The interaction module output is an RDF graph which is forwarded to the synchronization module input.

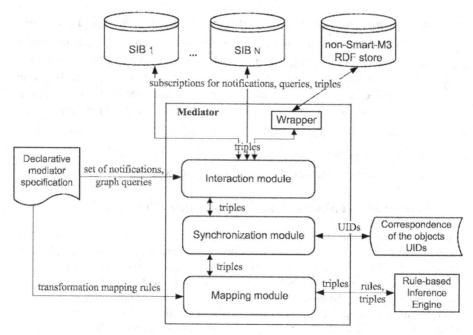

Fig. 3. Mediator architecture

The mediator internal memory is structured into the lookup table. The synchronization module generates and stores the lookup table which lists the unique identifiers of integrated entities. The lookup table is used to receive information on correspondence between unique identifiers of integrated entities in the different smart spaces without sending any extra queries to the data sources. This module also serves as a filter of RDF triples: it decides whether new entities should be generated or it has been done earlier. When the generation of new triples is needed, the source triples are forwarded to the mapping module input.

The mapping module is initiated by a set of transformation mapping rules specified in the logic programming language by an expert. This module uses a third-party inference engine (e.g. smodels) which receives a set of rules and facts. Today, the Smart-M3 platform is integrated with smodels, which allows to treat triples as facts [12]. Following the logical inference, resultant triples are forwarded to the interaction module in order to be inserted into the relevant data sources. Correspondence data on identifiers are passed to the synchronization module.

The basic mediator algorithm consists of the following steps.

1. Load a set of integration scenarios, including notifications, graph queries and transformation mapping rules.
2. Subscribe to notifications.
3. Wait for notification.
4. Receive notification.
5. Identify related graph query on the basis of notification, send the graph query to the source smart space.

6. Receive source triples the mapping rules should be applied to.
7. Consult with the lookup table and decide whether the integration is needed. If the entities have already been integrated and do not need to be updated, go to 3.
8. On the basis of triple load notification, identify related sets of mapping rules. Perform steps 9-13 for each set (i. e. for each target smart space).
9. Send source triples and mapping rules to the inference engine.
10. Receive resulting triples from the inference engine.
11. If generated data entities (represented by resulting triples) need to be bound with the entities that have been integrated earlier, send identifier correspondence query to the lookup table and update resulting triples.
12. Add data on generated entities to the lookup table.
13. Insert resulting triples in the target smart spaces.
14. Go to 3.

It is worth mentioning that possible error messages inserted in the smart space by integrated services enable an expert to specify an exception behavior of applications to be integrated. To that effect, a set of integration scenarios where an error message acts as an initiating notification should be defined. However, the automatic validity and satisfiability check of an expert's declarative specification is impossible in most cases. The specification errors may result in circularity, i.e. several integration scenarios initiating each other. Therefore, this approach implies that it is an expert who is fully responsible for semantic correctness of integration scenarios.

6   Related Work

Prior art [2] defined the generative communication model where common information is shared in a tuplespace and is further processed, providing deduced knowledge that otherwise cannot be available from a single source. Semantic web spaces use RDF triples as tuples and deduction with RDF reasoning capabilities [13]. Although the work on smart spaces [1, 8, 14, 16] assumes the possibility of communication between applications, it provides no concrete design rules and the discussion is limited with a generic architectural primitive: dedicated KPs implement the space mediation for applications.

The mediation principle for information systems is well-known [21] and recently is very demanded in ubiquitous environments with linked data integration [3, 17, 19]. Traditional use of mediator applies matching the existing instances (instance-to-instance exchange) or query reformulation (making a uniform view). Smart space applications integration needs, however, import/transformation/inference of knowledge from one space to another.

Declarative mediator specification is a known concept [15, 20]. We provided mediation formalism that supports semi-automatic implementation based on declarative ontological descriptions and rule-based mechanisms. The support is available partially in the recent Smart-M3 release: SmartSlog SDK for generating ontology libraries [9] and ssls tool for rule-based logic program reasoning [12].

The closest work is [7], where a mediator-based solution for a particular applications integration problem was presented: integrating the SmartScribo blogging service into SCS. It was successfully demonstrated at the 9th, 10th and 11th FRUCT Conferences. We use this problem and achieved experience as a reference example to derive generic mediation formalism for Smart-M3.

7   Conclusion

We have described the problem of interoperability of several smart space applications and have proposed our solution to this problem based on special software mediator. It should be noted that a mediator for smart spaces applications integration can be used for effective communication between the agents of one application. Obviously, if the logically separated agents are strongly connected to each other, then modifying and supporting the whole multi-agent application is quite difficult. The task of coordination between agents can be delegated to the mediator. In this case, an expert should specify the logical correspondence between the services provided by different applications. This approach allows an application changing on-the-fly its use cases. Also such multi-agent smart space applications are horizontally scalable and can be easily integrated with external services.

Further we plan to develop a universal mediator for automated integration of several Smart-M3 applications using the integration scenarios defined by an expert. The main research direction will be the development of declarative domain-specific language which best defines the smart space applications integration process.

Acknowledgment. Dmitry Korzun and Ivan Galov were supported by grant KA179 "Complex development of regional cooperation in the field of open ICT innovations" of Karelia ENPI CBC programme 2007-2013, which is co-funded by the European Union, the Russian Federation and the Republic of Finland. The authors are grateful to the Open Innovations Association FRUCT for its support and R&D infrastructure. We would like to thank Sergey Balandin and Alexey Kashevnik for their valuable comments.

References

1. Cook, D.J., Das, S.K.: How smart are our environments? An updated look at the state of the art. Pervasive and Mobile Computing 3(2), 53–73 (2007)
2. Gelernter, D.: Generative communication in Linda. ACM Trans. Program. Lang. Syst. 7, 80–112 (1985)
3. Gutierrez, C., Hurtado, C.A., Mendelzon, A.O., Perez, J.: Foundations of Semantic Web databases. J. Comput. Syst. Sci. 77(3), 520–541 (2011)
4. Honkola, J., Laine, H., Brown, R., Tyrkko, O.: Smart-M3 information sharing platform. In: Proc. IEEE Symp. Computers and Communications (ISCC 2010), pp. 1041–1046 (2010)
5. Horrocks, I.: Ontologies and the semantic web. Commun. ACM 51(12), 58–67 (2010)
6. Ivanov, P., Voigt, K.: Schema, Ontology and Metamodel Matching - Different, But Indeed the Same? In: Bellatreche, L., Mota Pinto, F. (eds.) MEDI 2011. LNCS, vol. 6918, pp. 18–30. Springer, Heidelberg (2011)

7. Korzun, D.G., Galov, I.V., Kashevnik, A.M., Shilov, N.G., Krinkin, K., Korolev, Y.: Integration of Smart-M3 Applications: Blogging in Smart Conference. In: Balandin, S., Koucheryavy, Y., Hu, H. (eds.) NEW2AN/ruSMART 2011. LNCS, vol. 6869, pp. 51–62. Springer, Heidelberg (2011)
8. Korzun, D., Balandin, S., Luukkala, L., Liuha, P., Gurtov, A.: Overview of Smart-M3 Principles for Application Development. In: Proc. Congress on Information Systems and Technologies (IS&IT 2011), Conf. Artificial Intelligence and Systems (AIS 2011), vol. 4, pp. 64–71. Physmathlit, Moscow (2011)
9. Korzun, D., Lomov, A., Vanag, P., Honkola, J., Balandin, S.: Multilingual Ontology Library Generator for Smart-M3 Information Sharing Platform. Int'l J. Advances in Intelligent Systems 4(3&4), 68–81 (2011)
10. Korzun, D., Galov, I., Balandin, S.: Proactive Personalized Mobile Mutli-Blogging Service on Smart-M3. In: Proc. 34th Int'l Conf. Information Technology Interfaces (2012)
11. Libkin, L., Sirangelo, C.: Data exchange and schema mappings in open and closed worlds. J. Comput. Syst. Sci. 77(3), 542–571 (2011)
12. Luukkala, V., Honkola, J.: Integration of an Answer Set Engine to Smart-M3. In: Balandin, S., Dunaytsev, R., Koucheryavy, Y. (eds.) ruSMART/NEW2AN 2010. LNCS, vol. 6294, pp. 92–101. Springer, Heidelberg (2010)
13. Nixon, L.J.B., Simperl, E., Krummenacher, R., Martinrecuerda, F.: Tuplespace-based computing for the semantic web: A survey of the state-of-the-art. Knowl. Eng. Rev. 23, 181–212 (2008)
14. Oliver, I., Boldyrev, S.: Operations on spaces of information. In: Proc. IEEE Int'l Conf. Semantic Computing (ICSC 2009), pp. 267–274. IEEE Comp. Soc. (2009)
15. Papakonstantinou, Y., Garcia-Molina, H., Ullman, J.: MedMaker: A Mediation System Based on Declarative Specifications (1996)
16. Prehofer, C., van Gurp, J., Stirbu, V., Satish, S., Tarkoma, S., di Flora, C., Liimatainen, P.P.: Practical Web-Based Smart Spaces. IEEE Pervasive Computing, 72–80 (2010)
17. Shvaiko, P., Euzenat, J.: Ontology matching: state of the art and future challenges. IEEE Transactions on Knowledge and Data Engineering (2012)
18. Smirnov, A., Kashevnik, A., Shilov, N., Balandin, S., Oliver, I., Boldyrev, S.: On-the-Fly Ontology Matching in Smart Spaces: A Multi-model Approach. In: Balandin, S., Dunaytsev, R., Koucheryavy, Y. (eds.) ruSMART/NEW2AN 2010. LNCS, vol. 6294, pp. 72–83. Springer, Heidelberg (2010)
19. Vidal, V.M.P., Macedo, J.A.F., Pinheiro, J.C., Casanova, M.A., Porto, F.: Query Processing in a Mediator Based Framework for Linked Data Integration. IJBDCN 7(2), 29–47 (2011)
20. Wache, H., Scholz, T., Stieghahn, H., Konig-Ries, B.: An Integration Method for the Specification of Rule-Oriented Mediators. In: Proc. DANTE 1999, pp. 109–112 (1999)
21. Wiederhold, G.: Mediators in the architecture of future information systems. IEEE Computer 25, 38–49 (1992)
22. Fagin, R., Kolaitis, P.G., Miller, R.J., Popa, L.: Data exchange: semantics and query answering. Theor. Comput. Sci. 336(1), 89–124 (2005)
23. Grosof, B., Horrocks, I., Volz, R., Decker, S.: Description Logic Programs: Combining Logic Programs with Description Logics. In: Proc. of WWW 2003, pp. 48–57 (2003)
24. Madhavan, J., Bernstein, P., Domingos, P., Halevy, A.: Representing and Reasoning about Mappings between Domain Models. In: Eighteenth National Conference on Artificial Intelligence (AAAI 2002), Edmonton, Canada, pp. 80–86 (2002)

Smart Logistic Service for Dynamic Ridesharing

Alexander Smirnov, Nikolay Shilov, Alexey Kashevnik, and Nikolay Teslya

St.Petersburg Institute for Informatics and Automation RAS (SPIIRAS), Russia
{smir,nick,alexey,teslya}@iias.spb.su

Abstract. The paper describes a service-based approach to dynamic ridesharing based on the smart space concept. Presented system allows planning and assisting a tourist for museum attending and finds a driver for reaching the museums. The paper applies an approach presented in FRUCT 11 to assisting tourists in a certain area (e.g., in a given city). Smart-M3 information platform is used as a smart space infrastructure for the presented approach. The service is based on the Smart-M3 ontology which is formed by ontology slices of user's mobile devices. The paper presents an algorithm for finding appropriate fellow-travelers for drivers as well as definition of acceptable pick-up and drop-off points for them.

Keywords: ridesharing, ontology, smart-m3, smart space, smart museum.

1 Introduction

Recently, the tourist business has become more and more popular. People travel around the world and visit museums and other places of interests. They have a restricted amount of time and usually would like to see many museums.

The paper describes a system developed for this purpose. The system can plan tourist's path and find a possibility for the tourist to reach the places of interest. The idea of ridesharing is used in the system to provide the tourists with the transportation. Ridesharing (also known as carpooling, lift-sharing and covoiturage), is a shared use of a car by the driver and one or more passengers, usually for commuting [1]. Dynamic ridesharing (also known as instant ridesharing, ad-hoc ridesharing, real-time ridesharing or dynamic carpooling) denotes a special implementation of a ridesharing service which enables a dynamic formation of carpools depending on the current situation. Typical features of this type of ridesharing are:

— arrangement of one-time trips instead of recurrent appointments for commuters;
— the usage of mobile phones for placing ridesharing requests and offers;
— automatic and instant matching of rides through a network service.

The first historical incidence of successful ridesharing was the tremendously popular yet short lived "Jitney Craze" beginning in 1914, when the US economy fell into recession with the outbreak of WWI, and some entrepreneurial vehicle owners in Los Angeles began to pickup streetcar passengers in exchange for a 'jitney' (the five cent streetcar fare).

S. Andreev et al. (Eds.): NEW2AN/ruSMART 2012, LNCS 7469, pp. 140–151, 2012.
© Springer-Verlag Berlin Heidelberg 2012

The second major period of rideshare participation, and the period most likely to be identified as the first instance of traditional ridesharing, was during the World War II (WWII). Opposite to the jitney era, the government encouraged ridesharing heavily during WWII as a method of conserving resources for the war effort. This period of ridesharing promotion was exceptionally unique, since it entailed an extensive and cooperative effort between the federal government and American oil companies.

The third period of interest in ridesharing picked up substantially with the Arab Oil Embargo in the fall of 1973 and 1979 oil crisis [2].

Nowadays, the next period of interest in ridesharing is expecting. It is associated with the intensive development of data processing, transfer technologies, and computing capacities, which can simplify the search for fellow travelers.

The following main schemes are used by people in different countries for searching for fellow-travelers:

— Search via public forums and other communities. For example: *eRideShare.com* [3], *PickupPal* [4], *Zimride* [5], *RideshareOnline* [6], *rideshare.511.org* [7], *CarJungle* [8]. The advertisements about trips are posted on a Web-site by users. This advertisement includes the start and end points, some information about people who post this ad, trip cost, time of the trip, etc;
— Search via private Web-services. People can get account in a private service only if they have an invitation. For example, *Zimride* service has a private interface for universities and companies;
— Search via special applications on mobile devices. With these applications users can edit their profiles, routes and search for fellow-travelers. The examples are *PickupPal* [4] and *Avego* [9];
— Search via agents (e.g. taxi companies);
— Pick-up points (not pre-arranged).

Software for mobile devices uses the client-server architecture. This architecture provides for the implementation of a centralized server and clients sending data processing requests to the server. Presented in the paper approach is based on the decentralized smart space infrastructure. This approach allows increasing the service stability and speed, as well as reducing the network's load.

The rest of the paper is structured as follows. Section II presents the Smart-M3 platform. Section III introduces the logistic service ontology used for enabling interoperability between different devices in the smart space. The algorithm for finding matching driver and passenger paths with two heuristics is given in Section IV. The system working scenario can be found in Section V. Main results are summarized in Conclusion.

2   Smart-M3 Platform

The open source Smart-M3 platform [10] has been used for the implementation of the presented ridesharing system. Usage of this platform makes it possible to significantly simplify further development of the system, include new information sources and services, and makes the system highly scalable. The key idea of this platform is that

the formed smart space is device, domain, and vendor independent. Smart-M3 assumes that devices and software entities can publish their embedded information for
other devices and software entities through simple, shared information brokers.
Information exchange in the smart space is implemented via HTTP using Uniform
Resource Identifier (URI) [11]. Semantic Web technologies have been applied for
decentralization purposes. In particular, ontologies are used to provide for semantic
interoperability.

The Smart-M3 platform consists of two main parts: information agents and kernel (Fig. 1) [12]. The kernel consists of two elements: Semantic Information Broker
(SIB) and data storage. Information agents are software entities installed on the
mobile device of the smart space user. These agents interact with SIB through the
Smart Space Access Protocol (SSAP) [13]. The SIB is the access point for receiving
the information to be stored, or retrieving the stored information. All this information is stored in the data storage as a graph that conforms to the rules of the Resource Description Framework (RDF) [14]. In accordance with these rules all
information is described by triples "Subject - Predicate - Object". More details
about Smart-M3 can be found in [13].

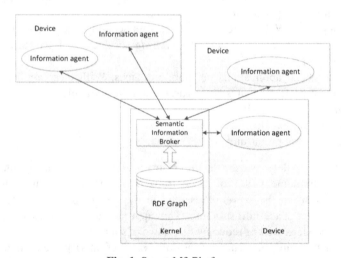

Fig. 1. Smart-M3 Platform

3   The Logistic Service Ontology

The logistic service ontology describes the domain area of ridesharing at the macro
level (Fig. 2).

The macro level ontology is based on integration of parts of the mobile devices'
ontologies. The logistics service ontology consists of three main parts: vehicles, actors
and paths. More details about the logistics service ontology can be found in [15].

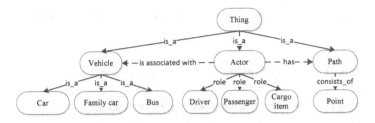

Fig. 2. Logistics service ontology on the macro level

3.1   Actors

The actors are: drivers, passengers and cargo items. All of them are associated with vehicles and have paths. For example, driver has his/her own car and several points defining his/her home, work and other locations. Passenger may prefer some vehicle type and has points of home, work, and other locations. Cargo items have size and vehicle type needed for their transportation.

The class actor consists of (Fig. 3):

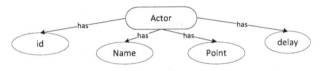

Fig. 3. Class "Actor"

— ID. Unique ID for each user;
— Name. First and last name of the user;
— Point. Path point belonging to the user's path (2 minimum: the start and the end);
— Delay. Maximal possible time of waiting in the meeting point.

The class "Driver" is a subclass of the class Actor and inherits all its properties with two own properties:

— Vehicle. Vehicle type;
— Detour. Maximal detour from the shortest path.

The class "Passenger" is a subclass of the class Actor and inherits all its properties with own property "Detour" the same as in the class "Driver".

The class "Cargo item" is a subclass of the class Actor and inherits all its properties with own property "size" defining the physical size of the cargo item.

3.2   Paths

For the path definition the set of points is used. This set is an ordered list of key points obtained as result of the shortest path searching algorithm (e.g., Dijkstra or A*). The class "Point" has the following structure (Fig. 4):

- previousPoint. Contains the previous path point. For the start point its value is "FALSE";
- Latitude;
- Longitude;
- driveByVehicle. If the point belongs to the passenger, it contains the driver who gives a ride to this passenger. If the passenger walks then its value is "FALSE";
- vacantseats. The number of vacant seats in vehicle in point;
- vacantItemPlace. The number of vacant places for cargo items;
- Date. Date, when the user will be at this point;
- Time. Time, when user will be at this point;
- Wait_time. How long the user will be waiting in this point.

Since the ontology in the smart space is represented in RDF standard, it looks like follows:

('user1',   'name',   'Name Surname')   - name of user1
('user1',   'is_a',   'Driver')   - user1 is a driver
('user1',   'vehicle',   'vehicle_type')   - user1 has this type of vehicle
etc.

In [15] the logistics service ontology is described in detail.

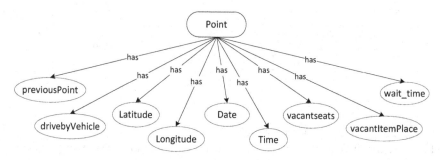

Fig. 4. Class "Point"

4   Algorithm for Finding Matching Driver and Passenger Paths

The problem of finding a matching path between the driver and the passenger in the ridesharing service can be formulated as follows: it is needed to determine the possibility of ridesharing between users, based on the information about their routes and restrictions set by users' services. The following algorithm describes the procedure of finding a matching path acceptable for the driver and the passenger in the presented ridesharing service.

Let A be the start point and B be the end point of the pedestrian's path. C is the start point and D is the end point of the driver's path. The shortest driver's path, which is found with the help of GIS, is indicated by the solid line (in generally, CD is not a straight line, it depends on the map of the region). Fig. 5 shows that the driver and pedestrian move almost in the same direction and in some parts of the routes the driver can give the pedestrian a ride. This situation is indicated in the figure by the dotted line (the CABD path) and it is the simplest situation, because the meeting points match with the start and end points of the pedestrian's path. A more difficult situation is searching for a meeting point when it belongs neither to the driver's shortest path nor to the pedestrian's one, but satisfies both the driver and the passenger. One of the possible situations is indicated in the figure by the dash-dot line with the meeting points E and F (the CEFD path). These points have to meet the following restrictions:

1. The distance between the start point of the passenger and his/her meeting point should be less than the maximum allowed detour of the passenger. This area is indicated in the figure by the dotted circle around point A.
2. The distance between the end point of the passenger and his/her drop-off point should be less than the maximum allowed detour of the passenger. This area is indicated in the figure by the dotted circle around point B.
3. The driver's detour should be less than the maximum allowed detour.

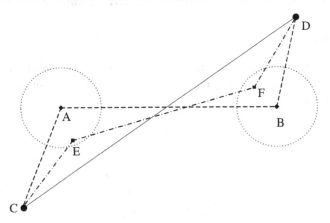

Fig. 5. The main idea of matching driver and passenger path search

The general scheme of the matching route searching algorithm will be follows:

```
FOR EACH driver DO
  FOR EACH passenger Do
    Find_mathing_path(driver.path,passenger.path); // ac-
cording to the above scheme
    constraint_checking();
    IF ALL constraints IS performed THEN
      set_passenger_for_driver();
  ENDFOR;
ENDFOR;
```

The goal functions for finding the meeting points are:

— Shortest total path (interesting for the driver);
— Minimal waiting time (interesting for the driver and passenger);
— Shortest distance between the passenger's start and end points and meeting points (interesting for the passenger).

As a result, the general task of matching paths has the exponential complexity, therefore, it is necessary to apply heuristics to reduce the task dimension.

4.1   Heuristics 1

Assumption: There is no need to calculate matching paths for all pairs of drivers and passengers. It is enough to build a set of candidate passengers for every driver.

$$(pp_1^x - dp_i^x)^2 + \left(pp_1^y - dp_i^y\right)^2 \leq (PDetour + DDetour)^2, \tag{1}$$

$$(pp_2^x - dp_{\prime i}^x)^2 + (pp_2^y - dp_{\prime i}^y)^2 \leq (PDetour + DDetour)^2, \tag{2}$$

where pp_1, pp_2 — the start and the end points of the passenger's path, dp_i — driver's path point i, *PDetour*, *DDetour* — detours of the driver and the passenger.

4.2   Heuristics 2

Assumption: There is no need to search through all possible combinations of meeting points. The following alternative sub-heuristics help to reduce the number of the possible combinations.

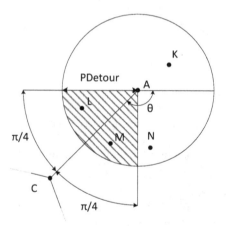

Fig. 6. The first sub-heuristics

The first sub-heuristics selects points of the sector from which the driver starts. Fig. 6 shows the situation when there is only one point ("C" point) meeting constraints (1) and (2). To determine the potential meeting points it is needed to calculate the angle (3) and select points in the area $\left[\theta - \frac{\pi}{4}, \theta + \frac{\pi}{4}\right]$ (points L and M in Fig. 6).

$$\theta = \text{arctg}\left(\frac{C^y - A^y}{C^x - A^x}\right) \tag{3}$$

Point A will always be within the list of the possible points as the passenger's start or end point. If there are more than one point meeting constraints (1) and (2), then the search area expands. This situation is shown in Fig. 7 with two points C and F meeting the constraints (1) and (2), and point N is also included in the expanded area.

The negative sides of this sub-heuristics are:

— selected points can be further than the driver's maximal detour;
— some of potential meeting points can be lost if an incorrect angle is chosen.

The second sub-heuristics (Fig. 8) selects meeting points in the intersections of the circles of radius PDetour around the passenger's start and end points with the circles of radius DDetour around the points of the driver's path. In this case all of the selected points are potentially reachable for both the driver and the passenger, with no need to determine the angle that restricts the selection area. The selection area can be expanded via increasing the number of the driver's path points meeting constraints (1) and (2).

Both sub-heuristics require the following constraints to work effectively:

— A large amount of drivers. Heuristics have strong limitations and filter out a lot of points. If there are no enough drivers, then the use of the heuristics will rarely produce positive result.
— A small value of DDetour. Heuristics will not be helpful with a large value of DDetour.
— Uniform distribution of roads on the map. The uneven distribution of roads (rivers, lakes, etc) leads to a lack of roads in some sectors, which could lead to the loss of possible meeting points due to the need to detour around the obstacles and to pick up the pedestrian on the other side.

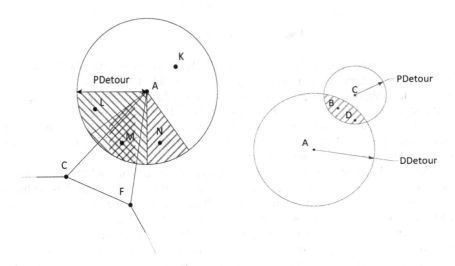

Fig. 7. The first sub-heuristics with two driver's points

Fig. 8. The second sub-heuristics

Both heuristics 1 and heuristics 2 are used in the logistics service prototype. Without using the heuristics, the system finds from 10 to 12 meeting points for each pair of driver and passenger and it needs to check all of 100–144 combinations to find the best one. With using the heuristics, the number of points is reduced to 8-9 points with 64–81 combinations for each pair of driver and passenger.

5   System Working Scenario

Common system working scenario is shown in Fig. 9. The mobile application is installed by all users of the service. This application collects the information about the user's agenda, preferences (Fig. 11, a), most frequent routes (Fig. 11, b), etc. with the agreement of the user. Also, the user can set additional constraints, for example, max. delay, max. detour, social interests, etc. (Fig. 11, b). This information is transferred into the smart space after the internal processing and depersonalization (only signs of information are transferred, not the raw information).

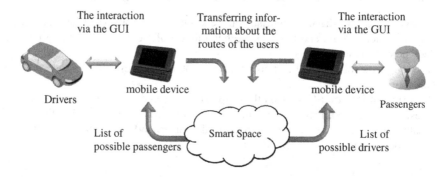

Fig. 9. System working scenario

During the execution of the above described algorithm the groups of fellow travelers are formed. Then, users interactively get the possible fellow travelers with their profiles, meeting points, meeting time, full recommendations about the route (Fig. 10, a-d) and if they have permission they can get the link to the external resources, e.g., social network page, which helps the user to make a decision. Sometimes, suggested driver can be a friend of the friend of the given passenger (this information can be useful for the user in decision making stage).

Also, the dynamic search is supported. Users can login into and logout from the smart space, change restrictions and then receive the list of fellow travelers within a short interval of time (almost in real-time). All this work is done by the smart space and users do not need to perform any actions to find fellow travelers.

For example, when the tourist is going to St. Petersburg an acceptable attending plan based on his/her preferences (how many days visitor is going to spend in St. Petersburg, his/her cultural preferences, and etc.) is formed. Fig. 12 presents the acceptable plan for a visitor. It consists of five museums: the Hermitage, Kunstkamera, the Museum of Karl May Gymnasium History, St. Isaac Cathedral, Dostoevsky museum. The system automatically finds drivers for the visitor to reach these museums.

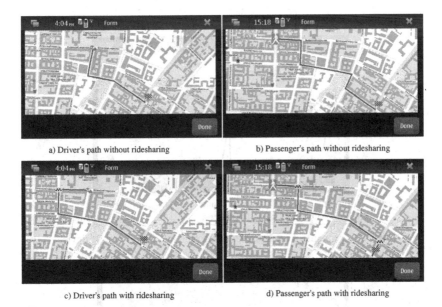

a) Driver's path without ridesharing

b) Passenger's path without ridesharing

c) Driver's path with ridesharing

d) Passenger's path with ridesharing

Fig. 10. Prototype screenshots (routes)

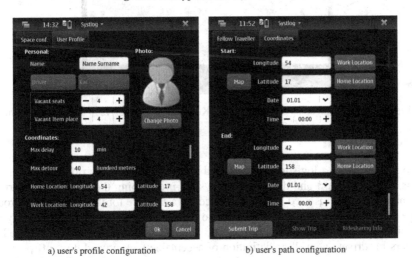

a) user's profile configuration

b) user's path configuration

Fig. 11. Prototype screenshots (user's routes and preferences)

From the Hermitage (label A in Fig. 12) to the Kunstkamera (label B in Fig. 12) visitor can walk (the distance is about 500 meters), but from Kunstkamera to the Museum of Karl May Gymnasium History it is reasonable to go by car. In this case the system finds the appropriate driver which goes in this direction in this time and picks visitor up.

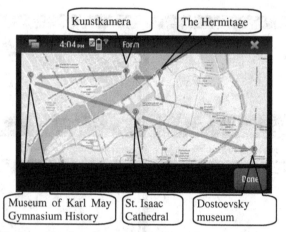

Fig. 12. A sample of museum attending plan in a visitor mobile device

Fig. 13. A driver picks up a tourist around Kunstkamera and drive to the next museum

6   Conclusion

The paper proposes a logistic service-based approach to dynamic ridesharing for museums visitors support and describes the system prototyping the approach. Smart-M3 information platform is used as a smart space infrastructure for the presented approach. Usage of this platform makes it possible to significantly increase the scalability and extensibility of the prototype system. The algorithm for finding appropriate fellow-travelers for drivers as well as definition of acceptable pick-up and drop-off points for them is presented in the paper. This algorithm can effectively find appropriate fellow-travelers for drivers. The presented heuristics help to reduce the time of search in more than 1.5 times.

Acknowledgment. The paper is a part of the research carried out within the ENPI project KA322 Development of cross-border e-tourism framework for the programme region (Smart e-Tourism); projects funded by grants # 12-07-00298-a, 11-07-00045-a, and 10-07-00368-a of the Russian Foundation for Basic Research; project # 213 of the research program "Intelligent information technologies, mathematical modeling, system analysis and automation" of the Russian Academy of Sciences; project 2.2 of the

Nano- & Information Technologies Branch of the Russian Academy of Sciences; and contract # 14.740.11.0357 of the Russian program "Research and Research-Human Resources for Innovating Russia in 2009-2013". Some parts of this work have been supported by Open Innovations Framework Program FRUCT – www.fruct.org.

References

1. Abrahamse, W., Keall, M.: Effectiveness of a web-based intervention to encourage car-pooling to work: A case study of Wellington, New Zealand. Transport Policy 21, 45–51 (2012)
2. Rideshare History & Statistics. MIT "Real-Time" Rideshare Research, http://ridesharechoices.scripts.mit.edu/home/histstats (last access June 05, 2012)
3. eRideShare.com., http://erideshare.com (last access June 05, 2012)
4. PickupPal, http://www.pickuppal.com (last access June 05, 2012)
5. Zimride, http://www.zimride.com (last access June 05, 2012)
6. RideshareOnline, http://www.rideshareonline.com (last access June 05, 2012)
7. Rideshare 511, http://rideshare.511.org (last access June 05, 2012)
8. CarJungle, http://www.carjungle.ru (last access June 05, 2012)
9. Avego, http://www.avego.com (last access June 05, 2012)
10. Smart-M3 at Sourceforge (2012), http://sourceforge.net/projects/smart-m3 (last access June 05, 2012)
11. Berners-Lee, T., Fielding, R., Masinter, L.: RFC 3986 – Uniform Resource Identifier (URI): Generic Syntax, http://tools.ietf.org/html/rfc3986 (last access June 05, 2012)
12. Smart-M3 at Wikipedia, http://en.wikipedia.org/wiki/Smart-M3 (last access June 05, 2012)
13. Honkola, J., Laine, H., Brown, R., Tyrkko, O.: Smart-M3 Information Sharing Platform. In: Proc. IEEE Symp. Computers and Communications (ISCC 2010), pp. 1041–1046. IEEE Comp. Soc. (June 2010)
14. Resource Description Framework (RDF). W3C standard, http://www.w3.org/RDF/ (last access June 05, 2012)
15. Smirnov, A., Kashevnik, A., Shilov, N., Paloheimo, H., Waris, H., Balandin, S.: Smart Space-Driven Sustainable Logistics: Ontology and Major Components. In: Balandin, S., Ovchinnikov, A. (eds.) Proceedings of the 8th Conference of Open Innovations Framework Program FRUCT, Lappeenranta, Finland, November 9-12, pp. 184–194 (2010)

A Methodological Approach to Quality of Future Context for Proactive Smart Systems

Yves Vanrompay[1] and Yolande Berbers[2]

[1] MAS Laboratory, Ecole Centrale Paris
Grande Voie des Vignes, F-92 295 Chatenay-Malabry, France
yves.vanrompay@ecp.fr
[2] Department of Computer Science, Katholieke Universiteit Leuven
Celestijnenlaan 200A, 3001 Heverlee, Belgium
yolande.berbers@cs.kuleuven.be

Abstract. Many current context-aware systems only react to the current situation and context changes as the occur. In order to anticipate to future situations and exhibit proactive behavior, these systems should also be aware of their future context. Since predicted context is uncertain and can be wrong, applications need to be able to assess the quality of the predicted context information. This allows applications to make a well-informed decision whether to act on the prediction or not. In this paper, we present prediction quality metrics to evaluate the probability of future situations. These metrics are integrated in a structured prediction component development methodology, which is illustrated by a health care application scenario. The metrics and the methodology address the needs of the developer aiming to build context-aware applications that realize proactive behavior with regard to past, present and future context.

Keywords: context prediction, quality of context, smart systems.

1 Introduction

In pervasive scenarios foreseen by ubiquitous computing, context awareness plays a central role. Context-aware systems are able to adapt themselves to their environment aiming at optimizing QoS for the user and resource consumption by taking into account and reasoning with context information. Many systems have been developed that are aware of the current context. However, these systems only react to changes in context information as they occur. As such they only take into account present (and possibly past) context information to act on. In many situations this is not sufficient since by only looking at the momentarily information it can already be too late to take appropriate actions and keep QoS at an acceptable level. Therefore applications should also take into account future context information. For example, instead of simply detecting a memory shortage as it occurs, a system should be able to predict the memory shortage in order to prevent it.

Instead of purely reactive context-aware systems, proactive context-aware systems are required that use past, present *and* future context information.

S. Andreev et al. (Eds.): NEW2AN/ruSMART 2012, LNCS 7469, pp. 152–163, 2012.

Future context information is predicted by learning of patterns and relations that are detected by analyzing the context history. We will argue for the importance of the integration of domain and prior knowledge to make the prediction process more efficient and transparent.

Proactive context-aware systems take actions based on predicted context information. The decision whether to act or not depends heavily on the quality of the predicted information. Since predictions are naturally uncertain and can be wrong, well-informed decision making taking into account the quality of the information is needed. While the cost of acting based on a wrong prediction can be limited to user annoyance or wasted system resources, more damage can be caused in the area of safety-critical systems. Therefore we propose a set of quality of future context (QoFC) metrics which allows the application to form a view on whether the prediction can be trusted enough to effectively be useable. As ubiquitous systems should be as unobtrusive for the user as possible, explicit user feedback should be kept to a minimum in the decision making process.

To support the developer in realizing *future context*-aware applications, we integrate the metrics in an overall structured methodology for developing concrete prediction components. All together, we aim to provide an efficient and transparent approach to context prediction. Efficiency is achieved by supporting the developer with a clear methodology. Transparency means that it is clear for the application what the quality is of a particular predicted context.

This paper is organized as follows. Section 2 introduces the concept of quality of future context together with a taxonomy of concrete quality metrics. We explain how these metrics can be concretely used by context-aware applications. Section 3 presents a step-by-step prediction component development methodology, illustrated by a case study from the health care domain. Section 4 gives an overview of related work, after which we draw conclusions and discuss future research directions.

2   Context Prediction Quality Metrics

Practical applications should not rely on context prediction results per se. The predicted value might be incorrect for several reasons. The given input might be erroneous or there might be several alternatives for the chosen value. As a consequence, the prediction results will not be accurate in all cases. Applications relying on these results should be able to assess the trust they can have that the prediction is correct. This allows the applications to decide whether they want to use the result, even if the probability of being incorrect is high.

In the following we define a set of Quality of Future Context (QoFC) metrics and explain how these metrics can assist an application in making decisions and take appropriate actions based on the quality of the predicted context information. We consider that QoFC is an objective notion when it is provided by a prediction component to an application. However, when it is effectively used by a specific application to determine its worth, it becomes subjective.

Quality of Context (QoC) is any inherent information that describes context information and can be used to determine the worth of the information for a specific application. This includes information about the provisioning process the information has undergone (history, age), but not estimations about future provisioning steps it might run through. [6]

The above QoC definition can equally be applied to QoFC. As such, QoFC metrics are objective metrics about the quality of the provided *future context* information. Applications can prioritize and weigh several QoFC metrics according to their specific requirements, turning these metrics into subjective ones.

2.1   Quality of Future Context Metrics

We propose a taxonomy of QoFC metrics used to assess the confidence one can have in the quality of both predictors in general and concrete predicted context values. As can be seen in Figure 1, QoFC metrics are divided into three categories.

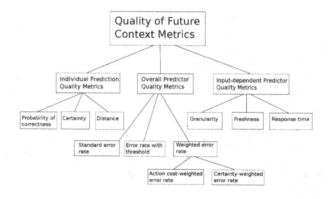

Fig. 1. Quality of Future Context metrics taxonomy

Individual Prediction Quality Metrics: Individual prediction quality metrics assess the confidence an application can have in individual, concrete context predictions of a predictor. For example, the probability of a predicted location tells something about the confidence one can have in that individual prediction, but says nothing about the general accuracy of the predictor over a whole dataset.

- *Probability of correctness P:* The probability of the most likely prediction, i.e. the prediction with the highest probability. In statistics, this corresponds to the confidence one has in a classification.
- *Certainty C:* This metric does not only take into account the probability of correctness P, but also m, the number of possible outcomes, i.e. predictions with a probability higher than zero. It is defined as follows: $c = \frac{p - \frac{1}{m}}{1 - \frac{1}{m}}$. For an extensive evaluation of this metric, we refer to previous work [1].

– *Distance metric D:* Suppose P_a is the probability of the most likely predicted value, and P_b is the probability of the second most likely predicted value. Then D is the difference between P_a and P_b: $P_a - P_b$.

Overall Predictor Quality Metrics: This second category of QoFC metrics, the overall predictor quality metrics, gives information about the overall accuracy of a given predictor. Several variants of the error rate (i.e. the number of erroneous predictions relative to the total number of predictions over a dataset) belong to this category.

– *Standard error rate E_s:* the number of incorrect predictions relative to the total number of predictions.
– *Error rate with threshold E_t:* the number of incorrect predictions relative to the number of total predictions, taking into account only the predictions with a probability of correctness above a threshold.
– *Certainty-weighted error rate E_c:* This error rate penalizes predictions with a high associated certainty c and which are wrong compared to those which have a low certainty c and are wrong by using the certainty values as weights in the error rate formula.
– *Action cost-weighted error rate E_{ac}:* This error rate penalizes predictions which are wrong and for which the actions taken based on the prediction are costly. A 3-dimensional cost matrix is associated with this error rate, expressing the cost of performing action a instead of the correct action b, given the future context.

Input-Dependent Predictor Quality Metrics: As a third category, we group quality metrics that can depend on the input needed by the predictor itself. For example, the freshness of the predicted context value can depend on the freshness of the data that is provided to the predictor to make its prediction. Also, the time needed for making a prediction can depend on the time that sensors need to provide the required information to the predictor, in addition to the time necessary to calculate the prediction itself.

– *Granularity G:* Granularity gives an indication on how fine-grained the predicted value is. E.g. an algorithm can be able to predict a user's location onto the street address level or the city level. The granularity of the predicted result can depend on the capabilities of the prediction algorithm itself, but also on the granularity of the input value(s) given to the algorithm, and thus on the kind of sensors that are available.
– *Freshness F:* Freshness gives an indication of the age of the predicted context information. The freshness can depend on the properties of the prediction algorithm, but more critical is the freshness of the context information on the basis of which the prediction is made. E.g. if all required input to make a prediction is only available once each hour, this will naturally influence the freshness of the prediction itself.
– *Response time T_r:* This is the time between the predictor receiving a prediction request and the predictor having calculated the predicted context

information together with the relevant QoFC metrics. Response time does not only depend on the computational complexity of the prediction algorithm, but also on the time needed to get the required context information to make the prediction.

It should be noted that the individual prediction quality metrics and the overall predictor quality metrics are only applicable to prediction algorithms that have a probabilistic outcome. Although a prediction is almost always probabilistic, i.e. a prediction is not absolutely guaranteed to be correct, there exist forecasting algorithms that do not give a probabilistic result. In that case, it is not clear in current research what kind of quality metrics that are transparent and understandable could be applied. However, probabilistic algorithms are the majority of prediction algorithms that are effectively applied for context prediction in current state-of-the-art, and thus this limitation is not an issue in most practical applications.

2.2 Use of QoFC Metrics by Applications

The quality metrics can be used by an application needing future context information at two points in time:

1. The quality metrics can assist an application in choosing the context predictor(s) most suited to the application requirements in case several predictors are available that provide the same context information (with different QoFC).
2. When an application is provided with a concrete predicted value, the associated QoFC metrics help in deciding whether to use the individual prediction or not.

1) Choosing the Appropriate Context Predictor. When an application requests a prediction of context information, several predictors can be available for the application. These predictors can be different algorithms, variants of the same algorithm (with different parameters) or being deployed on nodes with different resource characteristics, influencing e.g. response time. The predictors are all able to predict the same context information, but with different QoFC metrics. By taking into account the QoFC requirements requested by the application, a well-informed decision can be made to select one or more particular predictors which are available and which conform to the application QoFC requirements.

The QoFC metrics relevant to the decision process of choosing appropriate predictors are the input-dependent predictor quality metrics and the overall predictor quality metrics. Granularity, freshness and response time are quality metrics that act as preconditions on deciding whether the predictor is useful for the application or not. For example, a time-critical vehicle guidance application will need guarantees on a fast response time of a future location predictor. On the other hand, for an interactive museum visit application, response time is not that important, but granularity of the provided future location is of significance. Thus, the input-dependent predictor quality metrics are a first filtering mechanism for an application in selecting suitable predictors. In a second step the overall predictor quality metrics are taken into consideration. Since these metrics assess the overall quality of the predictor in terms of the number of correct predictions relative to the total number of predictions, they allow to discriminate between different predictors and to choose the best-performing one(s).

2) Assessing the Quality of Concrete Prediction Results. The second use of the proposed QoFC metrics is when the application has to decide whether to effectively utilize and act on a concrete predicted context value or not. Both the individual prediction quality metrics and overall predictor quality metrics play a role in the decision making. The metrics that are considered important for the application can be combined with developer-defined weights to obtain an overall confidence value in the predicted result.

For example:

$QoFC(pred) = w_1.P + w_2.C + w_3.E_c$, where $w_i = 1$, $0 < QoFC(pred) < 1$

The choice of the particular QoFC metrics included in the overall QoFC value and their associated weights depend on the application requirements. It should be noted that the QoFC metrics on themselves are objective metrics, as already mentioned. However, when they are effectively used by an application and/or combined with weights as in the formula above, the metrics become subjective. The decision whether to use the predicted context information does not only depend on the QoFC metrics mentioned above, but also on the cost of the actions performed as a result of the prediction. As such, both the QoFC metrics and the cost of the actions have to be taken into account and balanced.

The following rule D (adapted from [11]) balances the quality of the predicted context and the cost associated with deciding to act upon a wrong prediction:

$D : IF \, QoFC(pred) > A \, AND \, Cost(actions) < B \, THEN \, take actions$

By balancing the quality of the predicted context and the possible cost of an action, the application can decide whether to effectively use the predicted context or not.

3   Methodological Predictor Component Development

Reusable context and predictor plug-ins allow developers to separate the future context-awareness of an application from its functional logic. This section gives an overview of the steps a developer needs to take to create a predictor plug-in for the MUSIC context middleware [12]. The methodology consists of the following steps:

1. Identify the context type that will be predicted.
2. Identify the (context and domain) information that is required to make the prediction.
3. Identify context plug-ins for the different context types, take into account dependencies, and associate relevant domain knowledge to the context plug-ins.

4. Identify QoFC metrics applicable to the chosen prediction algorithm and relevant for the domain.
5. Implement the context model.
6. Define the XML descriptor of the plug-in and extend/implement the standard plug-in classes.

In the following, we describe each step in the development methodology, and apply it to a case study in the health care domain. A more detailed description of this case can be found in [2]. In hospital environments, nurses are equipped with a PDA on which they give in patient and other medical data. This data has to be synchronized as much as possible with a central server. Since there are only a few areas where (Bluetooth) network connectivity is available, and to prevent continuous polling for the network in order to save battery power, we predict the most likely time when the nurse will be at the location with network availability.

Step 1: Identify the predicted context type. The primary goal of the health care case is to efficiently synchronize patient data residing on the nurse's PDA with the central server. Since network availability is localized, the future context that needs to be predicted and will be provided by the predictor plug-in is the moment in time at which network connectivity can be expected to be available.

Step 2: Identify the information that is required to make the prediction. The predicted future context information is further analyzed to identify which context information is needed to make the prediction. In the health care case, we need to predict the expected time of network connectivity, which translates into predicting when the nurse will be at the location where the network is available. Firstly, the path the nurse follows through the hospital division depends on what kind of activity she/he is performing (e.g. bringing around food, measuring essential physical parameters, ...). Secondly, since the nurse follows a probable path visiting patient after patient (and thus location after location), the current location is of importance in predicting the series of future locations (including the location with Bluetooth activity). Thirdly, the time needed to reach the location with Bluetooth connectivity also depends on the activity, because for example medical treatment of patients will take longer on average than handing out the lunch. To summarize there are two relevant context types: current location and current activity, while the average time needed to perform the activity on a patient is domain-specific knowledge.

- *Current location:* the location of the nurse (i.e. the room she/he is in) can be inferred by placing RFID tags near the patient's bed or at the room entrance, which can be detected by an RFID-reader in the nurse's PDA.
- *Current activity:* as mentioned, the activity being performed has an influence both on the path typically being followed and on the time needed. The current activity is inferred by checking the kind of data the nurse is entering in the PDA (e.g. long daily patient treatment or just taking basic

physical parameters like temperature and blood pressure, or gathering lunch or dinner).

- *Time needed for the activity:* The average time needed for an activity is domain knowledge inferred by analyzing a dataset containing typical nurse's activities and the durations of these activities.

Additional domain knowledge includes the location of the Bluetooth access point. The resulting hierarchy of context types is illustrated in Figure 2:

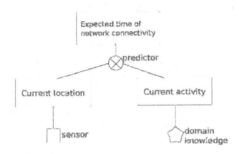

Fig. 2. Context type hierarchy for the nurse application scenario

The above Figure also gives information on the dependencies between the different context plug-ins (sensors, reasoners and predictors) that correspond to the context types, which is necessary information for step 3 below.

Step 3: Identify context plug-ins for the different context types, take into account dependencies, and associate relevant domain knowledge to the context plug-ins. In this step, a context plug-in is defined for each context type identified in the previous step. For each context plug-in that is identified, an existing implementation can be reused or a new one can be constructed. In most cases, context plug-ins for basic context information will be available. On the other hand, since predictor plug-ins use application-specific domain knowledge, they have to be implemented for each application. However, a set of predictor skeletons can be provided that implement popular context prediction algorithms. For example, part of the prediction process in the health care case is predicting future locations based on the current location. Predictor skeletons implementing standard location prediction algorithms like markov chains are provided as libraries to the developer.

For our case we identified the following context plug-ins:

- *Current location plug-in:* a sensor plug-in that reads RFID tags to provide the current location of the nurse.
- *Current activity plug-in:* A reasoner plug-in that infers the current activity based on the nurse's input of information on the PDA or the time of day.
- *Future locations plug-in:* a predictor plug-in that, based on the current activity and the current location, predicts a series of future locations representing

the most likely path the nurse will follow throughout the hospital division in performing the specific activity.

- *Expected time of network connectivity plug-in:* a predictor plug-in that predicts the expected moment of network connectivity by computing the time duration needed for the nurse to reach the location with Bluetooth activity. This predictor takes into account the path the nurse follows in performing her activities, and the average time duration needed to perform the activity to the individual patients that will be visited before reaching the Bluetooth range.

Figure 3 shows the dependencies between the different plug-ins which were identified.

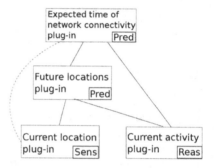

Fig. 3. Context plug-ins and dependencies for the nurse application scenario

Since the possibility must exist to check whether the predicted future context matches with what actually occurs, a dependency must be added between the predictor plug-in and the context plug-in that provides the current value of the predicted context type. This is denoted in the figure above by the red dashed line. For example, to check whether a predicted location is correct, the predictor needs input from the location sensor plug-in concerning the actual location at the predicted time. Also in this step, relevant static and dynamic domain knowledge identified in step 2 needs to be associated with the context plug-ins where the knowledge is effectively used.

Step 4: Identify relevant QoFC metrics. The QoFC metrics that were proposed in the previous section can be divided in two sets: those that are relevant for the predictor plug-ins general characteristics, and those that are relevant for individual concrete context predictions. The overall predictor quality metrics and the input-dependent predictor quality metrics give information on the overall quality of the predictor plug-in. As such, they are properties of the predictor plug-in and instantiated in the QoFCMetadata field of the plug-in. On the other hand, the individual prediction quality metrics give information on the quality of a concrete prediction. So these metrics belong to the context object that represents an individual prediction and they are instantiated in the metadata field of the ContextElement object.

For the expected time of network connectivity plug-in of the hospital use case the following QoFC metrics are considered relevant:

- *Probability of correctness:* This metric corresponds with the probability given by the markov chain of being at the location with Bluetooth connection within n time steps.
- *Standard error rate:* This metric gives the frequency of the predictor plug-in being correct in predicting the time of network connectivity availability.
- *Granularity:* This quality metric corresponds to the uncertainty of the exact time of network availability due to the variance in the duration of the activity currently performed by the nurse. This means the predictor can provide context information with different granularities which depend on the specific activity being carried out by the nurse.

Step 5: Implement the context model. Each provided and required context type needs to be modeled as an entity-scope pair with associated representation and QoFC metadata according to the MUSIC context model [13]. We illustrate this with the *current location* context type:

- Entity: The current location belongs to the user entity: *concepts.entities.user—self.*
- Scope: The scope is the location of the user: *concepts.scopes.environment.location.*
- Representation: The location is represented as a String, e.g. *room123.*
- QoFCMetadata: The only individual prediction quality metric identified as relevant in step 5 was the probability of correctness, which is represented as a double in the metadata field *ProbCorr.*

Step 6: Define the XML descriptor of the plug-in and extend the standard plug-in classes. The provided and required context types must be defined in the XML descriptor of the plug-in. Additionally, the symbolic name of the component, the class implementing the component and the provided interface must be specified to allow OSGi to dynamically activate and deactivate the predictor plug-in. The activate and deactivate methods are needed to start and stop the generation of context events and to initialize local variables and data structures. The *contextChanged* method handles context events that are received from other plug-ins. It analyzes the *ContextElement* encapsulated in the event to infer higher-level context information. Finally, everything is put in a JAR file for OSGi bundle packaging.

This section presented a step-by-step methodology that allows the developer to build context predictor plug-ins in a structured way. While parts of the methodology (e.g. step 6) are tailored to the MUSIC context middleware, we believe the approach can be applied in general for developing context prediction components in other systems or frameworks. Identification of the relevant context types, quality metrics and the dependencies between the different components, representing context information in an ontology and incorporating domain knowledge are universal requirements in developing context prediction components.

4   Related Work

Intensive research has been carried out in the domain of modeling and using quality of context information. Work by Buchholz et al. [3], Henricksen et al. [4] and Manzoor et al. [5] defined several quality metrics for context information. Buchholz et al. [3] argued on the importance of QoC for real-life applications to make effective use of provided context information. They defined QoC as any information that describes the quality of information that is used as context information. Authors presented an initial set of QoC parameters, being precision, trust-worthiness, probability of correctness, resolution and up-to-dateness. Other authors like Krause et al. [6], Sheikh et al. [7] and Abid et al. [8] have further added to these parameters. While the metrics proposed for QoC are also relevant for predicted context information, they are not sufficient. Due to the prediction step, additional uncertainty is added to the predicted context value. The metrics we presented contribute to quantifying this additional uncertainty. Boytsov et al. [9] mention as a major challenge concerning context prediction that there is a need for automated decision making which should be based on the quality of the predicted context. Also, Nurmi et al. [10] acknowledge the need for metrics in order to evaluate the quality of a context prediction. Prediction algorithms which have probabilistic outcomes, like the ones evaluated in our research, give a natural way in assessing their quality.

5   Conclusions

This paper presented some key issues towards achieving *future context* aware applications. Our main contribution is the introduction of a set of quality metrics for predicted context information, integrated in a development methodology for realizing in a structured and efficient way context predictor components. The methodology identifies the different context types involved, and the associated context plug-ins with their dependencies. Context information is modeled and quality information is added as metadata, both to the plug-ins as to the concrete context values. Future work includes the fusion of quality of context with QoFC information metrics, and the aggregation of different QoFC metrics in order to get a global view of the quality of the context. Also, instead of letting application developers decide when to effectively use the predictions based on the quality metric values, we have initial experiments on automatically learning when to take action or not using reinforcement learning techniques.

References

1. Vanrompay, Y., Mehlhase, S., Berbers, Y.: An effective quality measure for prediction of context information. In: Proceedings of the 7th IEEE Workshop on Context Modeling and Reasoning (CoMoRea) at the 8th IEEE Conference on Pervasive Computing and Communications (PerCom 2010), Mannheim, DE (March 2010)

2. Vanrompay, Y., Berbers, Y.: Efficient context prediction for decision making in pervasive health care environments: a case study. In: Supporting Real-Time Decision Making: The Role of Context in Decision Support on the Move. Annals of Information Systems, vol. 13, pp. 303–318. Springer (2010)

3. Buchholz, T., Schiffers, M.: Quality of context: What it is and why we need it. In: Proceedings of the 10th Workshop of the OpenView University Association: OVUA 2003 (2003)

4. Henricksen, K., Indulska, J.: Modelling and using imperfect context information. In: Proceedings of the Second IEEE Annual Conference on Pervasive Computing and Communications Workshops, PERCOMW 2004, Washington, DC, USA, pp. 33–37 (2004)

5. Manzoor, A., Truong, H.-L., Dustdar, S.: Using Quality of Context to Resolve Conflicts in Context-Aware Systems. In: Rothermel, K., Fritsch, D., Blochinger, W., Dürr, F. (eds.) QuaCon 2009. LNCS, vol. 5786, pp. 144–155. Springer, Heidelberg (2009)

6. Krause, M., Hochstatter, I.: Challenges in Modelling and Using Quality of Context (QoC). In: Magedanz, T., Karmouch, A., Pierre, S., Venieris, I.S. (eds.) MATA 2005. LNCS, vol. 3744, pp. 324–333. Springer, Heidelberg (2005)

7. Sheikh, K., Wegdam, M., van Sinderen, M.: Middleware support for quality of context in pervasive context-aware systems. In: Proceedings of the Fifth IEEE International Conference on Pervasive Computing and Communications Workshops, Washington, DC, USA, pp. 461–466 (2007)

8. Abid, Z., Chabridon, S., Conan, D.: A Framework for Quality of Context Management. In: Rothermel, K., Fritsch, D., Blochinger, W., Dürr, F. (eds.) QuaCon 2009. LNCS, vol. 5786, pp. 120–131. Springer, Heidelberg (2009)

9. Boytsov, A., Zaslavsky, A.: Context prediction in pervasive computing systems: Achievements and challenges. In: Supporting Real Time Decision-Making. Annals of Information Systems, vol. 13, pp. 35–63. Springer US (2011)

10. Nurmi, P., Martin, M., Flanagan, J.A.: Enabling proactiveness through context prediction. In: Workshop on Context Awareness for Proactive Systems, CAPS 2005 (2005)

11. Loke, S.W.: Facing uncertainty and consequence in context-aware systems: towards an argumentation approach (2004)

12. Paspallis, N., Rouvoy, R., Barone, P., Papadopoulos, G.A., Eliassen, F., Mamelli, A.: A Pluggable and Reconfigurable Architecture for a Context-Aware Enabling Middleware System. In: Meersman, R., Tari, Z. (eds.) OTM 2008, Part I. LNCS, vol. 5331, pp. 553–570. Springer, Heidelberg (2008)

13. Reichle, R., Wagner, M., Khan, M.U., Geihs, K., Lorenzo, J., Valla, M., Fra, C., Paspallis, N., Papadopoulos, G.A.: A Comprehensive Context Modeling Framework for Pervasive Computing Systems. In: Meier, R., Terzis, S. (eds.) DAIS 2008. LNCS, vol. 5053, pp. 281–295. Springer, Heidelberg (2008)

Integration of Advanced LTE Technology and MIMO Network Based on Adaptive Multi-beam Antennas

Natan Blaunstein[1] and Michael Borisovich Sergeev[2]

[1] University of Ben-Gurion, Beer Sheva, Israel
natanbcse@bgu.ac.il
[2] State University of Aerospace Instrumentation, Saint Petersburg, Russia
mbse@mail.ru

Abstract. In this work, we put a question on integration of the advanced LTE technology and multiple-input multiple-output (MIMO) system for multi-user (MU) multiple access deployment. We analyze such an integrated network based on multi-beam adaptive antennas application for the purposes of increase of spectral efficiency of such a network, rejection of inter-symbol and inter-user interference, as well as minimization of multiplicative noise effects, caused by multipath phenomena (e.g. fading phenomena), on information data sent to each subscriber located in the area of service of such an integrated network. Based on the framework of multi-parametric stochastic approach described "reaction" (or response) of each specific terrain channel, rural, sub-urban, and urban, it is shown that such a combination of LTE technology and MU-MIMO system, based on multi-beam adaptive antennas, can significantly increase capacity and spectral efficiency of each channel, and finally, for a whole system, during the multiple access deployment, as well as can fully mitigate multiplicative noises occurring in terrestrial built-up environments. The corresponding computer experiments and real measurements carried out for specific urban scenarios are presented to proof the proposed approach.

1 Introduction

The LTE (long term evolution) technologies (from Release 5 to Release 10) for the increasing demand of high mobile communication rates have recently received considerable attention both in applied applications and in the recent publications [1]-[6]. Specifically, the main goal of all releases, from 5 to 10, was to increase the spectral efficiency and the maximum data rate (e.g., capacity) of 3-G wireless networks using a new hardware a special digital signal processing techniques that recently have been developed. Development and implementation of advanced LTE technologies for modern 4-G mobile and stationary broadband communication networks require renewed investigation of propagation channel response for new practical scenarios, mostly for urban environments with dense building areas and complicated overlay profile of buildings.

Of course, the large capacity demands can be realized only by highly efficient and optimized network infrastructures. As was expected initially in deployments of LTE

S. Andreev et al. (Eds.): NEW2AN/ruSMART 2012, LNCS 7469, pp. 164–173, 2012.
© Springer-Verlag Berlin Heidelberg 2012

technologies, significant improvements in rejecting of inter-user and inter-carrier interference were expected with ongoing roll-out of OFDMA-based networks, such as WiMAX and 3GPP LTE. Unfortunately, these two standards do not fulfill the requirements for rejection all types of noises caused due to multipath fading occurring in the different scenarios of built-up channels and, therefore, under various "reactions" of each user channel. A relatively recent idea of extending the benefits of MIMO systems to multiuser access scenarios was proposed to obey a lot of worst effects by integrating multiuser (MU) scenarios with MIMO system and the advanced LTE technologies [7-13]. Here, the different MU-MIMO scenarios were studied in 3GPP standard integrated with regular LTE (releases 5 to 7) to LTE Advanced (releases 8 to 10) technology as a first steps towards the 4th generation (4-G) requirements defined by the International Telecommunication Union (ITU) – to achieve data rate via each user channel of 100Mbps in high-mobility applications and 1 Gbps for low-mobility applications such as LAN and/or WLAN access. For this purpose, a new European Project that currently was proposed and called EU FP7 project SAMURAI (Spectrum Aggregation and Multiuser MIMO: ReAL-World Impact), was investigated recently in [7] with special focus on practical implementations and deployment aspects. At the same time, as was shown in these works, despite the fact of increasing of channel (and then, the whole system) capacity, the problem to eliminate multipath fading effects of the propagation environment, as source of multiplicative noise, is still actual.

Therefore, in this work we study of how to mitigate propagation effects by usage of adaptive multibeam (or phased-array) antennas at both-end terminals of the MU-MIMO wireless communication network combined with the advanced LTE technologies.

Specially, Section 2 provides a brief overview on MU-MIMO system. In Section 3, the LTE-based technologies are briefly addressed. Section 4 presents short introduction in radio propagation for adapted multibeam antennas applications with the advantages and disadvantages of their deployment. In Section 5, we present some numerical and real experiments carried out recently for specific urban environments and show advantages of the integrated LTE and MU-MIMO system with respect to LTE combined with single-user (SU)-MIMO schemes. Simulations are based on LTE-release 8, following [14], as an example. Finally our conclusions are summarized in Section 6.

2   MU-MIMO Implementation

MIMO concept was introduced during recent decades as an effective technology based on spatial diversity surfacing from the usage of multiple antennas at both terminal ends of communication network to improve its reliability, increase spectral efficiency and achieving spatial separation of users for multi-user interference (MUI) elimination. Moreover, spatial diversity is also beneficial for MU cellular systems where spatial resources can be used to transmit information data to multiple users simultaneously. Recently, the MIMO transmission techniques and configurations were proposed to support radio access technologies, such as the WiMAX and LTE standards [1, 2, 4, 8].

The MU-MIMO was performed as a set of advanced MIMO techniques. This advanced technology exploits the availability of multiple independent user equipments (UEs) in order to enhance the communication capabilities of each individual UE (stationary or mobile). The canonical single user (SU) MIMO network is shown in Fig.1a as a comparison with that for multiple user (MU) MIMO network (see Fig. 1b).

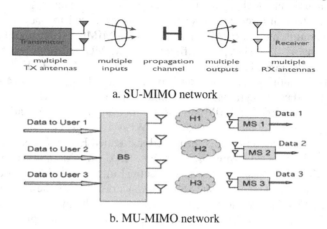

a. SU-MIMO network

b. MU-MIMO network

Fig. 1.

It is clearly seen that for each mobile subscriber (MS) in downlink communication, the base station (BS), as a multiple output antenna system send for each user with number i the corresponding data via the corresponding propagation channel with its response H_i, caused by multipath fading occurring in real communication environment (see below) that, finally, corrupt information data passing via each channel with noise.

In [7,14] was proposed some new precoding technique that reduces these corrupted effects in communication channels by getting feedback from each individual user equipment (UE), where receiver of each desired MS obtains the information about channel response H_i and the type of noise.

3   LTE-Based Technologies

As was mentioned above, the LTE were introduced to increase significantly the data rate in existing mobile WiMAX technologies. Thus, the IEEE 802.16e-2005 WiMAX system allows to achieve data rate from 25 Mbps (uplink) to 50Mbps (downlink), whereas 3GPP-LTE (E-UTRAN) system gives possibility to increase these numbers in twice [12, 13] with increase of grade of service (GoS) per twice too, and spectral efficiency – from 3.75 bits/sec/Hz (WiMAX)   to 5 bits/sec/Hz (LTE). The first advanced release of LTE is Release 8 that was aimed at defining the new OFDMA-based air-interface and introduced advanced SU-MIMO transmission scheme shown in Fig. 1a. The transmission mode from up to four antenna ports was supported (see fig. 2).

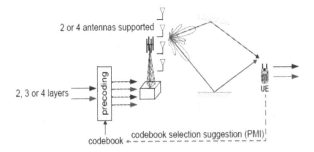

Fig. 2. Two or four transmitting antenna ports at BS in downlink spatial multiplexing

This LTE-Release 8 technology suggests existence from 2 to 4 layers, as simultaneous transmissions of separate data to multiple user equipments (UEs) on the same time-frequency resource, and usage of these separate layers to differentiate the transmission data for each desired user. Such a procedure can be achieved in downlink communication on the framework of 2-D OFDMA technique, as combination of TDD and FDD modes. For uplink communication, single carrier uses the FDMA technique. In both cases of communication between BS and UE, the adaptive modulation and coding techniques, such as QPSK/16QAM/64QAM, are used. The codebook via precoder matrix indicator can select and suggest on each channel response the corresponding noise occurring in it. Such an integration of different techniques allows achieving the date rate of 50Mbps and spectral efficiency of 2.5 bits/sec/Hz in uplink and correspondingly 100 Mbps and 5bits/sec/Hz in downlinks.

LTE-Release 9 technology was performed as an enhanced version of the Release 8.In this release, new support has been added for transmission modes utilizing two new virtual antenna ports with precoded UE-specific reference signals. CDMA technique was used to orthogonalize the transmission of these two virtual antenna ports, while nonorthogonal scrambling codes were introduced to support dual-layer transmission on each of the antenna ports [14]. This new dual-layer transmission mode was targeted for beamforming schemes and supports MU-MIMO transmission for up to 4 transmit antenna ports in downlink. As was mentioned in [14], a fully adaptive SU/MU-MIMO transmission mode was not supported in LTE Release 9, but was successfully adapted in Release 10, called the MU–MIMO LTE Advanced Technology [6, 12-14].

The Release 10, or LTE Advanced, technology combines in it all previous features of Release 8 and Release 9, but can be supported with up to 8 x 8 MIMO antenna ports configuration (see Fig. 3).

As is clearly seen from the sketched configuration, BS   downlink transmission is supported with 8 antennas, whereas the uplink multi-antennas transmission is with up to 4 antennas. Two codebooks support 6-layer mapping of the separate information data for each desired UE (up to 4 equipments). The layer corresponding de-map is utilized for differentiation of the 4-layered data to each desired UE. Consequently, the set of precoding codebooks can be extended for LTE Advanced depending of the configuration of MIMO system [13]. Thus, for configuration with 2 or 4 transmitting antennas, the codebook is the same as for previous releases, shown in Fig. 2. For configuration with 8 transmitting antennas (Fig. 3), a dual-codebook approach is used.

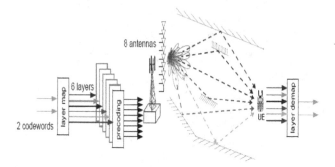

Fig. 3. Configuration of the advanced LTE-Release 10 network

The main problem that should be resolved in such an advanced MU/MIMO-LTE network is how to predict the channel response and the corresponding non-Gaussian noise in each user's communication link according to the "reaction" of the environment.

4   Propagation Conditions Modelling for Multibeam Antennas Application

The main goal of usage of multibeam antennas was to deploy in existing and advanced communication networks so-called spatial filtering of each desired signal. Usage of multiple beams concept with narrow beamwidth allows eliminating the inter-user interference (IUI), minimizing the multiplicative noise caused by fading phenomena occurring in the multipath communication channels, mostly for urban and sub-urban environments, and finally, tracking each mobile subscriber (MS) during his existence in area of service of multibeam BS antenna. Unfortunately, as was shown in [15-17], in each direction in azimuth and elevation domains for a current position of the desired subscriber the channel specific response should be introduced, and it was proposed to define this response by Ricean K-parameter of fading, as a ratio of the coherent (e.g. signal) and incoherent (e.g. noisy) components of the recorded signal. This parameter was evaluated for different terrestrial scenarios, rural, mixed residential, sub-urban and urban, and was shown that this parameter fully depends on features of built-up terrain, such as buildings' density, overlay profile, orientation and elevation with respect to BS and MS antenna, heights of the terminal antennas, and so on.

Thus, for the mixed residential area using results obtained in [17], we get the expression for K-factor along the radio path d between two terminal antennas:

$$
K = \frac{Ico}{Iinc}
$$

$$
= \frac{exp\left\{-\gamma_0 \cdot d \dfrac{\bar{h} - z_1}{z_2 - z_1}\right\} \left[\dfrac{sin\left(k \cdot z_1 z_2 / d\right)}{2\pi d}\right]^2}{\dfrac{\Gamma}{8\pi} \cdot \dfrac{\lambda \cdot l_h}{\lambda^2 + \left[2\pi l_h \gamma_0\right]^2} \cdot \dfrac{\lambda \cdot l_v}{\lambda^2 + \left[2\pi l_v \gamma_0 \left(\bar{h} - z_1\right)\right]^2} \cdot \dfrac{\left[\left(\lambda d / 4\pi^3\right)^2 + \left(z_2 - \bar{h}\right)^2\right]^{1/2}}{d^3}} \tag{1}
$$

For the urban and sub-urban environments following [17], we finally get:

$$K = \frac{Ico}{Iinc_1 + Iinc_2}$$

$$= \frac{exp\left\{-\gamma_0 d\frac{(z_1-\bar{h})}{(z_2-z_1)}\right\}\frac{sin^2(k \cdot z_1 z_2/d)}{4\pi^2 d^2}}{\dfrac{\Gamma \cdot \lambda \cdot l_v\left[(\lambda d/4\pi^3)+(z_2-\bar{h})^2\right]^{1/2}}{8\pi\left[\lambda^2+(2\pi l_v\gamma_0 \cdot(z_1-\bar{h}))^2\right]d^3} + \dfrac{\Gamma^2\lambda^3 l_v^2\left[(\lambda d/4\pi^3)+(z_2-\bar{h})^2\right]}{24\pi^2\left[\lambda^2+(2\pi \cdot l_v \cdot \gamma_0 \cdot(z_1-\bar{h}))^2\right]^2 d^3}} \tag{2}$$

Here, in formulas, $\gamma_0 = 2\bar{L}v/\pi$ is the density of the buildings' contours (in km^{-1}), v is the density of buildings in the area of service (in km^{-2}), $\bar{L}$ is the average length (or width) of buildings (in m), depending on its orientation with respect to terminal antenna, Γ is the absolute value of the reflection coefficient, l_v and l_h are the vertical and horizontal scales of coherency of the reflections from building walls, respectively (in m), $\bar{h}$ is the average buildings' height (in m), z_1 and z_2 are the heights of the MS and BS antenna, respectively.

5   Numerical Experiment for Specific Environment

In this section, we present some numerical experiment carried out for one of the the practical urban environment, such as the Ramat Gan market area in Israel shown in Fig. 4.

Fig. 4. Ramat Gan Built-up area; number of building floors are indicated by the corresponding color

The BS antenna height was $z_2 = 50m$. The outdoor desired MSs were located on the ground at the height of $z_{1i} = 2m$. The simulations were performed for the following parameters of propagation: $\bar{h} = 25m$, $\gamma_0 = 10km^{-1}$, $l_v = 1-2m$ [17].

Results of simulation of K-parameter of fading is shown in Figs. 5a and 5.b for the carrier frequency of f=900 MHz and 1,800 MHz respectively.

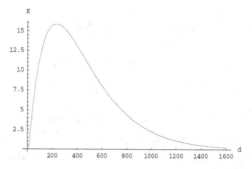

a. The Ricean *K*-factor vs. distance between BS and different MSs in Ramat-Gan market area for *f*=900 MHz

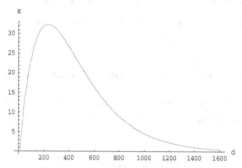

b. The same as in Fig. 5a, but for *f*=1.8 GHz

Fig. 5.

Based on these results, we can now for each MS with number i estimate signal to multiplicative noise ratio via K-factor that determines the losses and fading effects in each noisy communication channel for each desired UE. This procedure allows obtaining the channel spectral efficiency as function of K for various configurations of MIMO system consisting N outputs and M inputs.

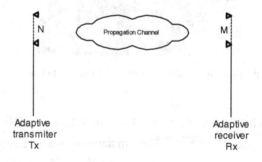

Fig. 6. Schematically presented MIMO antenna system

Thus, following [17], we can obtain for correlated and uncorrelated elements of the multibeam antenna the following formulas for spectral efficiency of the system:

— for correlated elements

$$\tilde{C}_{corr} = \log_2\left(1 + MN\left(K \cdot \frac{P}{N_{add}}\right)\middle/\left(K + \frac{P}{N_{add}}\right)\right) \text{ b/s/Hz} \qquad (3)$$

— for uncorrelated elements

$$\tilde{C}_{uncorr} = N\log_2\left(\left(K \cdot \frac{P}{N_{add}}\right)\middle/\left(K + \frac{P}{N_{add}}\right)\right) \quad b/s/Hz \qquad (4)$$

In both cases, the spectral efficiency depends not only on the number of output and input antenna elements, but also on the K-factor of fading. This dependence for different configurations of MIMO system is presented in Fig. 7.

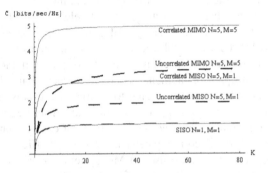

Fig. 7. Spectral efficiency of the MIMO and MISO systems vs. K-factor; as an example, SISO system is also presented

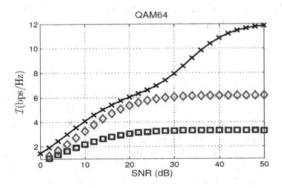

Fig. 8. Spectral efficiency in bps/Hz; □□□ indicates the SU-LTE-SISO, ◊◊◊ indicates the SU-LTE-SIMO, and +++ indicates MU-LTE-MIMO

It is clear seen that correlated antenna elements allow achieving better spectral efficiency compared with that for uncorrelated antenna elements.

Finally, we will introduce these data to SU-LTE-SISO integrated scheme (that is, for M=1, N=1), SU-LTE-MIMO (that is, for M=1, N=4), and for MU-LTE-MIMO (that is, for M=4, N=4) [antenna correlated elements]. The results of the numerical experiment are shown in Fig. 8, where K is presented, as in [17], as a signal-to-noise ratio.

It is clear seen that using the advanced LTE technology integrated with MU-MIMO system of various input-output antenna elements, it is possible to increase the spectral efficiency and the data rate in such an integrated 4-G network.

6   Conclusions

The framework for spectral efficiency (e.g. data capacity) analysis of the advanced LTE technology integrated with the heterogeneous SU-MIMO and MU-MIMO networks, which consist of the macro BS arranged by the multibeam antenna array of N elements and one or number UEs with array of M elements each, is investigated in this work. Different deployment strategies of antenna array configurations, SISO, MISO, SIMO, and MIMO were analyzed via the overall channel capacity or spectral efficiency Shannon's formulation accounting for the response of each communication channel on fading multipath phenomena occurring in each channel that can be evaluate via the Ricean K-factor, as the signal to multiplicative noise ratio. The corresponding computer experiment for one of the practical urban area shows that precise study of propagation environment and the "reaction" of each channel (for each individual UE) allows to estimate correctly capacity or spectral efficiency of each channel for different configurations of multi-user MIMO system integrated with the advanced LTE protocol standard. Finally, it can be found an effective configuration of MU-MIMO system that can satisfy increasing demand to spectral characteristics and capacity of future 4-G broadband mobile and personal networks. The main limiting factor observed during these simulations is the fading factor K as a response of each individual communication channel on data transmission.

References

1. Ghosh, A., Ratasuk, R., Mondal, B., et al.: LTE-advanced: next generation wireless broadband technology. IEEE Wireless Communicat. 17(3), 10–22 (2010)
2. Li, Q., Li, G., Lee, W., et al.: MIMO techniques in WiMAX and LTE: a features overview. IEEE Communic. Magazine 48(5), 86–92 (2010)
3. 3G Americas white paper, 3GPP Mobile Broadband Innovation Path to 4G: Release 9, Release 10 and Beyond: HSPA+, SAE/LTE and LTE-Advanced (February 2010)
4. Farajidana, A., et al.: 3GPP LTE downlink system performance. In: Proc. of the IEEE Global Communic. Conf. (Globecomm 2009), Honolulu, Hawaii, USA (November-December 2009)
5. 3GPP Technical Specification Group Radio Access Network; Evolved Universal Terrestrial Radio Access (E-UTRA), Physical Layer Procedures (Release 9), 3GPP TS36.213 V9.3.0 (June 2010)

6. 3GPP Technical Specification Group Radio Access Network; Evolved Universal Terrestrial Radio Access (E-UTRA), Further advancements for E-UTRA physical layer aspects (Release 9), 3GPP TS36.814 V9.0.0 (March 2010)
7. EU FP7 Project SAMURAI-Spectrum Aggregation and Multi-User MIMO: Real-World Impact, http://www.ict-samurai.eu/page1001.en.html
8. Kusume, K., et al.: System level performance of downlink MU-MIMO transmission for 3GPP LTE-advanced. In: Proc. of the IEEE Vehicular Technol. Conf.-Spring (VTC 2005), Ottawa, Canada (September 2010)
9. Ribeiro, C.B., Hugl, K., Lampinen, M., Kuusela, M.: Performance of linear muilti-user MIMO precoding in LTE system. In: Proc. of the 3rd Int. Symp. on Wireless Pervasive Computing (ISWPC 2008), Santorini, Greece, pp. 410–414 (May 2008)
10. Covavacs, I.Z., Ordonez, L.G., Navarro, M., et al.: Toward a reconfigurable MIMO downlink air interface and radio resource management: the SURFACE concept. IEEE Communic. Magazine 48(6), 22–29 (2010)
11. Spenser, Q.H., Swindlehurst, A.L., Haardt, M.: Zero-forcing methods for downlink spatial multiplexing in multi-user MIMO channels. IEEE Trans. on Signal Processing 52(2), 461–471 (2004)
12. 3GPP TSG RAN WG1 #62, Way Forward on Transmission Mode and DCI design for Rel-10 Enhanced Multiple Antenna Transmission, R1-105057, Madrid, Spain (August 2010)
13. 3GPP TSG RAN WG1 #62, Way Forward on 8 Tx Codebook fore Release 10 DL MIMO, R1-105011, Madrid, Spain (August 2010)
14. Duplicy, J.D., Badic, B., Balraj, R., et al.: MU-MIMO in LTE Systems. EURASIP Journal on Wireless Communic. and Networking 2011, Article 1D 496763, 13 pages
15. Yarkoni, N., Blaunstein, N., Katz, D.: Link budget and radio coverage design for various multipath urban communication links. Radio Science 42, 412–427 (2007)
16. Blaunstein, N., Yarkoni, N., Katz, D.: Spatial and temporal distribution of the VHF/UHF radio waves in built-up land communication links. IEEE Trans. Antennas and Propagation 54(8), 2345–2356 (2006)
17. Blaunstein, N., Christodoulou, C.: Radio Propagation and Adaptive Antennas for Wireless Communication Links: Terrestrial, Atmospheric and Ionosphere. Wiley, New Jersey (2007)

Feasibility Analysis of Dynamic Adjustment of TDD Configurations in Macro-Femto Heterogeneous LTE Networks

Alexey Khoryaev, Mikhail Shilov, Sergey Panteleev,
Andrey Chervyakov, and Artyom Lomayev

Intel Corporation, Nizhny Novgorod, Russia
{alexey.khoryaev,mikhail.shilov,sergey.panteleev,
andrey.chervyakov,artyom.lomayev}@intel.com

Abstract. The dynamic adaptation of TDD configurations to traffic conditions at low power base stations (femtocells and picocells) in heterogeneous LTE networks can be used to optimize the system performance. For instance, a significant performance improvement can be achieved in DL and UL packet throughputs. However, in practical multi-cell deployments the TDD configuration adaptation may result in opposite transmission directions in the neighboring cells and thus cause new types of inter-cell interference such as BS-BS and UE-UE. These interference types are often called DL-UL cross interference and may significantly deteriorate the overall system performance. For this reason all homogeneous Macro cell TDD deployments typically operate synchronously (i.e. use same TDD configurations). However the recent advances in the field of heterogeneous networks have reopened the dynamic TDD problem since low power small cells are rather isolated from each other and thus can be considered as candidates for dynamic TDD adaptation. In this paper we provide a feasibility analysis of dynamic TDD configuration adjustment in application to Macro-Femto LTE TDD networks.

Keywords: LTE, dynamic TDD, heterogeneous networks, femtocells, DL-UL interference management.

1 Introduction

The concept of heterogeneous network (HetNet) deployments is expected to be one of the major technological trends in the upcoming years and it promises to substantially boost performance of future broadband wireless networks. The HetNet is a multi-layered deployment which consists of macrocells (Macro cells) and low power small cells such as femtocells (Femto cells), picocells (Pico cells) or remote radio heads (RRHs) which are mainly aimed to increase the system capacity and operate in the coverage area of Macro cells. Multiple small cells are deployed at the user proximity and thus have improved channel link quality due to the reduced distance between the transmitter and receiver points. The increased cell densities lead to more efficient spectrum reuse and therefore substantially increase data rates and the amount of

S. Andreev et al. (Eds.): NEW2AN/ruSMART 2012, LNCS 7469, pp. 174–185, 2012.

traffic that can be processed by the network in a given area. The substantial benefits which HetNet deployments bring to LTE-Advanced systems come at the expense of new technical challenges associated with the multi-layered network architecture. The major issues are the increased amount of cell edge users and the high level of the inter-cell interference among different network layers (e.g. Macro and Femto cells). The comprehensive overview of HetNet challenges and interference coordination solutions adopted in LTE can be found in overview papers [1-3].

The LTE systems may operate using either frequency division duplexing (FDD) or time division duplex (TDD) mode. In FDD mode the amount of resources is equally divided between DL and UL transmission direction and thus the system may experience resource underutilization in scenarios with significant DL/UL traffic asymmetry. In TDD systems DL and UL transmission intervals are multiplexed in time and the proportion of the DL and UL resources may be adjusted to fit the traffic conditions semi-statically on a long-term basis or dynamically. In general case each cell may apply the dynamic TDD approach and adjust its UL-DL TDD configuration (i.e. proportion of DL and UL resources) to the instantaneous traffic conditions. However in homogeneous deployments using non-controlled dynamic adaptation of TDD configurations and thus non-aligned UL-DL configurations may have significant negative impact on the system performance. So in practical LTE TDD deployments fixed UL-DL configurations are synchronized across the network.

In this paper we address the scenario of using LTE TDD systems in application to HetNet deployments and in particular in application to Macro-Femto deployments. In typical HetNet deployments the amount of traffic as well as its direction (DL or UL) significantly varies across cells and in time. The traffic statistical characteristics in Macro and Femto cell may be significantly different. For instance Femto cells often serve only one user at the period of time and only one of the transmission directions (either UL or DL) is dominant. In other words, the significant DL and UL traffic imbalance may exist over a short period of time. In this case, if all Femto cells are enforced to use the same UL-DL configuration as Macro cells the traffic adaptation capabilities of Femto cells are significantly constrained. The straightforward solution to improve the user experience is to enable dynamic adaptation of TDD UL-DL configuration to the instantaneous traffic conditions. However, this may have some drawbacks in terms of the system performance that need to be carefully analyzed. In this paper we provide the results of the interference environment analysis in case of using dynamic adjustment of TDD configurations in Macro-Femto HetNet deployments.

The rest of this paper is organized as follows. In Section 2, we briefly review the concept of dynamic adaptation of TDD configurations in application to homogeneous Macro deployments. Section 3 provides an overview of investigated Macro-Femto deployment scenarios and describes the specific interference environment in case when Femto cells are allowed to use different UL-DL configurations. In Section 4, we present the results of the system level evaluation of DL and UL SINRs for Macro and Femto users. Finally in Section 5 we conclude on feasibility of dynamic UL-DL reconfiguration in Macro-Femto cell networks and highlight the directions for future work.

2   Dynamic Adjustment of TDD Configurations in Homogeneous Macrocell Deployments

The concept of dynamic TDD, i.e. adaptation of the UL-DL configuration to the traffic conditions, in broadband wireless networks was actively discussed in the past in application to homogeneous Macro cell deployments [4-5]. The main practical technical challenge of the dynamic adaptation of UL-DL configurations in this environment is the emergence of new types of inter-cell interference such as DL-UL interference between neighboring base stations (BS-BS/eNodeB-eNodeB) and user terminals (UE-UE) (see Fig. 1). The BS-BS interference is considered to be the most severe problem for Macro cells when dynamic UL-DL configurations are used. Assuming that neighboring cells have opposite transmission directions the BS-BS interference can completely distort the reception of the uplink signals. Typically the receive power of the useful uplink signal is far below the level of the interference signals coming from the neighboring base stations. There are two main factors that contribute to this effect. First of all, the base station transmit power (~46dBm) is about 20 dB higher comparing to the maximum transmit power of an UE (which power is restricted to 23dBm). Secondly, assuming that Macro base stations are installed above rooftops the BS-BS propagation conditions are typically line-of-sight (LOS). So the pathloss of BS-BS link is much less comparing to the BS-UE links. Another counterargument against using the dynamic UL-DL configuration in Macro cells is that the amount of traffic in DL and UL directions is rather balanced since typically Macro cell simultaneously serve a larger number of users that may have different demands it terms of DL and UL resources. In this sense the semi-static synchronous adjustment of UL-DL configuration in the whole network may be sufficient enough to effectively address average user demands.

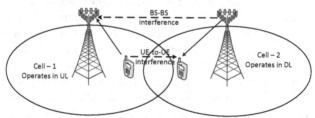

Fig. 1. Interference environment in Macro cells with dynamic TDD UL-DL configurations

The recent growth of HetNet deployments has reopened the problem of dynamic adaptation of the TDD UL-DL configurations in application to low power small cells and stimulated further research in this area. Recently in LTE Rel-11 the new study item on "Further enhancements to LTE TDD for DL-UL interference management and traffic adaptation" was initiated [6]. The main goal of this study item is to provide feasibility analysis and evaluate potential benefits from dynamic UL-DL reconfigurations in multi-cell HetNet deployment scenarios. In this paper we provide the results of our interference analysis that was conducted in the framework of LTE TDD SI [6] for the most feasible Macro-Femto deployment scenarios.

3   Macro-Femto Heterogeneous Deployment Scenario

Two Macro-Femto deployment scenarios which are of high practical interest for LTE TDD networks are investigated in this paper.

Adjacent Channel Macro-Femto Deployment Scenario. In this scenario all Femto cells operate in co-channel (same frequency channel) and all Macro cells operate in the adjacent frequency channel relative to the Femto cells. So due to the transmission on the adjacent carrier frequencies the inter-cell interference from Macro cells and Macro UEs on Femto cells and Femto UEs is substantially reduced. This scenario is mainly motivated by the adoption of carrier aggregation concepts where one base station may operate on a set of intra-band or inter-band carrier frequencies.

Co-channel Macro-Femto Deployment Scenario. In this scenario both Macro and Femto cells operate in co-channel and thus co-channel interference from Macro cells directly impacts Femto cells performance and vice versa.

The main focus of this paper is to analyze feasibility of applying dynamic UL-DL configurations in the LTE HetNet Macro-Femto network deployments. The feasibility is assessed by means of system level simulations. The mutual impact from Macro cells and Macro UEs on Femto cells and Femto UEs is evaluated. The DL and UL geometry SINRs are used as the main metrics of system performance and are separately measured for user terminals attached to Macro and Femto cells. The geometry SINR is defined as the ratio of the received signal power from useful link to the sum of noise power and total intercell interference power received from all interfering stations. For calculation of received signal power from each link the following factors are taken into account: large scale channel propagation characteristics (pathloss and shadow fading), antenna gains at the base station and mobile sides, antenna connector losses, and transmit signal power.

To analyze the Macro-Femto scenario the deployment and channel propagation models described in [7-8] are used. These models are often used for the assessment and modeling of the 3GPP LTE networks and are briefly reviewed in the next subsections. Note that the detailed assumptions on simulation parameters that were used in this study are aligned with the assumptions recommended for feasibility analysis by the 3GPP RAN4 WG and can be found in [8].

3.1   Dual Stripe Femtocell Deployment Model

For realistic modeling and evaluation of practical Femto deployments the dual stripe model (see Fig. 2) with two building blocks (each having six floors and two apartment rows on each floor) was adopted [7-8]. This model was adopted for system level evaluation and is commonly considered as one of the baseline models for evaluation of Femto cell deployments [7].

In this model Femto stations are uniformly distributed over apartments of two neighboring building blocks. Each Femto station has one associated UE which is located in the same apartment with the base station. The dual stripe model defines a rich set of propagation characteristics depending on the location of user terminal and

Femto station and aims to accurately represent the propagation characteristics for femtocell deployment scenario. The summary of path loss propagation characteristics for dual stripe model can be found in [7-8].

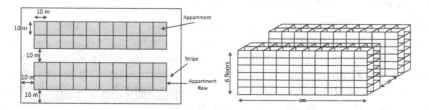

Fig. 2. Dual Stripe Femtocell Deployment Scenario

3.2   Macro-Femto Deployment Model

In this subsection a brief description of the system level model for evaluation of a single operator HetNet deployment where Macro and Femto stations operate in co-channel or in adjacent channels is provided. The illustration of this deployment scenario is shown in Fig. 3.

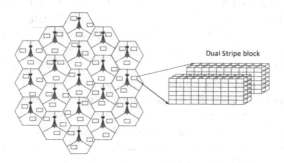

Fig. 3. Macro-Femto Deployment Scenario

For modeling of Macro cells a two-tier hexagonal cellular deployment with 19 three-sector base stations is used and a cell wrap-around mechanism is applied. To model Femto deployment one dual stripe block is randomly dropped within each sector of the Macro cell area. User terminals are divided into two groups: Macro UEs (MUE) and Femto UEs (FUE) which are associated to Macro and Femto cells, respectively. The 35% of Macro UEs are uniformly distributed inside the dual stripe buildings and the remaining 65% of Macro UEs are uniformly distributed outdoors over the Macro cell area. In each dual stripe block the Femto base stations are uniformly distributed over building floors and apartments. In total, there are twenty four activated Femto stations per two buildings of one dual stripe block which corresponds to Femto deployment ratio equal to 0.1. Each Femto cell is assumed to provide services to a closed subscriber group and in particular has only one attached Femto UE. Macro UEs do not have access to Femto stations and thus cannot be associated to them even if they have more favored channel propagation conditions to Femto cells comparing to Macro cells. Further details on the evaluation parameters adopted by 3GPP for feasibility study of dynamic UL-DL configuration in Femto cells can be found in [8].

3.3   Interference Environment in Macro-Femto Deployments

As discussed in Section 2 the mutual interference between Macro cells is rather severe so in both considered scenarios it is assumed that all Macro cells use the same UL-DL configuration which is semi-statically configured by the network so that there is no opposite transmission direction among Macro cells. It is also assumed that all deployed cells operate synchronously and that Femto cells can dynamically adjust their respective UL-DL configurations to satisfy their traffic demands. Similar to the Macro cell deployment described in Section 2 the dynamic adjustment of UL-DL configurations in Femto cells introduces new types of interference. This interference types may appear on certain subframes of LTE TDD frame structure due to overlapped transmission of DL and UL subframes in different Femto cells. In general the intercell interference environment will be expanded by four additional DL-UL interference types as listed in Table 1 and illustrated in Fig. 4.

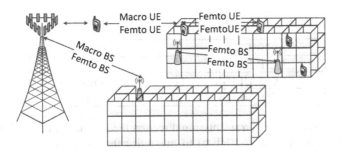

Fig. 4. DL-UL interference environment in Macro-Femto deployment model

Table 1. Intercell interference types in TDD networks with dynamic UL-DL configuration

Traditional intercell interference	New DL-UL intercell interference types
Macro BS ↔ Macro UE	Macro BS ↔ Femto BS
Femto BS ↔ Femto UE	Femto BS ↔ Femto BS
Macro BS ↔ Femto UE	Macro UE ↔ Femto UE
Femto BS ↔ Macro UE	Femto UE ↔ Femto UE

The impact of new DL-UL intercell interference environment on the system performance has not been well studied in the literature and especially in application to HetNet deployments. This factor was one of the main motivations to conduct the full comprehensive analysis of this interference environment in the framework of the LTE Rel-11 SI [6] and present the corresponding results in this paper.

4   Feasibility Analysis

In order to perform feasibility analysis of Macro-Femto deployment scenarios with dynamic UL-DL reconfiguration in Femto cells the system level simulations were used. All deployed Macro cells were assumed to operate synchronously and Femto cells could have asynchronous UL-DL configurations. The cross DL-UL interference in system level simulation was modeled by random assignment DL or UL transmission directions to Femto cells. The DL and UL SINRs of user terminals attached to Macro and Femto cells were analyzed and compared with the baseline scenario when all Macro and Femto cells operate synchronously (i.e. the transmission directions in all deployed cells are aligned in time). The system level analysis has been conducted for the co-channel and adjacent channel cases.

4.1   Adjacent Channel Macro-Femto Deployment Scenario

In the adjacent channel Macro-Femto deployment scenario, the impact from Macro cells on Femto cells is limited due to low coupling between different carrier frequencies. In current analysis the adjacent channel interference ratio (ACIR) model defined in [8] was used to take into account cross-carrier coupling (see Table 2).

Table 2. ACIR values applied for adjacent channel analysis

ACIR BS-BS	43dB
ACIR BS-UE	33dB
ACIR UE-BS	30dB
ACIR UE-UE	28dB

In general case in the DL transmission direction the Femto stations may transmit at the maximum power level (~20dBm) to maximize the coverage area. On the other hand, to minimize the DL interference to Macro UEs and other Femto UEs which may operate in the close proximity of the Femto station, Femto cells may decrease their output power [9]. The same mechanism may be applied to reduce the cross DL-UL interference problem. So to evaluate these effects the system level simulations were conducted for two cases of Femto cell transmit power: 1) maximum power level 20 dBm and 2) reduced power settings equal to -10 dBm.

In the UL transmission direction, the LTE uplink power control algorithm that fully or partially compensates pathloss and allows controlling the target SNR settings for each UE as well as intercell interference level was applied. The LTE open loop power control algorithm was used with the following settings of power control parameters: 1) Macro cells P_0 = -82dBm and α = 0.8, 2) Femto cells P_0 = -75dBm and α = 0.8.

When neighboring Femto cells have opposite transmission directions the coupling between them may significantly affect the UL SINR of Femto users. In Fig. 5 we present cumulative distribution functions (CDFs) of UL SINR metric for Macro and Femto UEs assuming that 50% of randomly selected Femto cells operate in DL and the remaining Femto cells operate in UL. These curves are compared with the baseline case when all Macro cells and Femto cells operate in UL. The analysis of the Femto UEs UL SINR was done for both DL and UL transmission directions in Macro cells.

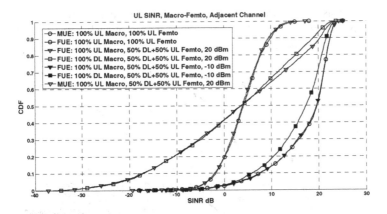

Fig. 5. UL SINR of Macro and Femto UEs in adjacent channel Macro-Femto deployment

The UL SINR distribution presented in Fig. 5 shows that for both cases of Femto station transmit power settings the UL SINR of Macro UEs does not change significantly when Femto stations adjust their transmission directions. On the other hand the UL SINR of Femto UEs significantly depends on the Femto station transmit power. For the maximum transmit power case the UL SINR of Femto UEs significantly degrades (becomes negative for about 40% of UEs) due to strong DL inter-cell interference from neighboring Femto cells and it becomes almost insensitive to the transmission direction of Macro cells. In case of using reduced transmit power level the UL SINR of Femto UEs does not depend on transmission directions in Femto cells but slightly degrades (< 3dB) if Macro cells are switched from UL to DL transmission direction.

The CDFs of DL SINR for Macro and Femto UEs are shown in Fig. 6. As it can be seen from the presented results the DL SINR of the Macro UE is insensitive to the transmission direction in Femto cells even when Femto stations transmit at the maximum power level. The DL SINR of Femto UEs depends on the Femto station power settings. For the case of using reduced power the DL SINR of Femto UEs is not sensitive to transmission directions in neighboring Femto cells and achieves practical SINR range. In case of using maximum transmit power, the DL SINR of Femto UEs is even improved comparing to the reference case. It can be explained by the reduced amount of Femto cells transmitting in DL and thus reduced level of dominant DL intercell interference. It can be also noticed that the change of transmission direction at the Macro stations from DL to UL is beneficial for Femto UEs and results in additional SINR improvements which is about 4-6 dB for the case of reduced Femto power.

Based on the presented results it can be concluded that adjacent channel Macro-Femto deployment scenario is feasible in terms of possibility to support dynamic UL-DL reconfigurations in Femto cells. The results of system level analysis have shown that significant DL-UL inter-cell interference issues may exist on Femto-Femto links if Femto cells operate at the maximum transmit power level. However, it was also demonstrated that this DL-UL interference problem can be avoided if Femto cells properly adjust their DL power settings. In the next section of this paper we continue feasibility analysis for the more challenging co-channel deployment scenario where the mutual impact of Macro and Femto cells is more noticeable due to transmission on the same carrier frequency.

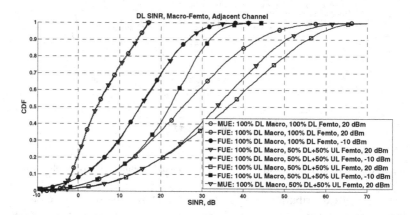

Fig. 6. DL SINR of Macro and Femto UEs in adjacent channel Macro-Femto deployment

4.2   Co-channel Macro-Femto Deployment Scenario

To avoid the DL-UL interference problem on Femto-Femto links the feasibility analysis of co-channel Macro-Femto deployment scenario is mainly provided for the case of reduced Femto station transmit power level (-10dBm). Following the same evaluation methodology as in previous section the measured UL SINR CDFs of Macro and Femto UEs are plotted in Fig. 7. Analyzing these results it can be concluded that the UL SINR of Macro UEs is not sensitive to the transmission direction of Femto cells. However the UL SINR of Femto UEs significantly depends on the transmission direction in Macro cells. For instance, the UL SINR of Femto UEs does not change if Macro cells operate in UL and Femto stations dynamically change their transmission directions from UL to DL. It means that if Macro cells operate in UL the UE-UE interference from Macro users is not a limiting factor and Femto stations may operate in either DL or UL transmission direction. On the other hand if Macro cells are switched to DL a significant degradation (~30dB) of Femto UE UL SINR is observed. This degradation prevents UL transmission in Femto cells during DL transmission in Macro cells.

The CDF curves for DL SINR of Macro and Femto UEs are shown in Fig. 8. From these results it can be seen that the DL SINR of Macro UEs is not sensitive to the transmission direction in Femto cells. The DL SINR performance of Macro UEs is limited by the DL inter-cell interference from neighboring Macro-cells.

The behavior of DL SINR for Femto UEs significantly depends on the transmission direction in Macro cells and Femto station transmission power settings. For instance the DL SINR of Femto UEs degrades significantly if all Macro cells operate in DL and Femto stations transmit power is reduced to -10dBm. To enable DL transmission in Femto cells the transmit power of Femto stations should be increased up to

20dBm. On the other hand the DL SINR performance of Femto cells is improved dramatically and practically feasible SINR ranges are demonstrated when Macro cells operate in UL and Femto stations apply reduced transmit power. This observation confirms that Femto station may use arbitrary transmission direction when Macro cells operate in UL.

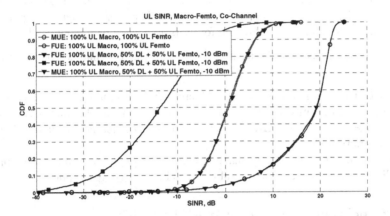

Fig. 7. UL SINR of Macro and Femto UEs in co-channel Macro-Femto deployment

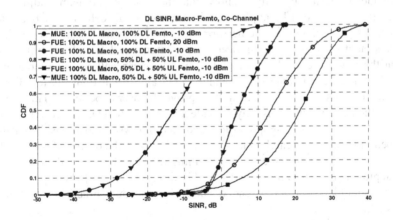

Fig. 8. DL SINR of Macro and Femto UEs in co-channel Macro-Femto deployment

4.3   Summary

In this section we summarize the results of the system level analysis for the two considered deployment scenarios when Macro and Femto stations operate in adjacent channel or in co-channel conditions.

The conducted system level analysis of the DL-UL interference in adjacent channel Macro-Femto deployment scenario has shown that:

- Femto cells can be considered as isolated cells in a sense that the DL and UL SINR performance of Femto UEs is insensitive to the transmission direction of the neighboring Femto cells if proper power adjustment mechanisms are applied;
- The performance of Macro UEs is insensitive to the transmission direction change in Femto cells;
- The sensitivity of Femto UEs to the DL-UL interference from the Femto and Macro sides depends on the Femto station transmit power settings. The practically feasible SINR values were observed for users associated with Femto cells.

The main conclusions from the feasibility study of Macro-Femto co-channel scenario are as follows:

- The DL inter-cell interference from Macro cells is a limiting factor that prevents adaptation of transmission direction for Femto stations in DL subframes of Macro cells. In other words if Macro stations transmit in DL then Femto station shall also use DL transmission.
- Femto stations can potentially change transmission direction if Macro cells operate in UL and interference management between Femto stations is applied.

5   Conclusions and Future Work Directions

The conducted feasibility analysis shows that the concept of dynamic adjustment of TDD configurations can be applied for low power small cells in practical HetNet deployments [8], [10-11]. So the performance of TDD systems can be optimized by the adjustment of TDD configurations to instantaneous DL and UL traffic conditions at the low power nodes deployed within the macrocell coverage area. For instance the packet throughput performance in DL and/or UL transmission directions can be increased substantially and thus improve the user experience. This ability to adjust to the traffic conditions may provide advantages to TDD spectrum over FDD in terms of the packet throughput at low and medium system loadings. To enable all these benefits further research is needed in the fields of the design of traffic adaptation and interference management algorithms that can be adopted in future LTE TDD deployments. The initial works in this direction [12-14] show promising results and should be further extended taking into account practical constraints of LTE systems.

References

1. Lopez-Perez, D., et al.: Enhanced intercell interference coordination challenges in heterogeneous networks. IEEE Wireless Communications 18(3), 22–30 (2011)
2. Damnjanovic, A., et al.: A survey on 3GPP heterogeneous networks. IEEE Wireless Communications 18(3), 10–21 (2011)
3. Yeh, S.-P., et al.: Capacity and coverage enhancement in heterogeneous networks. IEEE Wireless Communications 18(3), 32–38 (2011)
4. Li, J., et al.: Dynamic TDD and Fixed Cellular Networks. IEEE Comm. Letters 4(7), 218–220 (2000)

5. Holma, H., Povey, G.J.R., Toskala, A.: Evaluation of interference between uplink and downlink in UTRA/TDD. In: Proc. of IEEE Vehicular Technology Conference, vol. 5, pp. 2616–2620 (1999)
6. 3GPP RP-110450, Further Enhancements to LTE TDD for DL-UL Interference Management and Traffic Adaptation, CATT, Ericsson, ST-Ericsson (2011)
7. 3GPP TR 36.814, Further Advancements for E-UTRA, Physical Layer Aspects, v.9.0.0 (2010)
8. 3GPP R4-121077, Email discussion summary on the feasibility study for LTE TDD eIMTA, CATT (February 2012)
9. 3GPP TS 36.104, Evolved Universal Terrestrial Radio Access (E-UTRA); Base Station (BS) radio transmission and reception, v10.3.0 (2011)
10. 3GPP R4-120834, DL-UL interference analysis for single operator Macro-Femto deployment scenario in adjacent channel, Intel Corporation (February 2012)
11. 3GPP R4-121601, Evaluation results of feasibility study for dynamic TDD UL-DL configuration, CATT (March 2012)
12. 3GPP R1-121529, Performance analysis of DL-UL interference management and traffic adaptation in multi-cell Pico-Pico deployment scenario, Intel Corporation (March 2012)
13. 3GPP R1-121711, Evaluation of multi-cell scenarios for dynamic TDD traffic adaptation, Ericsson, ST-Ericsson (March 2012)
14. 3GPP R1-121111, Evaluation on TDD UL-DL reconfiguration for multiple out-door pico cell scenario, CATT (March 2012)

Performance Comparison of System Level Simulators for 3GPP LTE Uplink*

Mikhail Gerasimenko[1], Sergey Andreev[1], Yevgeni Koucheryavy[1],
Alexey Trushanin[2], Vyacheslav Shumilov[2], Michael Shashanov[2],
and Sergey Sosnin[2]

[1] Tampere University of Technology, Finland
{mikhail.gerasimenko,sergey.andreev}@tut.fi, yk@cs.tut.fi
[2] University of Nizhny Novgorod, Russia
{alexey.trushanin,vyacheslav.shumilov,mikhail.shashanov}@wcc.unn.ru,
sosninsd@gmail.com

Abstract. In this paper, we survey and classify various system level simulator (SLS) environments for advanced cellular networks. Our overview indicates two distinct groups of SLS tools with respect to their purpose and scope. Further, a typical example within either group is reviewed and compared against its counterpart. In particular, we contrast the Opnet modeler by OPNET Technologies against the LTE Uplink SLS by the University of Nizhny Novgorod (UNN). The target of this research is to assess several comprehensive network deployments with both simulators and compare their throughput-related results. Whereas the Opnet modeler is expected to have more powerful instruments for modeling the MAC and higher layer features, the UNN SLS has stronger focus on the accurate implementation of the Physical layer techniques. We therefore demonstrate the difference in performance by both tools, as well as benchmark against some theoretical predictions.

1 Introduction and Background

1.1 Motivation and Targets

Contemporary wireless communications technology is one of the most rapidly growing segments of the market. Novel solutions and features are being actively developed and deployed in this field. However, the analysis of modern wireless technologies requires significant resources. To facilitate performance assessment, system level simulator (SLS) tools are typically used. Their main purpose is to predict how the network will operate before actually running it in a real environment. Moreover, system level simulators can be used as network planning tools and for educational purposes.

There exist different criteria to classify the available SLS tools: efficiency, complexity, scalability, etc. However, we may broadly divide them into two

---

* Part of this work has been completed when Alexey Trushanin and Vyacheslav Shumilov were on a research visit at Tampere University of Technology.

S. Andreev et al. (Eds.): NEW2AN/ruSMART 2012, LNCS 7469, pp. 186–197, 2012.
© Springer-Verlag Berlin Heidelberg 2012

large classes: general-purpose and problem-oriented. Several examples of general-purpose SLS tools are shown in Table 1. It will be fair to say that the evaluations shown in Table 1 are based on the personal opinions and belief of the authors. More information about the mentioned simulators can be extracted from the respective websites (see references in Table 1).

Table 1. General-purpose SLS examples

Name	Detalization	Usability	Complexity
Opnet modeler 17.0 [1]	High	High	High
OMNet++ [2]	Medium	High	High
NS2 [3]	Low	Medium	Low
NS3 [4]	Medium	High	Medium
GNS [5]	High	Low	High
Hurricane 2 [6]	Low	Medium	High

General-purpose solutions are wider than their problem-oriented counterparts and may include different technologies as so-called "modules" inside one integral tool. The major disadvantage of the simulators belonging to this class is not very detailed (simplified) system structure. As the result, they are sometimes not enough accurate, or their performance cannot be verified. Another drawback is the complexity of their source code. General-purpose SLS tools are typically used for the educational purposes, that is, to study the basic principles behind a particular technology. They can also be applied in academic research [7] and Opnet modeler by OPNET Technologies is a typical representative within its class.

Problem-oriented SLS tools are specifically produced for the simulation of upcoming wireless technologies and are mainly developed by the network operators, vendors, or with the assistance of some third-party companies. Advantages of these simulators are: simplicity of usage, wider opportunities for calibration, and broader range of available statistics. However, the simulators belonging to this class are typically not intended for commercial use. One reason is that these SLS tools have smaller performance scope. They are normally exploited within companies for detailed research on specific technological issues and are therefore difficult to use in educational process. Particularly, problem-oriented SLS tools are important instruments exploited during the standardization process and the development of new technology. One typical example of this class of tools is the University of Nizhny Novgorod Long Term Evolution Uplink (UNN LTE UL) SLS.

In order to conclude on their precision, different simulators can be compared with each other and also against theoretical predictions. In particular, the most interesting statistics at the physical layer is related to throughput per user (or overall), spectral efficiency (center and cell-edge), signal to interference plus noise ratio (SINR), and block error rate (BLER). In this paper, the two simulators were compared to show how close general-purpose solutions are to their problem-oriented equivalents. In Subsection 1.2 of this work, the overview of the considered technology is presented to show the main direction and the scope of this research. In Section 2, the proposed SLS tools are briefly summarized and compared. Section 3 describes the considered scenarios and the analytical calculations for the simplest of them. In Section 4, the results are shown and discussed.

1.2   Technology Details

In this paper, the 3GPP Long Term Evolution (LTE) technology is considered. LTE is a cellular-based wireless network belonging to the 4th generation (4G), which development is ongoing for at least last four years. The main advantages of LTE are better flexibility and higher data rates. However, these features come with higher protocol complexity and more advanced techniques applied primarily at the physical layer. Below we briefly mention the basics of LTE topology and some physical layer features.

The architecture of an LTE network consists of two major components: evolved UMTS Terrestrial Radio Access Network (EUTRAN) and evolved packet core (EPC) [8]. The main elements of EUTRAN are user equipment (UE) and eNodeB. In this paper, we are mostly concentrated on the EUTRAN part, specifically, on the connections between UE and eNodeB. Particularly, we focus on Frequency Division Duplex (FDD), which means that we are considering the frame type 1 [9]. Each frame contains 10 time slots. In the frequency domain, one slot consists of a group of resource blocks (RB); every RB uses 12 subcarriers, each of which is 15 kHz wide. The length of a resource block in the time domain is exactly one slot [9].

One important feature of the physical layer is modulation and coding schemes (MCS). Changing radio environment cause variation in the SINR levels meaning that technology should support adaptation mechanisms to mitigate these fluctuations. Decreasing its MCS, the UE reduces the coding rate such that the SINR level needed for the successful data reception is also reduced by sacrificing maximum system throughput. There are 32 MCS sets [10] as per LTE release 10 and if adaptive MCS is enabled it can automatically adjust performance to the varying channel conditions.

2   Description of the Simulators

2.1   Opnet Modeler

OPNET Modeler 17.0 is a network simulator, which purpose generally is the modeling of different types of networks. The OPNET Modeler Wireless Suite presents plenty of capabilities and functions for various types of wireless networks including GSM, UMTS, IEEE 802.16 (WiMAX), and 3GPP LTE. Each network type is included as a separate module and is visible through graphical modules, called editors. The **Project Editor** defines the basic functionality of network planning. Every network element (node) in the Project Editor can be configured also in the **Node Editor**. Further, each block is represented as a finite state machine (FSM) that can be adjusted in the **Process Editor**. The typical work flow to create a new model is demonstrated in Figure 1.

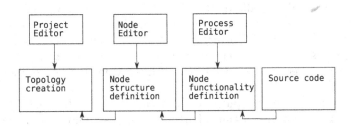

Fig. 1. Model creation workflow in Opnet modeler

Our module of interest is the LTE module, which includes different eNodeB models, UE models, and an EPC model. Moreover, it has a complex and wide structure with a number of limitations and parameters to configure. For the sake of this research, the most important component of the LTE module is the **Physical Layer Measurements**. It includes SINR measurement in both Downlink and Uplink directions, which are collected separately for each UE.

Other interesting statistics at the physical layer is related to the reference signal receive power (RSRP) and reference signal receive quality (RSRQ). A UE measures RSRP and RSRQ continuously (with a 5ms update interval) and independently for each eNodeB, which it can hear.

Continuing with physical layer measurements, we should also mention **Pathloss, Interference**, and **Multipath Fading** models supported by the module. For instance, the following pathloss models are available in the Opnet modeler [1]: Macrocell suburban based on COST231 Hata Urban model, Macrocell urban based on COST231 Hata Urban model, Microcell urban based on COST231 Walfish-Ikegami, Erceg suburban fixed, and shadow fading based on the lognormal distribution model.

The following ITU multipath models are supported [1]: Pedestrian A and B, as well as Vehicular A and B. The interference model is supported as follows [1]:

- Interference module detects time and frequency overlaps among different bursts.
- Interference is proportional to the amount of burst overlaps.
- Interference may cause burst drops for physical uplink shared channel (PUSCH) and physical downlink shared channel (PDSCH) bursts.
- Interference effects for control channels are based on a probability distribution function.
- Interference is computed among nodes using the same or different LTE physical layer profiles.
- The frequency attributes of a given LTE physical layer profile are accounted for when computing interference.

2.2   UNN LTE UL SLS

The UNN LTE UL SLS is an example of a problem-oriented SLS mentioned above and designed primarily to assess client cooperation techniques in mobile networks. Additionally, it allows for the evaluation of the advanced data transmission techniques. This SLS has been developed by the Wireless Competence Center (WCC) of the State University of Nizhny Novgorod and is based on the SLS used by Russian Evaluation Group within the IMT-Advanced standardization process intended for the evaluation of LTE-A as a 4th generation system candidate [11].

In contrast to the general-purpose SLS tools, the UNN LTE UL SLS has no support for higher-layer algorithms. Further, it has a simplified interface, flexible configuration tools, and convenient post-processing instruments. However, the major feature of the UNN LTE UL SLS is the possibility of detailed and accurate simulations of the technology in question and of the deployment scenario including layout, antenna configurations, and channel models.

The UNN LTE UL SLS supports simulation methodology typical for problem-oriented SLS tools and is actively used by the international standardization committees (e.g. 3GPP, IEEE, ITU, etc.), as well as equipment manufacturers. Particularly, hexagonal layout is assumed with base stations located in the centers of hexagonal cells and with UE uniformly and randomly distributed over the entire simulated area. Each base station has 3 sectors divided by different sectors antenna beams (see Figure 2 where UE belonging to different sectors are shown with different colors and markers).

Further, the eNodeB entities located in each sector perform power control, scheduling and resource management for the associated UE similarly to the real-world algorithms operated at the currently deployed base stations. Data packets are transmitted from UE to eNodeB and the success of each packet reception is modeled depending on the physical layer algorithms operation and signal propagation conditions: transmit power, path loss, channel models, interference from other transmitting UEs, etc. Then, system operation statistics is stored and analyzed further to derive the required SLS results.

The UNN LTE UL SLS structure is provided in Figure 3. The simulator has been developed within the framework of the common SLS platform. The basic principle of the common platform is the division of SLS functionality into two main blocks: system-independent part which is fully reusable between different problem-oriented SLS tools for different mobile radio-access networks or for different purposes, whereas system-dependent part is developed specifically for the system under consideration. The interface between system-independent and system-dependent parts is formalized and minimized to simplify porting of system-dependent part between different simulators.

Along with the auxiliary functionality, the system-independent part includes the deployment module responsible for the simulation of layout, antenna models, channel models, etc. and its own mathematical library optimized for and oriented at SLS purposes. System-dependent part was developed directly for the LTE-A Release 10 Uplink and in turn consists of two program modules:

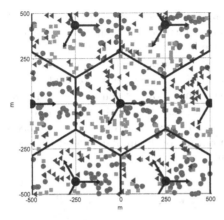

Fig. 2. Hexagonal layout used in the LTE UL SLS

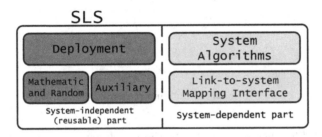

Fig. 3. Key components of the UNN LTE UL SLS

link-to-system (L2S) mapping interface and implementation of the main system control algorithms. L2S mapping interface is used for fast and accurate prediction of the packet transmission performance of each link avoiding the exact direct modeling of physical layer processing in the system. Simulated system control algorithms include power control and scheduling, mentioned above, and Hybrid Automatic Repeat reQuest (H-ARQ) algorithm.

To the moment of writing this paper, the basic functionality of the UNN LTE UL SLS is designed and implemented, whereas further development is also considered. In particular, there is a large field for considered problem-related improvements of the SLS. Enhancements of the program platform are also planned, primarily with respect to input (configuration) and output (Simulation Results Analyzer) tools.

3   Simulation Scenarios and Methodology

3.1   General Concepts

The central concept of this research is to firstly compare both simulators on a small-scale simplified scenario that can be also evaluated theoretically.

This ensures that both SLS tools give adequate and predictable results. Further, we consider scenarios with more advanced physical effects and evaluate the difference in performance by both tools.

Basically, some mismatch is expected due to the difference in signal power distribution dependent on the antenna pattern representation. Both simulators have flexible 3-D antenna pattern editors, which allow setting basic antenna parameters. As UE antenna is omni-directional, the core difference is in the BS antenna representation. The main antenna parameters are explained in Figure 4. All these parameters are configured for both simulators and have a significant impact on the network performance.

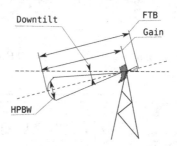

Fig. 4. BS antenna parameters

The second difference may be due to the interference representation. If both power control and adaptive MCS are disabled, interference representation differences will become more evident. Opnet is expected to show higher overall throughput, whereas UNN LTE UL SLS should give lower numbers due to wraparound. Technically, wrap-around carries the signals of border cells (as shown in Figure 5) into an opposite-side cell, thus border cells do not have any advantage in interference environment comparing to the center cells.

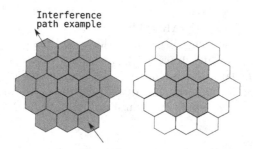

Fig. 5. Wrap around signal path example (left side) and realistic interference zone without wrap around (right side) in typical 19-cell scenario

3.2 Baseline Simulation Parameters and Calculations

Three scenarios are considered in this paper. Firstly, there is one-cell scenario with minimal physical effects. The purpose of this scenario is to compare both SLS tools against the theoretical predictions. Secondly, there is one-cell scenario with complete physical layer effects in order to highlight the difference between the SLS tools on a smaller scale. Thirdly, there is a typical 19-cells scenario, which is capturing the interference effects. Baseline parameters for all the three scenarios are summarized in the Table 2. Also some assumptions can be made for simplicity:

Assumption 1. No Hybrid automatic repeat request (HARQ).
Assumption 2. No sounding reference signals (SRS).
Assumption 3. No link adaptation and no channel-dependent scheduling.
Assumption 4. No penetration losses.
Assumption 5. The locations of users are chosen in a random fashion.

Table 2. Baseline simulation parameters

Inter-cell distance	500 m (288 m cell radius)
Number of users per cell	30 users per 3-sector cell (10 users per sector)
UE antenna configuration	omni-directional, antenna gain 0 dbi
BS antenna configuration	directional, 17 db in boresight gain
BS antenna HPBW	azimuth plane = 70 deg, polar plane = 15
BS antenna downtilting	12 deg
BS antenna FTB	20 db
Number of BS and MS antennae	1
MCS	7, 14, 21
Used bandwidth	10 MHz (1 MHz guard band)
Base frequency	1920 MHz
Used mode	FDD
Cyclic prefix	Normal
Scheduler type	Round robin scheduler (pure frequency division scheduling, FDS)

Some overhead is known to be presented in Opnet due to the explicit modeling of the control channels. Therefore, it should be taken into account in UNN SLS and in theoretical results. Basically, there is physical random access channel (PRACH) that takes 6 resource block pairs in one subframe out of 10 (each frame), decreasing the throughput by around 1.2 % on average (0.988 in (1)), and physical uplink control channel (PUCCH), that takes 1 additional resource block pair in each subframe. In Opnet modeler, the FDS is used meaning that resource blocks are equally divided between all the users in a sector. For this case, there exists a simplified method to calculate the physical-layer throughput. One should take a transport block size (TBS) value for a given number of RB (Table 7.1.7.2.1-1 in [11]) and divide it by the length of the subframe (equation

1) to obtain the throughput of one user. If there are 10 users in the cell, Opnet will allocate 5 RB for 9 of them and 4 RB for the last one. Therefore, the calculations must be presented for each of the 10 users separately. Summarizing, the theoretical throughput for one user can be defined as:

$$T = t \cdot 0.988/m, \tag{1}$$

where t is the TBS value, T is the theoretical value of the throughput, and m is the subframe length.

3.3 Scenarios Definition, Configuration, and Parameters

One-Cell with Minimal Physical Effects. For verification purposes, one hexagonal cell with 3 sectors and 10 users in each sector was considered. The limitations for this scenario are: no mobility (all users are static), no multipath, and no shadow fading. The related parameters are given in Table 3.

Table 3. Simulation parameters-I

Transmission power, per RB (per UE)	23 dbm
Transmission power, overall (per UE)	40 dbm
Path loss model	Free space

One-Cell with Complete Physical Effects. This scenario is similar to the previous one, but physical effects are taken into account and the system is not anymore static. The parameters for this scenario are shown in Table 4.

Table 4. Simulation parameters-II

Transmission power, per RB (per UE)	13 dbm
Transmission power, overall (per UE)	30 dbm
Path loss model	3GPP 25.996 Urban macrocell
Multipath model	ITU pedestrian A
User speed	3 km/h

19-cells Scenario. The third scenario is a typical 19-cells setup (see Figure 5) [11]. In each cell, there are 3 sectors and 10 users per sector. Hence, the total number of users per a simulation run is 570. The parameters are effectively the same as before, but the interference is accounted for explicitly.

4   Numerical Results, Comparison, and Conclusions

4.1   Theoretical and Simulation Results

Equation (1) was established for 1 user per sector. However, in case of 10 UE per sector, each of them has 5 RB per subframe on average. As such, the overall

Table 5. Theoretical throughput for ideal channel conditions

MCS	7	14	21
Average throughput per UE, Mbps	0.565	1.215	2.085
Overall throughput, Mbps	16.978	36.469	62.552

throughput is 16.978 Mbps per cell. Theoretical throughput for one user, with parameters defined in Subsection 3.3 is shown in Table 5.

In Tables 6, 7, 8, and in Figure 6 the values of throughput for all the three scenarios are shown. In the first scenario, one may see that the throughput is lower than theoretical for both SLS tools, whereas Opnet modeler demonstrates more optimistic results. Dramatic reduction of throughput in case of MCS = 21 can be explained by the higher BLER values for cell-edge (sector-edge) users.

Theoretical values can be achieved by accumulating the throughput and the traffic that was lost due to increased BLER. In equation (2), it is shown how to calculate loss S, where T is the initial throughout, B = BLER, and G = 1-BLER.

$$S = (T \cdot B)/G \tag{2}$$

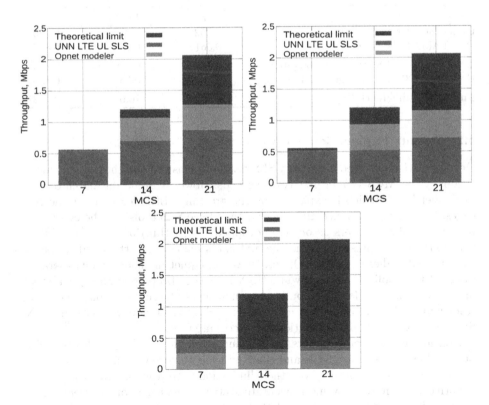

Fig. 6. Average throughput per user. Scenario 1 – top left, scenario 2 – top right, scenario 3 – bottom.

Table 6. 1st scenario results

MCS	7	14	21
Opnet Modeler average throughput	0.533	1.063	1.266
Opnet Modeler average BLER	0.058	0.128	0.401
UNN LTE UL SLS average throughput	0.543	0.701	0.869
UNN LTE UL SLS average BLER	0.033	0.420	0.570
Throughput difference (%)	1.841	34.055	31.359

Table 7. 2nd scenario results

MCS	7	14	21
Opnet Modeler average throughput	0.366	0.55	0.733
Opnet Modeler average BLER	0.370	0.557	0.651
UNN LTE UL SLS average throughput	0.513	0.523	0.721
UNN LTE UL SLS average BLER	0.110	0.570	0.650
Throughput difference (%)	28.655	6.727	1.637

Table 8. 3rd scenario results

MCS	7	14	21
Opnet Modeler average throughput	0.252	0.257	0.280
Opnet Modeler average BLER	0.561	0.789	0.867
UNN LTE UL SLS average throughput	0.484	0.306	0.355
UNN LTE UL SLS average BLER	0.160	0.750	0.830
Throughput difference (%)	47.934	16.013	21.127

4.2 Results Explanation and Conclusion

Our simulations showed that both SLS tools are close to the theoretical calculations, when MCS = 7 in first two scenarios. However, some mismatch occurs for higher MCS, which is explained by the growing BLER for users at the cell edges. Further performance differences between the SLS tools may be caused by unequal number of users associated (dropped) to (within) a sector and different interference representation in the first two scenarios. For the third scenario, we expected higher results for Opnet. However, Opnet modeler traffic generator works at the application layer, while UNN LTE UL SLS physical layer is always in the full-buffer mode. Thus, buffer may not always be full for Opnet modeler. Additionally, UE movement is modeled explicitly in Opnet, whereas in UNN SLS only the UE channel deviations are accounted for.

Considering the above factors, we conclude that the results are theoretically predictable for both SLS tools and that the tools themselves match with a reasonable degree of accuracy given their limitations. Additional simulations with adaptive MCS may show more accurate statistics, which is our current work. Nevertheless, we may already conclude that Opnet performance is good enough for less detailed research at higher layers.

References

1. Opnet website,
   http://www.opnet.com/solutions/network_rd/modeler_wireless.html
2. Omnet ++ website, http://www.omnetpp.org/
3. NS2 website, http://www.isi.edu/nsnam/ns/
4. NS3 website, http://www.nsnam.org/
5. GNS website, http://www.gns3.net/
6. Hurricane 2 website,
   http://www.packetstorm.com/psc/psc.nsf/site/Hurricane-II-software
7. Vassileva, N., Koucheryavy, Y., Arroyo, F.B.: Guard Capacity Implementation in Opnet Modeler WiMAX Suite. In: International Conference on Ultra Modern Telecommunications & Workshops, ICUMT 2009, pp. 1–6 (2009)
8. UTRA-UTRAN Long Term Evolution (LTE) and 3GPP System Architecture Evolution (SAE). Technical report, 3gpp
9. TS 36.211 V10.3.0 Evolved Universal Terrestrial Radio Access (E-UTRA); Physical Channels and Modulation (Release 10)
10. Evolved Universal Terrestrial Radio Access (E-UTRA); Physical layer procedures (Release 10)
11. Maltsev, A., Khoryaev, A., Maslennikov, R., Shilov, M., Bovykin, M., Morozov, G., Chervyakov, A., Pudeyev, A., Sergeyev, V., Davydov, A.: Analysis of IEEE 802.16m and 3GPP LTE Release 10 Technologies by Russian Evaluation Group for IMT-Advanced. In: International Congress on Ultra Modern Telecommunications and Control Systems and Workshops (ICUMT), pp. 901–908 (2010)

Performance of Multiflow Aggregation Scheme for HSDPA with Joint Intra-site Scheduling and in Presence of CQI Imperfections

Dmitry Petrov, Ilmari Repo, and Marko Lampinen

[1] Magister Solutions Ltd., Jyvaskyla, Finland
[2] Renesas Mobile Europe, Oulu, Finland
{dmitry.petrov,ilmari.repo}@magister.fi,
marko.lampinen@renesasmobile.com

Abstract. Lately several *Single Frequency Dual Cell* (SF-DC) techniques for *High Speed Downlink Packet Access* (HSDPA) were considered in the *Third Generation Partnership Project* (3GPP), which were aimed to enhance network performance for users at the cell edges. In this paper we are concentrating on intra-site multiflow aggregation scheme, where *User Equipment* (UE) can receive two independent transmissions from separate sectors. Furthermore, this paper focuses on the impact of *Channel Quality Indicator* (CQI) reporting interval and measurement imperfections on the system performance. Moreover, we present a joint scheduling algorithm, which selects users site-wide instead of independently in each sector. Simulation results for scenarios mentioned above are presented and analysed in the paper. This study shows that joint scheduling gives some extra gain compared to independent scheduling. In addition, impact of CQI errors and CQI update period are studied.

Keywords: HSDPA, SF-DC, 3GPP, multiflow, scheduling, CQI.

1 Introduction

Currently *High Speed Packet Access* (HSPA) is the leading mobile broadband technology globally with 451 networks commercially launched in 174 countries. Even though *Long Term Evolution* (LTE) networks are rapidly developing (already 49 commercial networks worldwide) the *Wideband Code Division Multiple Access* (WCDMA) networks still show constant subscriptions growth about 40% per year. Over 41% of commercial HSPA networks have already launched Evolved HSPA, also called HSPA+ [1]. According to the latest issued 3GPP Release 10, achievable peak data rates in downlink direction (HSDPA) are up to 168 Mbps. Deployment of LTE networks will still require time and considerable capital expenses from operators. At the same time already existing HSPA networks will need to accommodate huge mobile traffic growth. For that reason new techniques that can enhance 3G networks' performance are constantly examined in the 3GPP.

Good examples of such enhancements is *Dual-Cell HSDPA* (DC-HSDPA) feature first defined in 3GPP Release 8 [2]. It provided considerably better user

S. Andreev et al. (Eds.): NEW2AN/ruSMART 2012, LNCS 7469, pp. 198–207, 2012.

experience in two geographically overlapping cells in deployments when several 5 MHz paired spectrum allocations were available to an operator. Performance was improved especially at the cell edges where channel conditions were not favourable and existing techniques such as *Multiple Input and Multiple Output* (MIMO) could not be used.

At the same time, it is rather usual situation that only one 5 MHz frequency band is assigned to an operator in the specified geographical region. Thus multi-carrier multi-cell approach cannot be utilized. Nevertheless reception quality at the cell edges may degrade considerably. Multipoint HSDPA transmission, i.e. transmission from two different cells to a UE situated in *Soft* (SHO) or *Softer Handover* (SofterHO) region can be used to evolve system performance in *Single Frequency* (SF) deployments. Similar approach was firstly introduced for LTE Release 8 where it was called *Coordinated Multipoint* (CoMP) [3]. Adaptation of this technology into the HSDPA system was proposed as a study item at the 3GPP *Radio Access Network* (RAN) *Technical Specification Group* (TSG) meeting in December 2010 [4].

In this paper our own research is focused on so called intra-site aggregation scheme where UEs have two independent receiving chains and are capable to receive data blocks simultaneously and independently from two cells of one site. It is quite probable that this scheme will be included into the coming 3GPP Release 11 planned for the third quarter of 2012. Our aim is to go further than standard 3GPP simulation scenario [5] by evaluating the performance of the HSDPA system under two assumptions: firstly, when the standard *Proportional Fair* (PF) scheduling algorithm acting independently in every sector is substituted by a joint site-wise PF scheduler; secondly, when CQIs are reported less often and with errors. The study is done by means of quasi-static system-level simulator, in which the most essential physical parameters and radio resource management functions like channel fading, UE scheduling, etc. are modelled in details.

The rest of the paper is organized as follows: Section II contains the brief overview of SF multipoint concepts for HSDPA. In the following section the most important parameters and assumptions are introduced together with more detailed description of joint scheduling scheme and CQI imperfections. In Section IV simulation results are presented and analysed. Finally Section V concludes the article.

2   Multipoint Concepts for HSDPA

The idea of multipoint transmission is to perform spatial processing over several transmitting antennas from different cells. Two or more independent communication channels with different characteristics can be used to either improve the reliability of the transmission or to increase the capacity. Development of multipoint HSDPA techniques are motivated, firstly, by the problems with the reception quality at the cell edges, when users are far from the transmitter and being interfered strongly by the neighbouring cell. Secondly, by the load balancing function between the sectors, when UEs can utilize resources from less loaded cells.

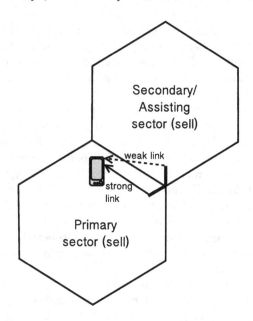

Fig. 1. Intra-site multipoint transmission scheme

The possibility of multipoint transmission is determined individually for every UE according to the difference in the received power between the two strongest NodeBs. This procedure is similar to one used to define the *HandOver* (HO) state of the UE. If the received powers of the two cells are close enough, the UE is informed to monitor the *High Speed Shared Control Channel* (HS-SCCH) on both of them. Transport blocks can be scheduled on *High Speed Downlink Shared Channel* (HS-DSCH) from secondary or even both sectors (Fig. 1). Two scenarios can be considered: intra-site case when UE falls into SofterHO coverage region and is served by only one NodeB; and inter-site case referred to SHO state when primary and secondary cells belong to different NodeBs. In this publication we study only intra-site scenario for two main reasons. Firstly, inter-site case has more impact of the network and require certain enhancements to *Radio Link Control* (RLC) and Iub interface flow control [6]. Secondly, practical realisation of proposed joint scheduling algorithm is more complicated when separate NodeBs are involved.

Now we will briefly introduce multipoint techniques-candidates, which were considered in the 3GPP for the SF deployments (for more detailed information see [5] and [7]):

- *High Speed Data-Discontinuous Transmission* (HS-DDTx) and *Fast cell switching* techniques are not really multipoint transmission schemes because they do not assume that UEs can receive data from several cells simultaneously. This is similar to single point data transmission and can be used with UEs having only one receiving antenna chain. UEs can get data blocks from only one cell in each *Transmission Time Interval* (TTI) whereas cells can be changed dynamically from TTI to TTI. Thus sectors are chosen in compliance with reported CQI values. The main difference between DDTx and Fast cell switching is that in the first scheme it is not required to switch off HS-SCCH reception from the secondary sector.

- *High Speed Single Frequency Network* (HS-SFN) technique allows UEs, which are in SofterHO state, to receive data blocks from two sectors simultaneously. Both of them transmit exactly the same data blocks to one UE. In such a way HS-SFN-capable UEs benefit from extended spatial diversity. Transmission from assisting cell adds additional power and also removes the strongest source of interference.
- *Single Frequency Dual Cell HSDPA* (SFDC-HSDPA) also called *Multiflow aggregation* technique originates from DC-HSDPA feature and gives use to two receiving chains of DC-HSDPA capable UEs in SF scenario. It can be understood as a spatial multiplexing MIMO scheme when transmitting antennas are situated in different sectors. Achievable gains in this scenario are the highest among all multipoint schemes presented above. It is quite probable that it will be adopted in 3GPP Release 11. For that reason we have selected it for evaluation in this paper.

3 Simulation Assumptions and Methodology

This study has been performed by means of proprietary comprehensive quasi-static system simulator, which models HSDPA with a slot resolution. It is widely used to support 3GPP standardization work [8,9]. The simulation tool enables detailed simulation of users in multiple cells with realistic traffic generation, propagation and fading.

Quasi-static simulations mean that UEs are stationary but both slow and fast fading are explicitly modeled. Fast fading is modeled as a function of time for each UE according to the *International Telecommunication Union* (ITU) channel profiles which are modified to chip level sampling [10]. Actual simulation area consists of 19 NodeBs which results into 57 hexagonal cells. Statistics are collected from all cells. Statistical confidence is obtained through running multiple drops, i.e. independent simulation iterations. In each iteration UE locations, fading, imbalance and other random variables are varied. The statistics are gathered and averaged over all drops. *Actual Value Interface* (AVI) mapping is used for mapping link level Eb/N0 values to frame error rates [11].

3.1 Scheduling Algorithms

Independent Scheduling. Basically, SFDC-HSDPA scheduling is considered to be independent. Practically it means that even multiflow capable UE is considered like a combination of two separate receivers connected to the different sectors and placed to one physical point. Nevertheless such UEs have one file buffer divided in two flows. Scheduling decisions are made depending on the whole load profile in the sector. Specifically, every TTI PF scheduler selects user with the highest priority $SP = \max_i(SP^i) = \max_i(\frac{R^i}{M^i})$ where R^i is the rate achievable by user i associated to this sector and M^i is its average rate [12]. The achievable rate R^i is defined directly from the channel quality measurements, i.e. from the CQI. The average rate refers to the PF metric and is calculated as moving average over a one TTI time window: $M^i(t+1) = (1-\alpha)M^i(t) + \alpha R^i$ where α is the forgetting factor. Thus PF scheduling achieves high throughput while maintaining proportional fairness among all users in the sector. Multiflow capable UEs are particular in that sense that there are two *Scheduling Priorities* (SP) associated with them: $SP_{10} = \frac{R_{10}}{M_{10}}$ at the primary sector and $SP_{02} = \frac{R_{02}}{M_{02}}$ at the assisting sector.

Moreover, In the multiflow case special prioritization is used. For each cell, two classes of UEs are defined during scheduling [13]:

- "Class A" UEs have this cell as serving, i.e. are connected via strongest link;
- "Class B" UEs do not have this cell as serving, i.e. are connected via the weak link.

Class A UE prioritization means that serving HS-DSCH cell UEs are always scheduled before multiflow UEs connected to this cell as to a secondary one. This rule is important in high load scenarios. In such a way loss in performance of regular UEs belonging to this serving cell is avoided.

Retransmissions are also taken into account in the scheduling algorithm. They always have highest priority over regular transmissions. It is achieved by simple adding of big enough positive number to the scheduling priority. Such approach always forces retransmissions to be scheduled first and at the same time keeps the order of UEs. This rule is fulfilled even in the case when retransmission is needed for Class B UE, while active Class A UE exists in the same sector.

Joint Intra-site Scheduling. In this paper we propose and evaluate a different scheduling algorithm based on site-wide user selection. Taking into account that in intra-site scenario scheduling decision are made inside one NodeB we can use PF algorithm to select users with maximum scheduling priorities from the list of all active users of the site. In the case when multiflow is switched off there will be almost no difference with regular independent scheduling algorithm because, again, UEs with the highest SP will be selected in every cell. The situation will change if there are multiflow capable users which can be scheduled simultaneously from both sectors. To take that into account we are considering three SPs for these kind of UEs:

- $SP_{10} = \frac{R_{10}}{M}$ and $SP_{02} = \frac{R_{02}}{M}$, where average rate $M = M_{10} + M_{02}$ takes into account whole traffic received by the UE;
- $SP_{12} = \frac{R_{12}}{M}$, where $R_{12} = R_{10} + R_{02}$ is the achievable rate in the case of aggregated transmission.

The conceptual scheme of joint scheduling can be seen on Fig. 2. First, all active users from three sectors are collected to one list, which is then sorted according

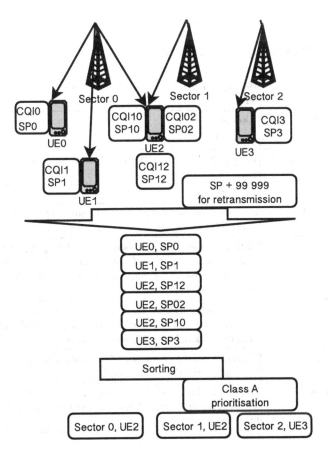

Fig. 2. Scheme of joint intra-site scheduling algorithm

to the scheduling priority value. After that we go through the sorted list and select one UE for every sector taking into account Class A UE prioritization. In joint scheduling the additional gain is expected from the better organized user selection mechanism. It allows to choose users in an optimal way out from three possible modes available for multiflow capable UEs: aggregated transmission, transmission only from primary sector, transmission only from the secondary sector. Simulation results can be found in the next section.

3.2 CQI Imperfections

CQI reports are sent from the UE to the NodeB every CQI reporting period. These values are used by the scheduling algorithm to select appropriate scheduling grants and to observe changes in the UE channel conditions. However CQI might be reported incorrectly and with varied period length. Thus it should be studied how these variations affect the performance of considered multipoint scheme. For this purpose, errors of 3 and 6 dB were simulated in addition to ideal case, and CQI reporting period from 3 TTIs to 12 TTIs is analyzed.

Table 1. Main simulation parameters

Feature / Parameters	Description / Value
Cell Layout	Hexagonal grid, 19 NodeBs, 3 sectors per NodeB with wrap-around.
Inter-site distance	1000 m
Carrier Frequency	2000 MHz
Number of UEs/cell	1, 2, 4, 8, 16, 32 (distributed uniformly across the system).
Receiver Type	Type 3i
Channel model	PA3; Fading across all pairs of antennas is completely uncorrelated.
CQI	CQI measurements are 3 slot delayed; CQI update period is 3, 6, 9, 12 TTIs; CQI estimation is ideal and with Gaussian errors of variance 3 and 6 dB; CQI decoding at NodeB is ideal.
Traffic	Bursty traffic model: File size - truncated lognormal distribution ($\mu = 11.736$, $\sigma = 0.0$); Inter-arrival time - exponential distribution (Mean = 5 sec.).
Scheduling	Proportional Fair: Independent; Intra-site joint; Forgetting factor 0.001.
MP-HSDPA UE capabilities	All MP-HSDPA UEs are capable of 15 SF 16 codes and 64QAM for each cell; Percentage of SFDC-HSDPA capable UEs is 100%.

3.3 Simulation Parameters

The main parameters used in the system simulation are summarized in Table 1. A hexagonal wrap-around multi-cell layout is utilized. Wrap-around is used to model the interference correctly also for outer cells. This is achieved by limiting UE placement inside the actual simulation area but replicating the cell transmissions around the whole simulation area to offer more realistic interference situation throughout the scenario. UE is also able to connect to the replicated cells, for example as a part of SHO active set. UEs are created according to a uniform spatial distribution over the whole area. This results into some cells being more heavily loaded while others can be even empty. 100% of the UEs are SFDC capable but multiflow is available only for user in SofterHO state. As it was shown in [7] considerable gain from multiflow feature can be achieved only with interference aware Type 3i receivers. Thus in our study all UEs use only this kind of receiver.

A bursty traffic generation model is assumed, which means that the UEs do not constantly have data in the transmission buffers. File inter-arrival time is modeled with exponential distribution. File size is also variable and follows

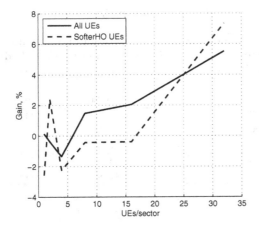

Fig. 3. Gain from joint intra-site scheduling for all and SofterHO UEs

log-normal distribution (see Table 1). The data available in the UE buffer is transmitted as fast as it is allowed by the NodeB. Those decisions are made according to the scheduling algorithms discussed above. Only pedestrian A channel with 3 km/h is studied in this paper.

4   Results Analysis

Simulation results are presented in this section. The performance is evaluated through mean user throughputs for both all and SofterHO UEs. Gain is calculated as the difference in throughputs divided by the throughput of baseline scenario. On Fig. 3 independent PF scheduling plays the role of baseline scenario. Joint scheduling is benchmarked against it. Actual values of throughput can be found in the Table 2. As Fig. 3 and Table 2 indicate there is a positive gain from joint scheduling in higher load scenarios. The gain increases and reaches the maximum value of about 6% in 32 UE/sector case. We see the explanation of this relatively small gain in several facts. Firstly, the ratio of multiflow capable UEs, i.e. percentage of UEs which are in SofterHO region is about 9% so it is difficult to expect their high influence on the overall system performance. Secondly, precise assessment of SPs requires more detailed knowledge of CQI. Currently it is impossible to distinguish situations when it is better, for example, to use aggregated instated of only one-flow transmissions removing the main source of interference for the neighbouring sector. Nevertheless the first step is the study of the influence of CQI imperfection on the system performance, which is also done in this paper.

Fig. 4 and 5 illustrate the degradation of system performance in the case of CQI imperfections. Legends in the figures refer to different cases so that:

- "Mflow off" means normal HSDPA operation with data transmission only from serving HS-DSCH cell;
- "Intra" equals to the case where UEs in SofterHO state can utilize HS-DSCH transmission from multiple cells (depending on scheduling decisions).

Table 2. Mean user throughputs

UEs / Sector	Independent scheduling, Mbit/sec	Joint scheduling, Mbit/sec	Independent scheduling, Mbit/sec	Joint scheduling, Mbit/sec
	All UEs		SofterHO UEs	
1	7.92	7.93	10.86	10.58
2	7.69	7.57	9.74	9.97
4	7.00	6.90	9.07	8.87
8	5.61	5.69	7.06	7.03
16	2.83	2.89	3.30	3.29
32	0.34	0.36	2.85	3.06

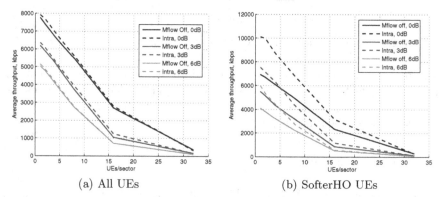

(a) All UEs   (b) SofterHO UEs

Fig. 4. Influence of CQI errors on average user throughput

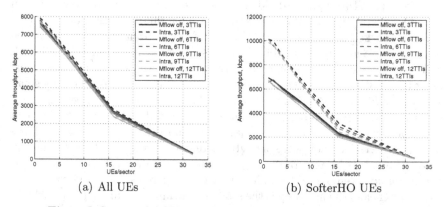

(a) All UEs   (b) SofterHO UEs

Fig. 5. Influence of CQI reporting period on average user throughput

Results are expected for both scenarios. With higher errors in CQI for all users (Fig.4(a)) average throughput goes down for baseline scenario without multiflow feature and for intra-site as well. As it can be noticed from Fig.4(b) for SofterHO UEs the gain from multiflow also becomes less significant. The influence of CQI reporting period is not so strong. However, from Fig.5(b) it follows that in high load scenarios delayed information on channel quality causes loss of the gain for SofterHO users.

5 Conclusion

This paper continues our study of multipoint HSDPA transmissions. It presents the results of our own simulations for one of the most practically interesting intra-site aggregation scheme. The effect of intra-site joint scheduling on system performance was studied and benchmarked against regular independent algorithm. In intra-site case moderate gain can be achieved by changing the scheduling algorithm. It is also shown how the precision of the channel quality feedback impacts on the efficiency of the multiflow feature.

Acknowledgement. This study is done in cooperation with Renesas Mobile Corporation. The authors would like to thank all of the co workers and colleagues for their comments and support.

References

1. Global mobile Suppliers Association (GSA), GSM/3G Market/Technology Update (February 7, 2012), http://www.gsacom.com/downloads/pdf/mobile_broadband_fact_sheet_070212.php4
2. Tapia, P., Lui, J., Karimli, Y., Feuerstein, M.J.: HSPA Perforamnce and Evolution: A Practical Persective. Wiley (2009)
3. Taoka, H., et al.: MIMO and CoMP in LTE-Advanced. NTT DOCOMO Technical Journal 12(2), 10–28 (2010)
4. HSDPA multipoint transmission, 3GPP Work Item Description RP-101439 (December 2010)
5. HSDPA Multipoint Transmission, 3GPP Technical Report (TR) 25.872, Release 11 (September 2011)
6. DL Scheduling, RLC and Flow Control assumption for Inter-NodeB Multi-Point Transmissions, 3GPP Contribution R1-110126 (January 2011)
7. Petrov, D., Repo, I., Lampinen, M.: Overview of Single Frequency Multipoint Transmission Concepts for HSDPA and Performance Evaluation of Intra-site Multiflow Aggregation Scheme. In: Proc. of IEEE VTC, Yokohama, Japan (May 2012)
8. System Performance Evaluation of SF-DC Inter-NodeB aggregation with Type 3 and Type 3i receivers in Pedestrian A channel, 3GPP contribution R1-112303 (August 2011)
9. System Performance Evaluation of SF-DC Inter-NodeB aggregation with Type 3 and Type 3i receivers in Vehicular A channel, 3GPP contribution R1-112304 (August 2011)
10. Hamalainen, S., Slanina, P., Hartman, M., Lappetelainen, A., Holma, H., Salonaho, O.: A Novel Interface Between Link and System Level Simulations. In: Proc. of ACTS 1997, Aalborg, Denmark (October 1997)
11. Guidelines for Evaluation of Radio Transmission Technologies for IMT-2000, ITU-R M.1225 Recommendation (1997)
12. Bu, T., Li, L., Ramjee, R.: Generalized Proportional Fair Scheduling in Third Generation Wireless Data Networks. In: IEEE INFOCOM (2006)
13. Simulation Framework for System Evaluation of Muti-point HSDPA, 3GPP Contribution R1-110563 (January 2011)

Modelling a Radio Admission Control Scheme for Video Telephony Service in Wireless Networks*

Irina A. Gudkova and Konstantin E. Samouylov

Telecommunication Systems Department,
Peoples' Friendship University of Russia,
Ordzhonikidze str. 3, 115419 Moscow, Russia
{igudkova,ksam}@sci.pfu.edu.ru
http://www.telesys.pfu.edu.ru

Abstract. Modern 4G wireless networks are multi-service networks, and an important issue is the development of an optimal radio admission control scheme (RAC) specific to various service types. The 3GPP recommendations for LTE and LTE-Advanced networks specify nine service classes that differ in terms of the priority level, bit-rate, and packet error loss. In this paper, we propose an RAC scheme for video telephony service that is provided in two modes: voice and video. Voice telephony users have a higher priority level than video telephony users; this is realised by the possibility of accepting voice calls owing to the degradation in video call quality. We analyse a model of a single cell with video telephony service, and we prove that the system state space is of an explicit form. This enables us to compute the performance measures, such as blocking probabilities, mean number of calls, and utilization factor.

Keywords: LTE, conversational voice, conversational video, radio admission control, blocking probability.

1 Introduction

Recently, 4G cellular wireless networks [1] have witnessed rapid development. LTE network deployment is inseparably linked with maintaining the quality of service (QoS) and enhancing customer base. Radio admission control (RAC) is a key factor affecting the QoS of modern wireless networks. RAC schemes are closely related to the types of services provided to customers. Table 1 summarizes nine service classes [2,3,4] defined for LTE networks, which differ in terms of the QoS Class Identifier (QCI) and priority level (Allocation and Retention Priority, ARP).

According to 3GPP TS 23.203 [4]: "The range of the ARP priority level is 1 to 9 with 1 as the highest level of priority. The pre-emption capability information

---

* This work was supported in part by the Russian Foundation for Basic Research (grants 10-07-00487-a and 12-07-00108), and by Rosobrazovanie (project no. 020619-1-174).

S. Andreev et al. (Eds.): NEW2AN/ruSMART 2012, LNCS 7469, pp. 208–215, 2012.

Table 1. Characteristics of LTE service classes [4]

QCI	Resource type	Priority level	Examples of services	
1		2	Conversational voice	
2	GBR	4	Conversational live streaming video	
3		3	Real time gaming	
4		5	Non-conversational buffered streaming video	
5		1	IMS signalling	
7		7	Voice, live streaming video, interactive gaming	
6	Non-GBR	6	Buffered streaming video, TCP-based applications	for Multimedia Priority Services subscribers
8		8	(e.g., www, e-mail, chat,	for premium subscribers
9		9	FTP, P2P file sharing, progressive video, etc.)	for non-privileged subscribers

defines whether a service data flow can get resources that were already assigned to another service data flow with a lower priority level. The pre-emption vulnerability information defines whether a service data flow can lose the resources assigned to it in order to admit a service data flow with a higher priority level. The resource type determines if dedicated network resources related to a service Guaranteed Bit Rate (GBR) value are permanently allocated."

With regard to video telephony service, 3GPP TS 23.401 [3] states the following: "The video telephony is one use case where it may be beneficial to use different ARP values for the same user. In this use case an operator could map voice to a higher ARP, and video to a lower ARP. In a congestion situation, e.g. cell edge, the base station can then drop the video data flow without affecting the voice data flow."

The 3GPP recommendations do not specify RAC schemes, and operators have to develop and select an optimal scheme accounting for the service level agreement. Researchers have proposed various admission control schemes [5,6] for different combinations of service classes.

The remainder of this paper is organized as follows. In Sect. 2, we propose a multi-rate model of an RAC scheme for two GBR services: conversational voice (QCI = 1) and conversational live streaming video (QCI = 2). In Sect. 3, we describe the main performance measures of the proposed model, which are illustrated via a numerical example. Finally, we conclude the paper in Sect. 4.

2 Mathematical Model

We consider a single cell with a total capacity of C bandwidth units (b.u.) supporting two GBR services: conversational voice (QCI = 1; voice, in short) and conversational live streaming video (QCI = 2; video, in short) (Fig. 1).

The voice service is provided on single guaranteed bit rate b_{voice} b.u., whereas the video service is a multi-rate service provided not only on guaranteed bit rate $b_{\text{video}}^{\text{gbr}}$ b.u. but also on maximum bit rate (MBR) $b_{\text{video}}^{\text{mbr}}$ b.u. The voice priority level is higher than the video priority level. First, this fact is realized by reserving capacity of C_1 b.u. for voice calls. Second, admission control is achieved such that a new voice call is accepted by the so-called pre-emption owing to the lack of reserved bandwidth C_1. Pre-empting refers to the reallocation of bandwidth occupied by MBR video calls, i.e., degradation of the bit rate from MBR to GBR. Without loss of generality and based on close-to-reality data, we assume this MBR video call number to be one, i.e., $\left\lceil \frac{b_{\text{voice}}}{b_{\text{video}}^{\text{mbr}} - b_{\text{video}}^{\text{gbr}}} \right\rceil = 1$. Let all arrival rates λ_{voice} and λ_{video} be Poisson distributed, and let the call durations be exponentially distributed with means $\frac{1}{\mu_{\text{voice}}}$ and $\frac{1}{\mu_{\text{video}}}$.

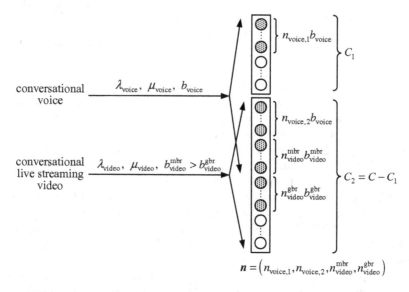

$$n = \left(n_{\text{voice},1}, n_{\text{voice},2}, n_{\text{video}}^{\text{mbr}}, n_{\text{video}}^{\text{gbr}} \right)$$

Fig. 1. Multi-rate model of RAC scheme for two GBR services: conversational voice (QCI = 1) and conversational video (QCI = 2)

Given the above considerations, when a new voice call arrives, three scenarios are possible.

1. The voice call will be accepted with bandwidth b_{voice} allocated in reserved capacity C_1, which is possible if the call finds capacity C_1 having at least b_{voice} b.u. free.
2. The voice call will be accepted with bandwidth b_{voice} allocated in capacity $C_2 := C - C_1$, which is possible if the call finds capacity C_1 having less than b_{voice} b.u. free and at least one MBR video call.
3. The voice call will be blocked without any after-effect on the corresponding Poisson process arrival rate.

Similarly, when a new video call arrives, three scenarios are possible.

1. The video call will be accepted with maximum bit rate b_{video}^{mbr}, which is possible if the call finds capacity C_2 having at least b_{video}^{mbr} b.u. free.
2. The video call will be accepted with guaranteed bit rate b_{video}^{gbr}, which is possible if the call finds capacity C_2 having less than b_{video}^{mbr} b.u. free and greater than or equal to b_{video}^{gbr} b.u. free.
3. The video call will be blocked without any after-effect on the corresponding Poisson process arrival rate.

Let us introduce the following notations:

$n_{voice,1}$ – number of voice calls allocating bandwidth in reserved capacity C_1;
$n_{voice,2}$ – number of voice calls allocating bandwidth in capacity C_2;
n_{video}^{mbr} – number of MBR video calls;
n_{video}^{gbr} – number of GBR video calls.

We denote the state of the system as $\boldsymbol{n} := \left(n_{voice,1}, n_{voice,2}, n_{video}^{mbr}, n_{video}^{gbr} \right)$ and the system state space as $\mathcal{X}$. Let $b_1(\boldsymbol{n}) := n_{voice,1} b_{voice}$ and $b_2(\boldsymbol{n}) := $
$:= n_{voice,2} b_{voice} + n_{video}^{mbr} b_{video}^{mbr} + n_{video}^{gbr} b_{video}^{gbr}$ be the bandwidth allocated in capacities C_1 and C_2, respectively, when the system is in state $\boldsymbol{n}$. Obviously, owing to limited bandwidth, the system state space $\mathcal{X}$ satisfies the relation

$$\mathcal{X} \subseteq \mathcal{X}_1 := \{ \boldsymbol{n} \geq \boldsymbol{0} : \ b_1(\boldsymbol{n}) \leq C_1, \ b_2(\boldsymbol{n}) \leq C_2 \}. \tag{1}$$

To form state space $\mathcal{X}$, we use the so-called access functions [7] $\boldsymbol{f}(\boldsymbol{n}) :=$
$:= \left(f_{voice,1}(\boldsymbol{n}), f_{voice,2}(\boldsymbol{n}), f_{video}^{mbr}(\boldsymbol{n}), f_{video}^{gbr}(\boldsymbol{n}) \right)$ that will be equal to 1 if a new arriving call is accepted, or 0, otherwise. The conditions corresponding to 1 refer to scenarios 1 and 2 for the rules of accepting voice and video calls described above

$$\mathcal{X} := \{\boldsymbol{0}\} \cup \{\boldsymbol{n} \in \mathcal{X}_1 : \ f_{voice,1}(\boldsymbol{n} - \boldsymbol{e}_1) = 1 \ \vee \ f_{voice,2}(\boldsymbol{n} - \boldsymbol{e}_2) = 1 \ \vee$$
$$\vee \ f_{video}^{mbr}(\boldsymbol{n} - \boldsymbol{e}_3) = 1 \ \vee \ f_{video}^{gbr}(\boldsymbol{n} - \boldsymbol{e}_4) = 1\}, \tag{2}$$

where $\boldsymbol{e}_i$ is the unit vector and

$$f_{voice,1}(\boldsymbol{n}) := \begin{cases} 1, & \boldsymbol{n} \in \mathcal{X}_1 : \ b_1(\boldsymbol{n}) \leq C_1 - b_{voice}, \\ 0, & \text{otherwise}, \end{cases} \tag{3}$$

$$f_{voice,2}(\boldsymbol{n}) := \begin{cases} 1, & \boldsymbol{n} \in \mathcal{X}_1 : \ b_1(\boldsymbol{n}) > C_1 - b_{voice}, \ n_{video}^{mbr} \geq \left\lceil \frac{b_{voice}}{b_{video}^{mbr} - b_{video}^{gbr}} \right\rceil, \\ 0, & \text{otherwise}, \end{cases} \tag{4}$$

$$f_{video}^{mbr}(\boldsymbol{n}) := \begin{cases} 1, & \boldsymbol{n} \in \mathcal{X}_1 : \ b_2(\boldsymbol{n}) \leq C_2 - b_{video}^{mbr}, \\ 0, & \text{otherwise}, \end{cases} \tag{5}$$

$$f_{video}^{gbr}(\boldsymbol{n}) := \begin{cases} 1, & \boldsymbol{n} \in \mathcal{X}_1 : \ C_2 - b_{video}^{mbr} < b_2(\boldsymbol{n}) \leq C_2 - b_{video}^{gbr}, \\ 0, & \text{otherwise}. \end{cases} \tag{6}$$

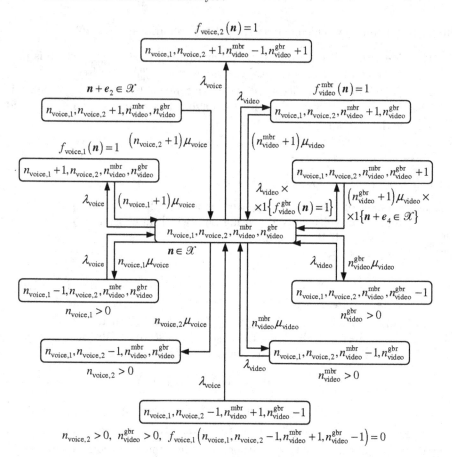

Fig. 2. State transition diagram

The process representing the system states is described by the state transition diagram shown in Fig. 2. It seems only possible to determine the system probability distribution $p(n)$, $n \in \mathcal{X}$, by means of numerical methods for solving systems of linear equations.

3   Performance Measures and Numerical Example

Let us introduce blocking sets $\mathcal{B}_{\text{voice}}$ and $\mathcal{B}_{\text{video}}$ as follows:

$$\mathcal{B}_{\text{voice}} := \left\{ n \in \mathcal{X} : \ b_1(n) > C_1 - b_{\text{voice}}, \ n_{\text{video}}^{\text{mbr}} \left(b_{\text{video}}^{\text{mbr}} - b_{\text{video}}^{\text{gbr}} \right) < b_{\text{voice}} \right\},$$

(7)

$$\mathcal{B}_{\text{video}} := \left\{ n \in \mathcal{X} : \ b_2(n) > C_2 - b_{\text{video}}^{\text{gbr}} \right\}.$$

(8)

Having found the probability distribution $p(n)$, $n \in \mathcal{X}$, of the multi-rate model of the RAC scheme for voice and video services, one may compute its performance measures, notable blocking probabilities B_{voice} and B_{video}, mean number of calls M_{voice} and M_{video}, and utilization factor UTIL:

$$B_{\text{voice}} = \sum_{n \in \mathcal{B}_{\text{voice}}} p(n), \tag{9}$$

$$B_{\text{video}} = \sum_{n \in \mathcal{B}_{\text{video}}} p(n), \tag{10}$$

$$M_{\text{voice}} = \sum_{n \in \mathcal{X}} (n_{\text{voice},1} + n_{\text{voice},2}) \cdot p(n), \tag{11}$$

$$M_{\text{video}} = \sum_{n \in \mathcal{X}} \left(n_{\text{video}}^{\text{mbr}} + n_{\text{video}}^{\text{gbr}}\right) \cdot p(n), \tag{12}$$

$$\text{UTIL} = \frac{1}{C} \sum_{n \in \mathcal{X}} [b_1(n) + b_2(n)] \cdot p(n). \tag{13}$$

Computing the probability distribution involves the inconvenient formation of system state space (2) with the aid of access functions $f(n)$. The system state space $\mathcal{X}$ has been proved to include only those states $n \in \mathcal{X}_1$ that satisfy the inequality

$$\mathcal{X} = \{n \in \mathcal{X}_1 : n_{\text{voice},2} \leq N_{\text{voice},2}\}, \tag{14}$$

where

$$N_{\text{voice},2} = \sum_{i=0}^{\left\lfloor \frac{C_2}{b_{\text{voice}}} \right\rfloor} \left\lfloor \frac{C_2 - i \cdot b_{\text{voice}}}{b_{\text{video}}^{\text{mbr}}} \right\rfloor. \tag{15}$$

Note that the constraint $n_{\text{voice},2} \leq N_{\text{voice},2}$ is related to the pre-emption process. The variable (15) is the maximum number of voice calls allocating bandwidth in capacity C_2. Remember that owing to pre-emption, a new voice call allocates bandwidth in capacity C_2 only if there is at least one MBR video call. Thus, the dimension of $\mathcal{X}$ can be computed as follows

$$|\mathcal{X}| = \sum_{n_{\text{voice},1}=0}^{\left\lfloor \frac{C_1}{b_{\text{voice}}} \right\rfloor} \sum_{n_{\text{voice},2}=0}^{N_{\text{voice},2}} \sum_{n_{\text{video}}^{\text{mbr}}=0}^{\left\lfloor \frac{C_2 - n_{\text{voice},2} b_{\text{voice}}}{b_{\text{video}}^{\text{mbr}}} \right\rfloor} \left(\left\lfloor \frac{C_2 - n_{\text{voice},2} b_{\text{voice}} - n_{\text{video}}^{\text{mbr}} b_{\text{video}}^{\text{mbr}}}{b_{\text{video}}^{\text{gbr}}} \right\rfloor + 1\right). \tag{16}$$

We present an example of a single cell supporting conversational voice and conversational video services to illustrate the performance measures defined above. Let us consider a cell with a total capacity of $C = 3000$ Kbps; let $C_1 = 500, 1000, 1500$ Kbps be the capacity reserved for voice service. The bit rates are $b_{\text{voice}} = 100$ Kbps, $b_{\text{video}}^{\text{mbr}} = 500$ Kbps, and $b_{\text{video}}^{\text{gbr}} = 300$ Kbps. Assume that the arrival rates and mean call durations are equal for both services. By changing the offered load $\frac{\lambda_{\text{voice}}}{\mu_{\text{voice}}} + \frac{\lambda_{\text{video}}}{\mu_{\text{video}}}$ from 0 to 10, we compute the performance measures (9), (10), (13). The results are plotted in Fig. 3.

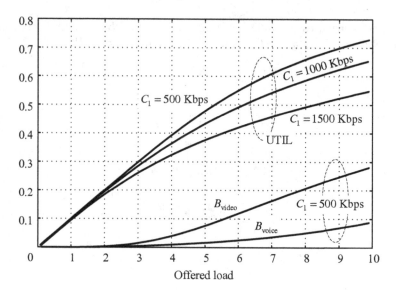

Fig. 3. Performance measures

4 Conclusion

In this paper, we addressed an admission control problem for a multi-service LTE radio network, and we presented a multi-rate model for two guaranteed bit rate services: voice and video telephony. The RAC scheme is based on video telephony quality degradation from high to standard definition. The scheme assumes that a voice telephony user can pre-empt high definition video telephony users. An interesting task for future studies is the analysis of various admission control schemes not only for GBR services but also for Non-GBR services [8], such as P2P file sharing.

Acknowledgments. The authors are grateful to B.Sc. graduate Ivan Matvey-chuk for performing the numerical experiment.

References

1. Stasiak, M., Glabowski, M., Wisniewski, A., Zwierzykowski, P.: Modelling and Dimensioning of Mobile Wireless Networks: from GSM to LTE. Johns Wiley & Sons Ltd., Chichester (2010)
2. 3GPP 36.300: Evolved Universal Terrestrial Radio Access (E-UTRA) and Evolved Universal Terrestrial Radio Access Network (E-UTRAN); Overall Description; Stage 2 (Release 11) (2012)
3. 3GPP 23.401: General Packet Radio Service (GPRS) Enhancements for Evolved Universal Terrestrial Radio Access Network (E-UTRAN) Access (Release 11) (2012)
4. 3GPP 23.203: Policy and Charging Control Architecture (Release 11) (2012)

5. Chowdhury, M.Z., Yeong, M.J., Haas, Z.J.: Call Admission Control based on Adaptive Bandwidth Allocation for Multi-Class Services in Wireless Networks. In: Proc. of the 1st International Conference on Information and Communication Technology Convergence, Jeju Island, Korea, pp. 358–361 (2010)
6. Qian, M., Huang, Y., Shi, J., Yuan, Y., Tian, L., Dutkiewicz, E.: A Novel Radio Admission Control Scheme for Multiclass Services in LTE Systems. In: Proc. of the 7th IEEE Global Telecommunications Conference, Honolulu, Hawaii, USA, pp. 1–6 (2009)
7. Basharin, G.P., Samouylov, K.E., Yarkina, N.V., Gudkova, I.A.: A New Stage in Mathematical Teletraffic Theory. Automation and Remote Control 70(12), 1954–1964 (2009)
8. Gudkova, I.A., Samouylov, K.E.: Approximating Performance Measures of a Triple Play Loss Network Model. In: Balandin, S., Koucheryavy, Y., Hu, H. (eds.) NEW2AN/ruSMART 2011. LNCS, vol. 6869, pp. 360–369. Springer, Heidelberg (2011)

Multi-point Cooperative Fountain Codes Multicast for LTE Cellular System[*]

Wei Liu[1], Yueyun Chen[1], and Yudong Yao[2]

[1] School of Computer & Communication Engineering,
University of Science & Technology, Beijing, China
chenyy@ustb.edu.cn
[2] Department of Electrical and Computer Engineering,
Stevens Institute of Technology, Hoboken, NJ, USA
yyao@stevens.edu

Abstract. Cooperative transmission is an effective technique to improve transmission performance. In this paper, we propose a novel multi-point cooperative fountain codes multicast (CF-Multicast) transmission strategy. In the proposed strategy, any user can be selected to act as a relay for its neighbors in a multicast group. Then we introduce two schemes, R-CUT and O-CUT, for selecting cooperative user terminals (CUTs) to decrease the number of cooperative users in order to consider both decoding efficiency and energy consumption, and compare their performance. Simulation results show that the proposed schemes of R-CUT and O-CUT are able to provide tradeoff between transmission efficiency and system energy consumption, and O-CUT scheme has better transmission performance compared with R-CUT scheme.

Keywords: LTE, multicast, cooperative communication, fountain codes.

1   Introduction

Multicast, as an effective transmission mode for improving resource efficiency, will be adopted in the future LTE systems [1]. In a multicast group, more than one terminal can receive the same packets from an eNodeB. However wireless multicast performance is degraded significantly due to fading channels compared with wireline multicast. Cooperative communications is attracting a lot of attentions because of its advantages in many areas, including power consumption, BER, and cover region. In cooperative communications, more than one nodes to work cooperatively to transmit information from a source to the destinations [2][3]. Some relay strategies, such as amplify and forward (AF) and decode and forward (DF) have been proposed in [4][5][6].

---

[*] The authors are supported by National Science Foundation of Beijing under Grants 4102041, and by The National Science and Technology Key Programs under Grants 2011ZX03003-002-03 and 2011ZX03004-002-02.

S. Andreev et al. (Eds.): NEW2AN/ruSMART 2012, LNCS 7469, pp. 216–224, 2012.
© Springer-Verlag Berlin Heidelberg 2012

In recent years, a new category of cooperative approaches, which integrated cooperative relay and coding techniques [7], such as fountain codes proposed by Luby [8], are proposed. In using fountain code, the receiver will not stop receiving data until the decoding is successful, which implies that the code rate varies according to the instantaneous channel condition. Using fountain code over a two-hop transmission from the source to the partner and from the partner to the destination, [9] proposed a coding framework, in which the partner assisted the source as a secondary antenna during the collaborative phase. In [10], the authors proposed using fountain coding in three-node cooperative relay networks to provide reliable data transmissions with the focus on relay strategies. However, multiple relay nodes introduce mutual interference, which can potentially degrade the transmission performance. The author in [11] proposed an amplitude modulation scheme to address the above issue, by scaling the output signal at each node according to their unique identifier. The reparation for data received at the destination from several transmitters is thus achieved.

In this paper, we introduce a novel strategy of multi-point cooperative multicast based on fountain codes in LTE cellular networks. In the strategy, any user terminal in a cell or adjacent cells can act as a relay for its neighbors. We introduce two schemes to decrease the number of cooperative users in order to reduce the total energy consumption in the system. We focus on the performance of the average number of packets transmitted by eNodeB and the total system energy consumption in order to examine the tradeoff between the transmission efficiency and energy consumption. The performance of the schemes are analyzed, and it is shown that the overall throughput is improved using the proposed schemes.

2 System Model

In a LTE cellular cell, one multicast group with several user terminals (UTs) is serviced by an eNodeB. The UTs are in a cell or adjacent cells. The cooperative relay multicast system scenario is shown in Fig.1.

A strategy of cooperative fountain code multicast (CF-Multicast) is modeled as follows.

The unit energy consumption of eNodeB (BS) for sending one packet is P_B. We assume that there are M UTs in a multicast group, which are independently identically distributed (i.i.d) in the area. The BS transmits K original information symbols encoded by fountain codes to all the M UTs. Each UT will decode the K symbols successfully from fountain codes as soon as it receives enough fountain codes. We set the number of symbol needed to complete decoding as N, where $N = K(1+\varepsilon)$, ε is the decoding overhead. We suppose that the channels from the BS to all the UTs are modeled as erasure channels. The erasure probabilities is denoted as P_{ei} for ith UT, $i=1,2, \dots, M$. The average number of encoded packets N_{ri} needed to be sent to UT_i is

$$\overline{N_{ri}} = \frac{N}{1-P_{ei}} = \frac{M \times (1+\varepsilon)}{1-P_{ei}} \tag{1}$$

Any of the UTs can act as a relay to assist other UTs which are not far away from it, namely, the UT can directly forward packet transmitted by BS to its neighbors. This means that any UTs can not only receive the encoding packets from the BS but also from its neighbors, which called cooperative UTs (CUTs). For simplicity, we assume that the channels from the UT to its CUTs are rayleigh fading channels. Each CUT has a cooperative range r_i, which is determined by its transmit power P_i, and the unit energy consumption of UT for sending one packet is P_U. The total system energy consumption is

$$E_A = E_B + E_C = \overline{N_B} P_B + E_U \tag{2}$$

where $\overline{N_B}$ is the average number of packets sent by BS, E_U is the total energy consumption of all CUTs, and P_B is the unit energy consumption of BS for sending one packet.

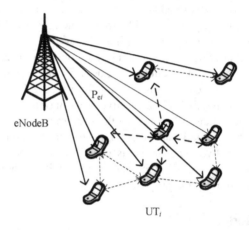

P_{ei}

eNodeB

UT_i

Fig. 1. Cooperative multicast scenario

3   Transmission Scheme

In a conventional multicast transmission system, each of the M UTs only receives fountain code packets directly from BS. In erasure channel condition, the average number of encoded packets N_{ri} needed to be sent to UT_i is expessed in (1), and the average number of packets sent by BS is

$$\overline{N_B} = \max_{i, i=\{1,2,...,M\}} \left\{ \overline{N_{ri}} \right\} \tag{3}$$

None of the UTs is working as relay, and $E_U = 0$. The total system energy consumption is

$$E_A = \overline{N_B} P_B \tag{4}$$

In the multi-point cooperative multicast, any UT can work as a relay for its neighbor UTs. Each UT has a cooperation range (CoR), and the radius r_i of its CoR is determined by the transmit power of UT. During the transmission, one UT can cooperate with its neighbor UTs which are inside its CoR. In this case, diversity gains are achieved due to BS and CUTs.

Assuming that the number of CUTs of UT_i is T, expressed as CUT_{i1}, CUT_{i2},...,CUT_{iT}, the channel erasure probability from BS to CUT_{ij} (i=1,2,...,M; j=1,2,...,T) is P_{eij}. The probability of correctly receiving a packet of UT_i is

$$P_{ri}^c = 1 - P_{ei} \prod_{j=1}^{T} P_{eij} \tag{5}$$

The average number of encoded packets N_{ri} received by UT_i is:

$$\overline{N_{ri}^c} = \frac{N}{P_{ri}^c} = \frac{N}{1 - P_{ei} \prod_{j=1}^{T} P_{eij}} \tag{6}$$

The average number of packets send by BS is

$$\overline{N_B^c} = \max_{i,i=\{1,2,...,M\}} \left\{ \overline{N_{ri}^c} \right\} \tag{7}$$

Thus the total system energy consumption is

$$E_A^c = \overline{N_B^c} P_B + \sum_{i=1}^{M} \overline{N_{ri}^c} P_U \tag{8}$$

where PU is the unit energy consumption of UT for sending one packet. For simplicity, we don't consider the energy of UTs for receiving packets.

Note the scenario that UTs are uniform and densely distributed in an area. If all UTs participate in cooperative communication at the same time, one UT may have several CUTs, and the CUTs may transmit the same packets for one UT, which results in wasting resource and increases the energy consuming of UTs and the system. Therefore, how to determine the CUTs in the cooperative multicast transmission is an important issue.

In the following, we develop two schemes to address this issue.

3.1 Multi-point Cooperation Scheme - R-CUT

D UTs are randomly and periodically selected by BS as CUTs (R-CUT) to cooperate for one of their neighbor, where $D<M$, and the channel erasure probability from each UT to BS is different. The probability for each UT to be randomly selected as CUT is

$$P_c = \frac{D}{M} \tag{9}$$

The probability P_{cit} of UTi cooperated by neighbor $t=\{0, 1, ..., T\}$ can be expressed as

$$P_{cit} = C_T^t P_c^t (1-P_c)^{T-t} \tag{10}$$

The probability of correctly receiving a packet for UTi is

$$P_{ri}^{c1} = \sum_{t=0}^{T} P_{cit} (1 - P_{ei} \prod_{j=1}^{t} P_{eij})$$

(11)

where P_{esj} is the channel erasure probabilities of jth CUTs. The average number of encoded packets received by UTi is

$$\overline{N_{ri}^{c1}} = \frac{N}{P_{ri}^{c1}} = \frac{N}{\sum_{t=0}^{T} P_{cit} (1 - P_{ei} \prod_{j=1}^{t} P_{eij})}$$

(12)

The average number of packets sent by BS is

$$\overline{N_{B}^{c1}} = \max_{i, i=\{1,2,...,M\}} \left\{ \overline{N_{ri}^{c1}} \right\}$$

(13)

Assuming BS select D CUTs randomly at every time after sending H packets. Then the total system energy consumption is

$$E_A^{c1} = \overline{N_B^c} P_B + \sum_{i=1}^{M} \overline{N_{ri}^c} \frac{H}{\overline{N_B^c}} P_U$$

(14)

Comparing $\overline{N_B^c}$ in (7) and $\overline{N_B^{c1}}$ in (13), it can be observed that the average number of packets sent by BS in R-CUT scheme is higher than that in CF-multicast. However, by comparing (8) and (14), the scheme of R-CUT can decrease the total system energy consumption.

R-CUT scheme is easy to implement for cooperation because BS selects CUTs without knowing channel state information (CSI). However, one or more UTs with bad channel conditions may be selected as relays, which have little effect for improving system performance but wasting their energy.

3.2   Multi-point Cooperation Scheme - O-CUT

In order to utilize the good channels between BS and UTs, BS optimally selects some CUTs (O-CUT) by setting a SINR threshold δ (SINR can be evaluated by BS from uplink ACK), namely BS periodically selects CUTs with SINR value larger than δ. A simple method to determine δ is for BS to sort the UTs SINR in descending order, and select the first D UTs as CUTs, which can make sure that enough UTs can be chosen to act as CUTs. The average number of encoded packets $\overline{N_{ri}^{c2}}$ received by UT$_i$ similar to $\overline{N_{ri}^{c1}}$, is expressed as

$$\overline{N_{ri}^{c2}} = \frac{N}{P_{ri}^{c2}} = \frac{N}{\sum_{t=0}^{T} P_{cit} (1 - P_{ei} \prod_{j=1}^{t} P_{ebj})}$$

(15)

where P_{ebj} is the erasure probability of jth selected CUTs. The average number of packets sent by BS is

$$\overline{N_B^{c2}} = \max_{i,i=\{1,2,...,M\}} \left\{ \overline{N_{ri}^{c2}} \right\} \tag{16}$$

Assuming $P_{ebj} < P_{ei}$ ($i=D+1,...,M$; $j=1,...,D$), the total system energy consumption is

$$E_A^{c2} = \overline{E_A^{c2}} = \overline{N_B^c} P_B + \sum_{i=1}^{D} \overline{N_{ri}^{c2}} P_U \tag{17}$$

Comparing (12) and (15), we see that O-CUT can gain better decoding efficiency than R-CUT. Besides, it decreases the energy consumption of BS.

Compared with traditional relay method, our scheme do not select one or more terminals as fixed relay, but we allow any of UE terminal in multicast group can work as a relay while it has a better channel condition. That's give full play to the superiority of UE terminals location, and the selection of relay mode take the balance between efficiency and energy consumption into consideration, it's convenient to adjust collaborative strategy.

4   Numerical Analysis

In order to estimate the performance of the proposed strategy of CF-multicast, simulation is formed in Matlab according to Fig.1. The simulation parameters are as follows. LT codes [8] are used as Fountain codes. M UTs are randomly (i.i.d) in the area. The channel condition is Rayleigh fading, and the erasure probabilities of channels from UTs to BS are 0.1, 0.15, 0.2, 0.25 and 0.3, etc. Each CUT can reach T neighbor UTs, and each UT can be assisted by at most T CUTs, The unit energy consumption of BS for sending one packet is 0.1 mWh and the unit energy consumption of ith UT for sending one packet is 0.01mWh.

The average number of packets (ANP) sent by BS reflects decoding efficiency. For a given number of original information symbols, the fewer ANP means that the decoding efficiency is higher. Fig.2 shows ANP comparing between without cooperation multicast and CF-multicast strategy for different numbers of UTs. In the Figure, $K=100$; $M=5$, 10, 15, 20; $T=2$. The ANP of CF-multicast is significantly fewer than the case without cooperation. Therefore, the proposed CF-multicast strategy can increase the system throughput by decreasing the number of fountain code data packets transmitted.

Fig.3 shows the variation of ANP with the number of cooperative neighbors. The number of cooperative neighbors for each UT to reach from 1 to 4, and $M=20$. For the strategy and two CUTs selection schemes, with more cooperative neighbors for one UT, the system will have higher decoding efficiency, namely fewer encoding packets transmitted in the channel from the BS to UTs and higher multicast throughput from BS. Besides, the generally decoding performance of CF-multicast is better than R-CUT and O-CUT, for it has fewest number of sending packets to complete decoding.

In Fig.4 and Fig.5, we compare conventional Fountain code multicast, the proposed basic CF-multicast strategy, and the proposed schemes of R-CUT and O-CUT based on the basic CF-multicast strategy. We consider $M=10$; $K=20$, 50, 80, 100, 120;

T=2; and D=4. It can be observed that the basic CF-multicast strategy has the least number of transmit packets from BS at the largest system energy consumption; both R-CUT and O-CUT achieve the tradeoff between transmission efficiency and system energy consumption. Because the CUTs with better channel conditions are selected periodically in O-CUT, they can provide better relay performance than R-CUT. Notice that, if channels states are stationary, the CUTs which have better channel conditions will serve as relays all the time, and it's unfair to these CUTs for their larger energy consumptions.

Different from the traditional relay method, the proposed schemes allow any of the neighbor UEs in multicast group to act as a relay when it has the best channel condition in a limited area. Taking CUTs' location into consideration, fewer CUTs are selected to sever as relays for more neighbor users. So, the proposed scheme O-CUT can provide a better tradeoff between decoding efficiency and system energy consumption while the system achieves better transmission performance.

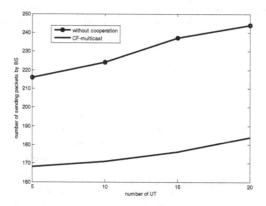

Fig. 2. ANP performance of the proposed scheme

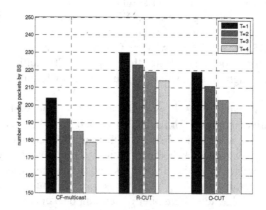

Fig. 3. Decoding efficiency

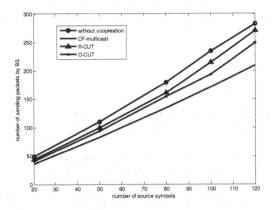

Fig. 4. ANP performance with proposed schemes

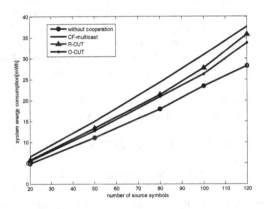

Fig. 5. System energy consumption comparison

5   Conclusion

In this paper, we propose a novel cooperative transmission multicast strategy, CF-multicast, based on fountain codes. The proposed strategy allows any UT in the network to act as a relay for its neighbor UTs. It is able to decrease the average number of transmitted packets from BS to UTs at the cost of energy consumption, thus the system throughput can be increased. Further, we propose two schemes, R-CUT and O-CUT, in which only some UTs are selected as CUTs base on random or optimization methods. The proposed schemes are able to provide tradeoff between decoding efficiency and system energy consumption. The scheme of O-CUT achieves better transmission performance compared with R-CUT.

Our paper proposes a general framework for cooperative multicast based on fountain codes, and provides performance analysis and simulation results for implementations using this framework. In the future, some other issues in the proposed strategy should be studied further, such as selecting cooperative receivers not only based on the SINR simply but also considering the other factors, such as fairness in UT energy consumption.

References

1. 3GPP TS 36.300, Evolved Universal Terrestrial Radio Access (E-UTRA) and Evolved Universal Terrestrial Radio Access Network (E-UTRAN): Overall description, pp. 28–35 (2008)
2. Zhang, S., Lau, V.K.N.: Multi-Relay Selection Design and Analysis for Multi-Stream Cooperative Communications. IEEE Transactions on Wireless Communications 10(4), 1082–1089 (2011)
3. Laneman, J.N., Tse, D.N.C., Womell, G.W.: Cooperative Diversity in Wireless Networks: Efficient Protocols and Outage Behavior. IEEE Transactions on Information Theory 50, 3062–3080 (2004)
4. Kramer, G., Gastpar, M.: Cooperative strategies and capacity theorems for relay networks. IEEE Transactions on Information Theory 51(9), 3037–3063 (2005)
5. Sendonaris, A., Erkip, E., Aazhang, B.: User cooperation diversity Part I: System description. IEEE Transactions on Communications 51(11), 1927–1938 (2003)
6. Sendonaris, A., Erkip, E., Aazhang, B.: User cooperation diversity Part II: Implementation aspects and performance. IEEE Transactions on Communications 51(11), 1927–1938 (2003)
7. Molish, A.F., Mehtra, N.B., Yedidia, J.S., Zhang, J.: Performance of fountain codes in collaborative relay networks. IEEE Transactions on Wireless Communications 6(11), 4108–4119 (2007)
8. Luby, M.: LT Codes. In: The 43rd Annual IEEE Symposium on Foundations of Computer Science, pp. 271–282 (2002)
9. Castura, J., Mao, Y.Y.: Rateless coding and relay networks. IEEE Signal Processing Magazine 24(5), 27–35 (2007)
10. Zhang, Y.Y., Ma, Y., Tafazolli, R.: Improved coded cooperation schemes for wireless communication. In: 6th International Symposium on WiOPT, pp. 533–538 (April 2008)
11. Kurniawan, E., Sun, S., Yen, K.: Transmission strategy of fountain code in coopera-tive networks with multiple relay nodes. In: IEEE Personal, Indoor and Mobile Radio Communications, pp. 1108–1112 (2009)

Clustering for Indoor and Dense MANETs

Luís Conceição and Marilia Curado

Dept. Informatics Engineering, Centre for Informatics and Systems,
University of Coimbra
{lamc,marilia}@dei.uc.pt

Abstract. Clustering is the most widely used performance solution for Mobile Ad Hoc Networks (MANETs), enabling their scalability for a large number of mobile nodes. The design of clustering schemes is quite complex, due to the highly dynamic topology of such networks. A numerous variety of clustering schemes have been proposed in the literature, focusing different characteristics and objectives. In this work, a new clustering scheme, designed for large cooperative environments, is proposed, namely Clustering for Indoor and Dense MANETs (CIDNET). CIDNET was evaluated featuring its stability, amount of clustered nodes and network load. Results demonstrate high and constant levels of network stability.

Keywords: MANET, distributed clustering, cooperative work, stability, indoor environment.

1 Introduction

MANETs are autonomous systems, capable of self deployment and maintenance, not requiring infrastructure support for their operation. As a result, the topology of such networks is very dynamic, especially due to the unpredictable behaviour of the nodes involved. In this context, numerous clustering schemes were developed, following different approaches and objectives, such as stability, low maintenance overhead or energy efficiency. Each one attempts to obtain the best efficiency by varying the characteristics of the system, like the usage of cluster-heads and gateways, the maximum hop distance between nodes and the location awareness. However, there are very few clustering schemes which provide a fully distributed cluster structure with no clusterheads.

In recent years, a wide growth of wireless systems has been noticed. Wireless technologies are present in consumer applications, medical, industrial, public services, transports and much more. Therefore, there is a high demand for accurate positioning in wireless networks, either for indoor or outdoor environments. Concerning the nature of the application, different types of location are needed, which can be characterized as physical location, symbolic location, absolute location and relative location.

Currently, there are many wireless location technologies, such as Radio Frequency (RF) based (WLAN, Bluetooth, ZigBee, RFID), Infrared (IR), Ultrasound, and Global Positioning System (GPS). However, each technology has

S. Andreev et al. (Eds.): NEW2AN/ruSMART 2012, LNCS 7469, pp. 225–236, 2012.

its advantages and disadvantages, and environment scope. No single technology is applicable to all services and circumstances. Recent studies have focused on developing indoor location systems, since GPS offers a good solution but for outdoor environments.

There are many solutions using location awareness designed to improve a wide diversity of goals, e.g. [1] and [2]. However, to the best of our knowledge, none focusing only the improvement of clustering in MANETs has been proposed. In this work, a new clustering scheme is proposed, named as Clustering for Indoor and Dense MANETs (CIDNET), aiming to improve the stability of the cluster structure. CIDNET takes advantage of node location information in order to provide a more efficient cluster creation and management, ultimately leading to a stabler network. The remaining of this document is organized as follows. Section 2 discusses the related work, covering some of most significant clustering schemes and location sensing solutions. Section 3 describes the CIDNET clustering scheme. Section 4 performs the evaluation of CIDNET and, finally, Section 5 concludes the article.

2   Related Work

Currently, there is a wide range of clustering and location algorithms, aiming at different objectives and scopes. This section exploits provides an overall description of the most popular algorithms in both areas, identifying their main characteristics and technologies.

2.1   Clustering

Clustering algorithms can be classified according to different characteristics and objectives [3]. One of the common features in clustering schemes is the utilization of clusterheads (CH) and most of the proposed schemes rely on centralized nodes to manage the clusters structure. The utilization of gateway (GW) nodes is also another important characteristic that is present in the majority of clustering schemes. Other properties of clustering schemes concern the single-hop or multi-hop environments, the multi-homing (MH) support, embedded routing capabilities and location awareness.

Combining the possible characteristics, each proposed clustering scheme attempts to accomplish a specific objective. The Stable Clustering Algorithm (SCA) [4] aims at supporting large MANETs containing nodes moving at high speeds by reducing re-clustering operations and stabilizing the network. To meet these requirements, the algorithm is based on the quick adaptation to the changes of the network topology and reduction of clusterhead reelections. A weight-based clustering scheme, named Distributed Weighted Clustering Algorithm (DWCA), was proposed with the objective to extend the lifetime of the network, by creating a distributed clustering structure [5]. The election of clusterheads is based on the weight value of nodes, which is calculated according to their number of neighbors, speed and energy. The Enhanced Performance Clustering Algorithm

(EPCA) [6] is also a weight based clustering solution. Once more, the weight parameters are only taken into account for the selection of the clusterhead. The Trust-related and Energy-concerned Distributed MANET Clustering (TEDMC) [7] is also a scheme driven by energy concerns. TEDMC considers that the most important nodes are the clusterheads, and therefore it elects them according to their trust level and residual energy. There are also clustering schemes capable of performing route discovery, such as the On-Demand Clustering Routing Protocol (OCRP) and On-Demand Routing-based Clustering (ORC) [8,9]. These schemes are capable of building cluster structures and routing paths on-demand. In these schemes, only the nodes that are necessary to satisfy a routing path are bounded to the cluster structure. The On-Demand Group Mobility-Based Clustering with Guest Node [10] provides a solution with the main purpose of building a cluster structure capable of supporting several types of routing protocols with identical efficiency. SALSA [11] is also a fully distributed scheme, aiming to provide stability at a reduced maintenance overhead. It utilizes two new mechanisms to accomplish its goal, the clustering balancing mechanism and a best clustering metric, which evaluates the most suitable cluster to join based on its connectivity.

2.2 Location Sensing Systems

In recent years, a wide growth of wireless systems has been noticed. Wireless technologies are present in consumer applications, medical, industrial, public services, transports and much more. Therefore, there is a high demand for accurate positioning in wireless networks, either for indoor or outdoor environments. Concerning the nature of the application, different types of location are needed, which can be characterized as physical location, symbolic location, absolute location and relative location. Physical location is expressed in coordinates, identifying a point on a map. Symbolic location refers to a location in natural language, such as a coffee shop, office, etc. Absolute location uses a global shared database system, which references all located objects. Finally, relative location is usually based on the proximity of devices, e.g. known reference points, providing an environment-dependent location. The latest is the most common used paradigm.

The main challenge of location estimation relies on the radio propagation interferences, due to severe multipath, low probability of a Line of Sight (LOS) path, reflecting surfaces, and environment dynamic characteristics, such as building restructuring and moving objects. There are three main techniques to model radio propagation: trilateration, fingerprinting and proximity.

- **Trilateration** - The process consists on determining radial distance, obtained by the received signal, from three or more different points. It can be used on most RF based technologies by measuring radio propagation characteristics, thus calculating distances from two different points.
- **Fingerprinting** - Algorithms first collect features (fingerprints) of a scene and then estimate the location of devices, by matching (or partially matching)

real time (online) measurements with fingerprints. Most of these algorithms define location fingerprints based on Received Signal Strength (RSS) values, previously obtained (offline). Thus, the fingerprinting technique must occur in two stages: the offline gathering of fingerprints, where multiple measurements of known locations are stored in a database, and a online location estimation, which obtains the most suitable match from the database. The major challenge of this technique is the dynamic environments, since building layouts and arranjement of objects are likely to change, thus affecting RSS measurements.

– **Proximity** - Algorithms determine symbolic locations. Typically, it relies on the installed base stations, each classified to be in a known position. When a mobile device is detected by the Base Station (BS) antenna, it is considered to be located in its coverage radius. Moreover, when multiple antennas detect a device (overlapping), is is considered to be located in the BS with the strongest signal, whereas the RSS value is typically used.

There are many proposed wireless location solutions, using different technologies, scopes and with different accuracies. The Active Badge [12] system was a pioneer contribution in location sensing systems and source of inspiration to many following projects. The main goal of this solution is the ability to locate persons or objects inside public buildings like hospitals. Each person wears a badge, which emits an IR signal within every 10 seconds. The sensors placed at known positions are responsible to receive the unique identifiers and relay these to the location manager software. Emitted signals are reflected by surrounding materials and therefore are not directional when used inside small rooms. Bahl *et al.* [13] proposed an WLAN indoor location tracking system called RADAR. In this work, two main types of approaches are employed to determine user location: empirical model and radio propagation model. The first depends on a database that consists of previously measured signal strength of points, recording user orientation and signal strength for each BS. In the second approach, authors adopted the Floor Attenuation Factor (FAF) and Wall Attenuation Factor (WAF) models [14], taking into consideration the number of obstructions walls and material types between the user and the BS. Raghavan *et al.* [15] proposed an location system, for indoor environments, suitable to any technology that provides Receiver Signal Strength Indication (RSSI) values, such as Bluetooth and WLAN. However, since it is designed to locate robots, the authors chose to use Bluetooth, as power consumption is significantly lower than WLAN, despite of providing a higher data rate. The method can provide more accurate results, however at a higher processing cost, by discarding the points with a low error, and repeating the computation process to the remaining. LANDMARC [16] is an indoor location sensing system using active Radio Frequency Identification (RFID), aiming to locate objects. The infrastructure consists of RFID readers, active RFID tags and a management server. All objects must be tagged with an active tag. Active tags are also deployed across the scenario, acting as reference tags, aiding the location process with a low installation cost. The main disadvantage of this approach resides in the sequential scan of all reading ranges, which takes about one minute per cycle. Cheng [17] proposed a room-based location technology using ZigBee wireless technology. Two ZigBee nodes are placed inside each room, one at the door, with the antenna pointing inwards the room and adjusted within 1.5 meters, and a second in a unspecific wall, adjusted within 10 meters. When the user tag passes the door or room and the secondary node senses the user tag, it can be certain that the user is in that specific room. The Bat system [18] present an location approach based on ultrasound. Each person or object carries a device called Bat that periodically sends an ultrasonic signal. Receivers are placed to fixed positions at the ceiling of rooms, and connected to a wireless network. Analyzing the arriving times, provided by several receiving units, the core management system calculates the position of devices. This project shows that ultrasound provides an high precision location sensing, however ultrasound is highly vulnerable to interferences. Moreover, the installation cost of this system can be very high, which difficult the extension of the system for large areas. Cricket [19] is another ultrasound based location system. In contrast with the Bat system, mobile devices are responsible to determine the location by themselves, ensuring privacy to users. Also, instead of receivers, beacons are placed in the ceiling, which periodically send radio and ultrasonic signals. Using multiple signals from different beacons, the mobile device calculates the current position.

3 Clustering for Indoor and Dense MANETs (CIDNET)

There is a large variety of clustering schemes in literature, with different mechanisms and objectives, aiming to build a suitable hierarchical structure in order to provide an efficient routing in MANETs. Despite the goal of the majority of schemes, which aims at the impromptu deployment of wireless networks in remote environments, clustering can also be an asset in common scenarios, where network infrastructures are present. Typical WLAN network infrastructures do not efficiently support a large quantity of associated nodes, becoming overloaded and consequently unresponsive. The utilization of ad-hoc networks in this environments would be a solution to address this issue.

CIDNET is a fully distributed clustering scheme designed for dense cooperative environments, where existent network infrastructures are not sufficient. This proposal offers a clustering solution for ad-hoc networks, utilizing surrounding network infrastructure as context information to ease cluster management. CIDNET uses the existent Access Points (APs) as a proximity location reference, in order to facilitate cluster creation and management. As studied in the previous section, location sensing systems are complex, particularly when concerning trilateration and fingerprinting. Thus, CIDNET is based on proximity location, relying on APs to determine the location information for the entire network. Nodes scan for WLAN Service Set Identifier (SSID) broadcasts and create clusters according to that information. This mechanism is more efficient since all nodes in-range to an AP are instantly assigned to a cluster (according to the SSID string) and an initial waiting period for cluster creation is not necessary. Furthermore, the proposal implements some mechanisms of the SALSA algorithm, namely the automatic clustering balancing and most suitable joining cluster determination.

3.1 Location Sensing and Dissemination

In CIDNET there are two distinct types of nodes. The nodes that are in-range with at least one AP, named as anchor nodes, and the nodes that are distant and not capable of receiving SSID broadcasts, named as blind nodes.

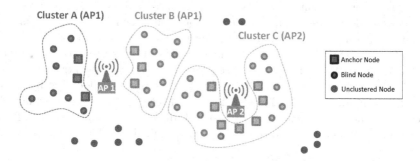

Fig. 1. CIDNET Clustering Example

Anchor nodes have the additional responsibility of creating clusters, based on the nearby SSID, and inform its blind neighbours that there is an AP nearby. Upon receiving this information, blind nodes decide whether to join the cluster associated with that AP and continue to broadcast the received information to its neighbours, until a configurable TTL reaches 0. Figure 1 shows a possible clustering scenario of CIDNET. As depicted, only anchor nodes (represented as a square dots) have connectivity to APs. Blind nodes (round dots) receive broadcast of the APs and join the clusters. To be noted that multiple clusters can be associated to the same AP, like Cluster A and Cluster B. In the best case scenario, each broadcast would be associated with one single cluster, however this situation is not always possible. Looking at Figure 1, Cluster A and B are associated with one specific SSID AP broadcast and they cannot be merged into one single cluster, since their nodes do not have connectivity. Addicionally, there are nodes that remain completely isolated and cannot be associated to the cluster structure. This situation occurs when nodes do not have any connectivity, even through neighbour nodes, to APs or AP broadcast messages did not reach them due to TTL expiry.

3.2 Node States

Nodes can be in one of three distinct states, namely *Unclustered*, *Clustered* and *Clustered-GW*.The *Unclustered* state typically represents a temporary role, as the node is waiting to be assigned to a cluster. In this state, when the node discover at least one APs it becomes an anchor node and creates a cluster. On the other hand, if the node receives an AP broadcast, it assumes the blind node role and decides, based on the received broadcasts, what is the best cluster to join. Nodes in the *Clustered* state usually represent the majority of nodes on the network, either anchor or blind nodes, whereas all in-range nodes must belong to its cluster. Thus, the communication with foreign nodes (i.e. nodes assigned to a different cluster) is performed through gateway nodes. Finally, the *Clustered-GW* state is assigned to nodes that have in-range foreign nodes, i.e. they must have direct connectivity with at least one different cluster. Thus, they are responsible of forwarding inter-cluster maintenance messages and typically are located on the edge of clusters.

State Transitions. The *Unclustered* state occurs on two different situations:

1. Node isolation (geographic position) - in this case the node does not have any in-range neighbour nodes or AP's, therefore cannot create or be assigned to a cluster
2. Node isolation (TTL expiry) - in this case, nodes have connectivity to neighbour nodes that may be clustered, however due to the TTL expiry of AP broadcasts, they cannot be associated with a cluster. This mechanism is necessary as it prevents the creation of very large clusters, leading to a higher instability of the network

Unclustered to Clustered. This state occurs when a node becomes aware of an AP broadcast or it has direct connectivity with an AP. In the first situation, the node has to evaluate which is the best cluster to join (if received distinct and multiple AP broadcasts) based on the number of received broadcasts per AP. This is, the more received broadcasts announcing an AP, the more neighbour nodes associated with that same AP, leading to a better connection and ultimately to a better stability. In the second situation, upon detecting an AP, nodes automatically create a cluster and broadcast a message announcing its neighbours the presence of an AP.

Unclustered to Clustered-GW. This transition is very similar to the previous, but with one difference. When a node becomes clustered, it is considered a gateway if it has direct connectivity with neighbour nodes belonging to different clusters.

Clustered to Clustered-GW. This transition occurs when a node becomes aware of clusters, excluding its own.

Clustered-GW to Clustered. Whenever a clustered gateway node loses connection with all its foreign clusters, it automatically transits to a normal clustered state.

Clustered/Clustered-GW to Unclustered. A node becomes unclustered when willingly disconnects from the network or loses connection with all its neighbor nodes. When this situation occurs, it is necessary to verify the consistency of the cluster, i.e. guarantee that all home nodes can communicate with each other.

4 Simulation Evaluation

To examine the effectiveness of CIDNET, a simulation was performed using the OPNET Modeler [20]. Therefore, the main purpose of this simulation evaluation is to assess the stability and low overhead capabilities of the proposal. To accomplish this objective, a set of different simulation environments, featuring the network size and speed of nodes, were defined.

4.1 Environment and Parameters

The scenarios utilized to evaluate CIDNET were selected in such a way that they represent, as much as possible, realistic scenarios. In this specification the evaluation parameters can be divided in two groups, the fixed-value and variable-value parameters, according to whether their value changes for different simulation scenarios (Table 1).

Table 1. Simulation parameters

Fixed-value parameters	
Simulator	OPNET Modeler 16.0
Field Size (m^2)	500 × 500
Node mobility algorithm	Random Waypoint Model
Pause time (s)	50
Transmission range (m)	150
WLAN IEEE Standard	802.11b (11 Mbps)
Simulation time (s)	900
Number of APs	49
AP Broadcast TTL	5
Variable-value parameters	
Network size (number of nodes)	80; 160; 240; 320; 400
Node maximum speed (m/s)	0; 1; 2

CIDNET relies on APs to the creation and position of clusters. Since APs will determine the position of clusters, it would be desirable to evenly scatter them across the deployment scenario. Thus, to a first validation of CIDNET, all scenarios will contain 49 APs (7 × 7), placed in a grid fashion. The parameters that most influence the scalability of the network are the network size (number of nodes) and the maximum speed that nodes can achieve. This simulation study aims to evaluate areas with the existence of network infrastructures, e.g. an university campus. Thus, a random model for mobility, namely the Random Waypoint, was chosen to simulate the movement of people. Also, the average speed on foot of humans does not exceed 2 m/s, which was considered as a good maximum movement speed. Each simulation execution was repeated 30 times, assigning to each a distinct seed value.

4.2 Results

This section presents the obtained results from the CIDNET simulation. As previously mentioned, CIDNET is a completely new algorithm, implementing some features of the SALSA scheme. For that reason and since both CIDNET and SALSA are full distributed algorithms (i.e. do not use clusterheads), the discussion of the following results will be conducted according to the results obtained in SALSA.

Number of Clustered Nodes. This metric provides the number of nodes that are associated with the cluster structure.

Table 2. Amount of clustered nodes (in percentage)

(a) CIDNET

Network Size \ Speed (m/s)	0	1	2
80	90.54	86.10	85.51
160	91.68	87.56	86.89
240	92.98	88.97	87.59
320	94.51	90.30	88.90
400	96.10	93.31	92.01

(b) SALSA

Network Size \ Speed (m/s)	0	1	2
80	88.91	90.54	90.69
160	91.21	92.88	91.54
240	93.28	92.95	90.89
320	94.55	92.10	89.28
400	95.29	91.50	87.21

Table 2a shows the percentage of clustered nodes for the different network sizes and node speeds in CIDNET. The percentage of clustered nodes for large networks is slightly higher than for smaller networks. This occurrence is due to the node density increase, i.e. the probability of a node being communication in-range with another is greater for networks with more nodes. In SALSA (Table 2b) the percentage of clustered nodes generally increases for larger networks, with the exception of scenario of nodes moving at the maximum speed of 2 m/s and larger than 240 nodes. This fact occurs due to the high density of nodes in the network and shows that SALSA is becoming overloaded, thus not being able to cluster such a large quantity of nodes.

Cluster Stability. The stability of clusters can be measured according to the amount of time that nodes belong to a cluster, without suffering re-clustering operations.

For this analysis, a cluster stability metric is utilized, which defines a stability time (ST), from which nodes are considered to be stable (1).

$$ST = k \times \frac{r \times p}{v \times d} \tag{1}$$

where r is the transmission range of nodes, p is the pause time, v the average of node speed (mean value of minimum and maximum speed), d the density of nodes (number of nodes per Km^2) and finally, k represents an arbitrary constant, equal in all simulation executions, enabling the transformation of the ratio to a real execution time.

The stability metric ST provides a mechanism of determining the amount of nodes that were stable during the simulation for a period greater than the ST

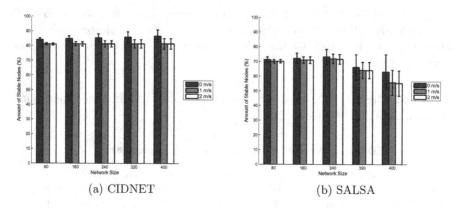

(a) CIDNET   (b) SALSA

Fig. 2. Amount of stable nodes (in percentage)

value. Figures 2a and 2b show the number of stable nodes per network size and their percentage, at different node speeds.

As a quick first analysis of the Figures, CIDNET clearly outperforms SALSA in terms of stability. CIDNET presents higher stability levels for all scenarios. Furthermore, the percentage of stable nodes is almost constant for all network sizes, whereas SALSA shows a significant decline for network sizes greater than 240 nodes. As previously seen, the amount of clustered nodes in SALSA also decreased at node speed of 2 m/s for network sizes greater than 240 nodes. Once again, this fact occurs due to the high density of the network, leading to instability. CIDNET however, is capable of overcoming this issue, as it always forces clusters to be near APs, providing a stabler network. There is, however some small drawbacks. In CIDNET the number of clustered nodes for dynamic nodes (1 m/s and 2 m/s) is slightly lower in smaller networks, as it can be seen in Table 2.

Network Load. The network load represents the received and transmitted traffic in the network. This metric translates the overall weight of the network, including the clustering control overhead.

Table 3. Average Network Load (Kbit/s)

(a) CIDNET

Network Size	Speed (m/s) 0	1	2
80	47,11	47,89	48,08
160	56,56	57,34	57,53
240	68,54	69,32	69,51
320	78,00	78,78	78,97
400	87,12	87,90	88,09

(b) SALSA

Network Size	Speed (m/s) 0	1	2
80	10,30	13,25	14,21
160	12,98	14,21	15,74
240	21,59	24,12	24,54
320	36,87	38,13	38,90
400	47,12	48,50	49,12

Table 3a and Table 3b show the average network load, for different velocities and network sizes, for CIDNET and SALSA, respectively. CIDNET handles clustering with a significant higher overhead. This is mainly due to the broadcasts announcing the presence of Access Points (APs). As previously described, anchor nodes broadcast messages announcing the presence of APs, and consequently, blind nodes follow the broadcasts until a configurable TTL expires. Naturally, if the TTL value is lower, CIDNET will present lower overhead levels. On the other hand, this value cannot be too small, otherwise there would be many unclustered nodes. Despite presenting a significant higher overhead, CIDNET can outperform SALSA in the amount of clustered nodes and, most importantly, in stability.

5   Conclusion

This article proposed CIDNET, a new clustering scheme aiming to improve the stability of the network, in order to provide reliable and large cooperative environments. This clustering scheme employs a context aware paradigm, utilizing the existent network infrastructure as a location reference to improve the stability and management of the cluster structure. Evaluation results shown that CIDNET outperforms SALSA in both the amount of clustered nodes and stability, despite of using a higher management overhead. Nevertheless, CIDNET never overcomes an average of 90 kbits/s in the entire network, which is pretty reasonable.

Acknowledgments. This work was supported by FCT project MORFEU (PTDC / EEA-CRO / 108348 / 2008). The authors would like to thank the OPNET University Program for the licenses provided for the OPNET Modeler-Wireless Suite R.

References

1. Kim, J.-I., Song, J.-Y., Cheol Hwang, Y.: Location-based routing algorithm using clustering in the manet. In: Future Generation Communication and Networking (FGCN 2007), vol. 2, pp. 527–531 (December 2007)
2. Wang, T., Yang, Z.: A location-aware-based data clustering algorithm in wireless sensor networks. In: 11th IEEE Singapore International Conference on Communication Systems, ICCS 2008, pp. 1–5 (November 2008)
3. Yu, J., Chong, P.: A survey of clustering schemes for mobile ad hoc networks. IEEE Communications Surveys & Tutorials 7(1), 32–48 (2005)
4. Tolba, F., Magoni, D., Lorenz, P.: A stable clustering algorithm for highly mobile ad hoc networks. In: Second International Conference on Systems and Networks Communications, ICSNC 2007, p. 11 (August 2007)
5. Choi, W., Woo, M.: A distributed weighted clustering algorithm for mobile ad hoc networks. In: International Conference on Internet and Web Applications and Services/Advanced International Conference on Telecommunications, AICT-ICIW 2006, p. 73 (February 2006)

6. Zoican, R.: An enhanced performance clustering algorithm for manet. In: MELE-CON 2010 - 2010 15th IEEE Mediterranean Electrotechnical Conference, pp. 1269–1272 (April 2010)
7. Qiang, Z., Ying, Z., Zheng-Hu, G.: A trust-related and energy-concerned distributed manet clustering design. In: 3rd International Conference on Intelligent System and Knowledge Engineering, ISKE 2008, vol. 1, pp. 146–151 (November 2008)
8. Huang, C., Zhang, Y., Jia, X., Shi, W., Cheng, Y., Zhou, H.: An on-demand clustering mechanism for hierarchical routing protocol in ad hoc networks. In: International Conference on Wireless Communications, Networking and Mobile Computing, WiCOM 2006, pp. 1–6 (September 2006)
9. Hsu, C.-H., Feng, K.-T.: On-demand routing-based clustering protocol for mobile ad hoc networks. In: IEEE 18th International Symposium on Personal, Indoor and Mobile Radio Communications, PIMRC 2007, pp. 1–5 (September 2007)
10. Dana, A., Yadegari, A., Salahi, A., Faramehr, S., Khosravi, H.: A new scheme for on-demand group mobility clustering in mobile ad hoc networks. In: 10th International Conference on Advanced Communication Technology, ICACT 2008, vol. 2, pp. 1370–1375 (February 2008)
11. Conceição, L., Curado, M.: In: Frey, H., Li, X., Ruehrup, S. (eds.): Ad-hoc, Mobile, and Wireless Networks
12. Want, R., Hopper, A., Falcão, V., Gibbons, J.: The active badge location system. ACM Trans. Inf. Syst. 10, 91–102 (1992), http://doi.acm.org/10.1145/128756.128759
13. Bahl, P., Padmanabhan, V.: Radar: an in-building rf-based user location and tracking system. In: Proceedings of the IEEE Nineteenth Annual Joint Conference of the IEEE Computer and Communications Societies, INFOCOM 2000, vol. 2, pp. 775–784 (2000)
14. Seidel, S., Rappaport, T.: 914 mhz path loss prediction models for indoor wireless communications in multifloored buildings. IEEE Transactions on Antennas and Propagation 40(2), 207–217 (1992)
15. Raghavan, A., Ananthapadmanaban, H., Sivamurugan, M., Ravindran, B.: Accurate mobile robot localization in indoor environments using bluetooth. In: 2010 IEEE International Conference on Robotics and Automation, ICRA, pp. 4391–4396 (May 2010)
16. Ni, L., Liu, Y., Lau, Y.C., Patil, A.: Landmarc: indoor location sensing using active rfid. In: Proceedings of the First IEEE International Conference on Pervasive Computing and Communications, PerCom 2003, pp. 407–415 (March 2003)
17. Cheng, Y.-M.: Using zigbee and room-based location technology to constructing an indoor location-based service platform. In: Fifth International Conference on Intelligent Information Hiding and Multimedia Signal Processing, IIH-MSP 2009, pp. 803–806 (September 2009)
18. Harter, A., Hopper, A., Steggles, P., Ward, A., Webster, P.: The anatomy of a context-aware application. Wirel. Netw. 8, 187–197 (2002), http://dx.doi.org/10.1023/A:1013767926256
19. Smith, A., Balakrishnan, H., Goraczko, M., Priyantha, N.: Tracking moving devices with the cricket location system. In: Proceedings of the 2nd ACM International Conference on Mobile Systems, Applications, and Services (MobiSys 2004), pp. 190–202. ACM Press (2004)
20. OPNET, Opnet simulator (1986), http://www.opnet.com/

Energy-Efficient Heuristics for Multihop Routing in User-Centric Environments

Antonio Junior[1], Rute Sofia[1], and António Costa[2]

[1] SITI, University Lusófona
{antonio.junior,rute.sofia}@ulusofona.pt
[2] ALGORITMI, University of Minho
costa@di.uminho.pt

Abstract. This paper proposes and validates routing metrics focused on improving energy-efficiency of multihop approaches in heterogeneous wireless environments. The validation is carried out through discrete event simulations and for the specific case of AODV.

Keywords: multihop routing, energy-efficiency, user-centric networks.

1 Introduction

The most recent advances in wireless technologies and enabled devices is leading to an expansion of wireless architectures where nodes are often carried or owned by regular Internet users. These environments, user-centric in nature, are often coined with the term *user-centric networks* [15,13]. Examples of such environments can be a network formed on-the-fly after a disaster of some nature or even a municipality network where some nodes are based on end-user devices (through Internet access sharing).

In user-centric environments, there is a behavior that we can root on social network theory, and where spontaneity is a main characteristic, due to the fact that individuals (humans) are not only carriers but also the decision makers for the operation of nodes that form the topology. Adding to the variability e.g. due to node movement, for instance, another key aspect is that some devices are multimedia capable with strong limitations in terms of energy capabilities.

Albeit being often spontaneously deployed, user-centric wireless environments rely on traditional multihop routing approaches. Multihop routing has been extensively analyzed and optimized in terms of resource management, but in terms of energy-efficiency there is a lack of a thorough analysis in particular in what concerns user-centric environments. On the other hand, there is considerable related work in the fields of energy-efficiency and energy-awareness for sensor networks, where nodes are normally considered to be *homogeneous* in terms of energy capability. In contrast, in user-centric networks, nodes are expected to be *heterogeneous* in terms of energy.

In previous work [5,6] we have discussed the potential of current energy-aware routing approaches for wireless networks, and whether or not they may make

S. Andreev et al. (Eds.): NEW2AN/ruSMART 2012, LNCS 7469, pp. 237–247, 2012.

sense when applied to routing in user-centric environments. We have also proposed a few initial concepts that could assist in making multihop routing more efficient in terms of energy-awareness, without necessarily having to change operational aspects of the underlying algorithms, or protocols.

Following such line of thought, this paper proposes and validates two routing metrics to improve network lifetime based on current multihop approaches. The main goal of this work is focused on making current multihop routing approaches, i.e., *shortest-path* routing approaches for wireless networks, more adequate to be applicable in user-centric environments. We evaluate the proposed heuristics through discrete event simulations having as particular case the *Ad-Hoc On-demand Distance Vector protocol (AODV)* [10] and as performance goal improving the overall network lifetime.

The rest of this paper is organized as follows. Section 2 describes related work focused on multihop energy-efficiency. Section 3 goes over notions concerning energy-awareness in multihop routing. Section 4 describes our proposed energy-awareness heuristics. Then, in section 5, we present the performance evaluation based on simulations with statistically rigorous results. Conclusions and future work are presented in section 6.

2   Related Work

A few approaches [18,8] have surveyed multihop proposals focused on energy-efficiency, considering both the energy spent when nodes are engaged in active communication or passive communication (e.g., in idle mode). Such work has as underlying scenarios homogeneous environments, and several proposals combine different energy-aware metrics to maximize the network lifetime.

Attempting to make multihop routing adaptive, some proposals [11,1,17] have explored new metrics having in mind different types of optimization, e.g., reduction of energy spent across a path or avoiding nodes with low residual energy, on the global network.

C. K. Toh provides a relevant overview [14] of different routing properties to consider in multihop routing. One of them is efficient utilization of battery capacity. In this work, the author also addresses the performance of power efficiency in ad-hoc mobile networks by analyzing four approaches which have as common goal to select an optimal path, being the optimum the minimization of the total power required on the network and also the maximization of the lifetime of all nodes in the network.

The cost function of the *Maximum Residual Packet Capacity* (MRPC) protocol [9] comprises a node perspective parameter (battery of the node) and a link perspective parameter (packet transmission energy in a link) across the link between nodes. MRPC identifies the capacity of a node not just by its residual battery energy, but also by the expected energy spent in reliably forwarding a packet over a specific link. However, such formulation is more adequate to scenarios where the link transmission cost depends on the physical distance between nodes and on the link error rates. Hence, the approach does not consider energy-awareness as a primary resource of the network.

A recent work [4] proposes a multi-objective prediction approach to optimize three main aspects of network operation: minimize average end-to-end delay, maximize network energy lifetime, and maximize packet delivery ratio. The authors have as main purpose to make multihop routing more flexible in terms of the three mentioned parameters. In terms of energy they consider a composite energy cost based upon transmission power and residual energy. The energy cost is then computed based on a time series prediction model and the authors show that proactive routing (based on the *Optimized Link State Routing (OLSR)* protocol) benefits from their proposal. In contrast to their work, our heuristics are not based on prediction and instead on a more realistic assessment of the energy consumption and thus our approach is expected to incur on a lower operational cost, as less changes to the protocols are expected.

Finally, we highlight that The Internet Engineering Task Force (IETF) Working Group *Routing Over Low Power and Lossy Networks* (ROLL) is currently discussing multihop metrics tailored to energy-efficiency [16].

3   Energy Awareness in Multihop Routing

This section provides a few notions concerning energy awareness in multihop routing.

A *node i* represents a wireless heterogeneous device with a single or with multiple network interfaces. An edge interconnecting two nodes i and j is represented as *link* (i, j) with a cost $e_{(i,j)}$ which is a measure of energy expenditure. Such cost e can be obtained from a single node perspective (source or destination); from the link perspective; from a global network utilization perspective. From a single node perspective, there are three main modes of energy expenditure which depend on the node status. A node is in *Transmission mode* when transmitting information. Hence, *Transmit Power (Tx Power)* for a node corresponds to the amount of energy (in Joules) spent when the node transmits a unit (bit) of information. A node is in *Receive mode* if it is receiving data. Hence, *Reception Power (Rx Power)* for a node corresponds to the amount of energy (in Joules) spent when the node receives a unit of information. Particularly for the case of 802.11, there are two additional states a node may be at. When the node is still passively listening to the shared medium (*overhearing*), is said to be in *Idle mode* which consumes energy. A fourth mode can be considered, *Sleep mode*. In this mode, the networking capabilities are shut down but the node still consumes little power.

Another relevant parameter to consider from an energy-awareness perspective is a node's degree, N_i as the surrounding nodes impact heavily not only the transmission channel, but also the rate of energy consumption. We use the node degree definition where N_i corresponds to the number of neighbors that a node i has at an instant in time. From an energy-awareness perspective, more relevant than the number of neighbors, is the history of variation of N_i through time as it can assist us in estimating potential fluctuations of energy levels - a faster or slower potential for energy consumption.

The most relevant energy-aware routing metrics for user-centric environments are the residual energy and drain rate of a node. The *Residual Energy (RE)* of a node i, $RE(i)$ [12] is defined as the amount of energy units that the battery of node i has at an instant in time. The *Drain Rate (DR)* of a node i, $DR(i)$ [7] is defined as the amount of energy being spent by node i through time, due to the activities the node is performing. $DR(i)$, can be computed by applying an *Exponential Weighted Moving Average (EWMA)*. The DR alone simply provides a way to measure energy being spent by a node i at an instant in time. Garcia-Lunes et. al. have considered the ratio between RE and DR, $C(i)$, as the estimated lifetime for node i, having in mind scenarios where nodes are homogeneous in terms of energy capability. Our heuristics proposed in the section 4 are based on the notion of $C(i)$.

A *node's lifetime* corresponds to the period of time since a node becomes active until the node is said to be dead, i.e., from a network perspective, the node ceases to exist.

Network lifetime is often associated in related literature to the time period that elapses since a topology becomes active, until a first node dies. In contrast, in our work we follow the definition where *network lifetime is associated to the time period since a topology becomes active, until the topology becomes disconnected, from a destination reachability perspective*. In other words, we consider that the energy lifetime covers the time a topology is stable enough to reach at least one of the destinations present at some instant in time.

4   Proposed Heuristics

This section is an overview on our metrics. Our belief is that they can assist multihop routing in becoming more energy-efficient without increasing a significant cost in terms of protocol implementation and operation.

4.1   Heuristic 1: Energy-Awareness Ranking of a Node Based on Idle Time Periods

Based on the notion that a node still spends energy per bit when in idle mode, this first proposal explores the fact that nodes may be in idle mode for a long time. Nodes that have been in idle mode for a long period of time in the past and that still have a reasonable large estimated lifetime are, in our opinion, better candidates to be elements in a shortest-path.

Hence In this first heuristic we take into consideration the periods over time where i is in idle mode. In other words, over time we estimate how much of its lifetime has node i been in idle mode, to then provide an estimate towards the node's future energy expenditure, as this will for sure impact the node's lifetime. Such periods are the ones that are the most expensive to i in terms of energy. So we consider the total period in idle time, t_{idle} over the full lifetime expected for a specific node, which is given by the sum of the elapsed time period T with the estimated lifetime of the node, as provided in equation 1.

$$E_1(i) = \frac{T - t_{idle}}{T \times C(i)} \qquad (1)$$

E_1 is therefore a node weight which provides a ranking in terms of the node robustness, from an energy perspective, and having as goal to optimize the network lifetime. Hence, the smaller $E_1(i)$ *is*, the more likelihood a node has to be part of a path.

4.2 Heuristic 2: Energy-Awareness Ranking of Node Based on Idle Time Periods and Node Degree History

Based on the previous line of thought, we consider a new parameter in terms of impact of energy expenditure of a node, namely, the node degree. Surrounding nodes impact the conditions of the wireless media and as such, the node degree history, in particular the variability of the node degree is one additional aspect that may impact node lifetime.

Hence, still following a simplistic approach, we consider ways to combine the history of the node degree with E_1, having derived as a first approach E_2, provided in equation 2.

$$E_2(i) = \frac{(T - t_{idle}) \times N_i}{T \times C(i)} \qquad (2)$$

For instance, let us assume that node i has, at a specific instant in time, a lifetime that is rather large. If the node has an history of having a low number of neighbors around as happens in the case of less dense networks, then in contrast to a node that has the same lifetime but a larger number of nodes around, we can decide on which node to opt. Deciding for a node that has a higher node degree implies having more alternate paths being the flip-side to this the possibility of seeing an abrupt change in the time left until the node exhausts energy. Opting for a node with a lower node degree may provide more robustness at the cost of having less alternate paths. Depending on the situation of the nodes around (e.g. movement; shorter lifetimes), there is some variability associated which impacts the node lifetime, and the network energy lifetime.

The node degree history, N_i, is provided by an *Exponential Moving Average (EMA)* as provided in equation 3.

$$N_i = \alpha \times N_{i_{t-1}} + (1 - \alpha) \times N'_{i_t} \qquad (3)$$

4.3 Brief Analysis of the Impact of the Different Parameters

To better explain how the different parameters may impact E_1 and E_2, Table 1 depicts the described parameters and the respective results in terms of E_1 and E_2. For each parameter we consider two extreme values, "low" and "high".

The outcome in terms of E_1 and E_2 is ranked as a potential candidate, low potential for being a candidate to the path, and avoiding if possible (*avoid*).

For instance, in line 1 we consider a potential case where a node has been most of its lifetime in idle mode and yet it has a large estimated lifetime. This is likely to occur if the node is in fact isolated or if its energy is being regularly increased (e.g. large battery capacity or AC power on). If the node has been most of its lifetime in idle mode than the E_1 value is strongly dependent on the value of $C(i)$. Therefore, a node that has a large $C(i)$ is expected to be chosen when compared with a node with a small $C(i)$. The situation where a node has been most of its time in idle mode but has a small estimated lifetime is considered in line 2. The result is that E_1 tends to 1 and therefore the node has low potential to be part of a path, when compared to nodes that have smaller E_1 values.

Table 1. Ranking the node cost

t_{idle}	$C(i)$	E_1	N_i	E_2
high	high	candidate	high	low potential
			low	candidate
high	low	low potential	high	low potential
			low	good potential
low	high	good potential	high	good potential
			low	candidate
low	low	avoid	high	avoid
			low	low avoid

The impact of t_{idle} in E_1 becomes less significant for the cases where t_{idle} is small, assuming that the node has enough energy (large value for $C(i)$ cf. line 3).

To analyze E_2 following this line of thought, Table 1 considers also N_i, i.e., situations where a node has a few or some relevant number of neighbors in range. A large number of neighbors through time implies that the node may not be an adequate candidate as loaded networks may exhaust nodes faster. The node degree seems only to become less prominent for situations where the estimated node lifetime is large enough.

5   Performance Evaluation

This section provides a performance evaluation for E_1 and E_2 based on NS-2 (version 2.34) simulations. Scenarios have been modeled as close as possible to reality. We have considered the NS-2 default physical layer, two-ray ground propagation model and DCF (Distributed Coordination Function) for MAC layer with 802.11g parameters. We have used the energy consumption model provided by [3].

In terms of topology we considered 25 nodes distributed across a square with an area of 1400m x 1400m, 5 nodes per row and per column. Hence, node degree varies as 2, 3, 4 according to the position of each node in the square.

We then considered a simple model for *Voice over IP (VoIP)*, where calls follow a Poisson model, and where each flow is based on *Constant Bit Rate (CBR)*, average packet size of 40 bytes, inter-packet times of 0.02 seconds. Sources and destinations are randomly selected from the available nodes. Then, we consider 5, 10, and 15 flows as a way to represent three different load levels. The simulation time has been set to 100 seconds.

In these first set of experiments, all of the nodes are static, as what is relevant to us is to understand how the network behaves in terms of energy consumption. Hence, each node has been modeled to have different levels of energy parameters in order to represent heterogeneous devices.

Albeit applicable to any shortest-path approach, in this paper the evaluation of the heuristics is performed based on AODV, being the reason simply the fact that this is still a first step towards a thorough evaluation of the heuristics in terms of its global applicability. For AODV we considered the native NS-2 module, here referenced as *AODV-native*. The *AODV-native* considers hop count as the metric to compute a shortest-path. Moreover, the original $C(i)$ has been developed to be applied to Dynamic Source Routing (DSR) protocol. The original specification of $C(i)$ therefore selects a best path based on a *min-max* approach, where the best path is the one that has the lowest bottleneck in terms of energy. So, we adapted the protocol to select the path in a min-max way as the original specification of the $C(i)$. We refer to this implementation as *AODV-minmax-Ci*. Moreover, *AODV-SP-E1* and *AODV-SP-E2* represent AODV running our two heuristics E_1 and E_2, respectively.

The results extracted intend to analyze benefits in terms of *network lifetime* as defined in section 3. Even though we analyze benefits in terms of network lifetime, we also want to understand the impact of the metrics in the overall network performance. For that, we consider three additional aspects: i) *average estimated node lifetime*, i.e., the estimated node lifetime, $C(i)$, across all nodes in the network; ii) *average end-to-end delay*, the time a packet takes between source and destination computed per destination and then averaged across all destinations; (iii) *throughput*, the average number of bytes reaching destination nodes, computed first per destination and then averaged across all destinations measured in Kbps.

To generate statistical sound results we relied on Akaroa2 [2] tool. All results have been computed within a 95% confidence interval.

5.1   Network Lifetime

In a first setting, based on the parameters described, all of the 25 nodes of the topology have been set with initial energy levels picked up randomly between 30 Joules and 180 Joules. Figure 1(a) shows the average network lifetime for the different approaches. The X-axis represents the number of flows, while the Y-axis provides the network lifetime in seconds.

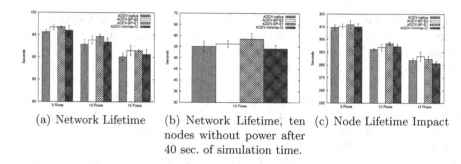

(a) Network Lifetime   (b) Network Lifetime, ten   (c) Node Lifetime Impact
nodes without power after
40 sec. of simulation time.

Fig. 1. Node and Network Lifetime Analysis

Even though the topology is simple in the sense that the nodes are equally placed, traffic is CBR based, and the network never becomes truly congested, from Figure 1(a) it can be observed that the behavior of *AODV-native* is constant, resulting in the lowest network lifetime.

AODV-SP-E1 and *AODV-SP-E2* do improve network lifetime, outperforming AODV. However, as the time-scale reflects seconds, it is not possible to state, based on this simple case, that the proposed heuristics provide an advantageous benefit when compared to the native behavior. For this concrete simulation scenario, Table 2 (col. 2, 3 and 4) shows that heuristics E_1 and E_2, when applied to AODV, seem to improve the overall network lifetime circa 1.5%. Moreover, it is also not possible to understand if an heuristic is truly better than the other.

Table 2. Network Lifetime Improvement

Approach	5 flows	10 flows	15 flows	exhausted network - 10 flows
AODV-SP-E1	1.20%	1.87%	1.55%	6.06%
AODV-SP-E2	1.15%	1.05%	1.53%	2.02%
AODV-minmax-Ci	0.38%	0.59%	0.61%	-1.87%

To better analyze if our heuristics behave coherently across different scenarios, we run a simulation with ten flows (average network load) configured as before, but where now 10 nodes have an energy level which has been randomly picked to be exhausted after 40 seconds of simulation time, being the results depicted in Figure 1(b). The intention is to create a topology where there is more variability in terms of nodes (and path) availability. From Table 2 (col. 5) the first conclusion we can draw relates to the fact that again *AODV-SP-E1* and *AODV-SP-E2* provide the best results, being *AODV-SP-E1* increases the network lifetime the most.

The E_1 and E_2 differ only in the application of the node degree. In the scenarios simulated, there is not a relevant benefit in applying the node degree, but this relates to the fact that the chosen topology does not provide adequate variability in terms of node degree, to reflect an adequate difference. This is an aspect we intend to explore in future work.

5.2   Node Lifetime Impact

Our approach is meant to improve network lifetime but nonetheless it is relevant
to understand and to ensure that new metrics do not negatively impact node
lifetime and other network operation parameters, while improving network lifetime. Hence, for the initial setting of 25 nodes described in section 5.1, we have
also analyzed how the metrics proposed affect node lifetime, and the results are
illustrated in Figure 1(c).

A first observation based on the results achieved is that the proposed heuristics do not impact negatively the node lifetime and in fact slightly improve the
behavior when compared to *AODV-native*. For a low load, *AODV-SP-E1* outperforms all other approaches. When the number of flows increases, however,
AODV-SP-E2 is the heuristic that provides a better node lifetime at the expense of more variability. We believe this may be related to the node degree
impact and the way we model such impact (product), which may be more severe
than expected.

5.3   End-to-End Delay and Throughput

As our main goal is to extend network lifetime without penalizing the end-to-end
delay and throughput. Figure 2(a) shows the average end-to-end delay of the E_1
and E_2 heuristics, *AODV-native* and *AODV-minmax-Ci*. It is represented by
the average end-to-end delay in seconds and by number of flows according to the
degree of load in the network.

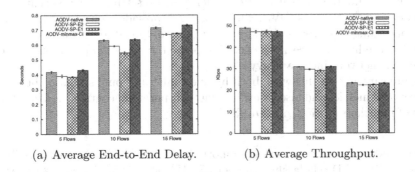

(a) Average End-to-End Delay.   (b) Average Throughput.

Fig. 2. Delay and Throughput Analysis

According to the results, our E_1 and E_2 heuristics result, for AODV, in a
lower end-to-end delay. There is a slight gain due to node ranking favoring more
robust paths by selecting a good path regarding energy resources and also for
delay constraints. The heuristics seem to provide AODV with lower end-to-end
delays and across all scenarios *AODV-SP-E2* is more stable than *AODV-SP-E1*
in terms of gain. The reason for E_2 to be more stable relates to the application
of the node degree. Our simulated environment corresponds to a static topology

where node degree does not vary. Over time, node degree smooths out variability in E_2 and overall is the heuristic that seems to result in more stability. *AODV-minmax-Ci* is the approach that provides additional delay. This is not surprising, as $C(i)$ has been developed as a metric for source-based routing.

We have then analyzed throughput impact and Figure 2(b) shows the average throughput represented in Kbps for different number of flows. There is a slight decrease of all of the approaches in comparison to *AODV-native* which we believe to be due to the fact that when in the presence of multiple shortest-paths computed at different instants in time between a source and a destination, AODV will select the first one available. While the three other approaches will always opt for a more robust path, independently of the previous selection of AODV. This is a hypothesis that we intend to further analyze by testing future, more variable scenarios.

6   Conclusions and Future Work

Out of the possible energy-efficiency aspects, choosing paths having in mind optimal network lifetime is an aspect that introduces more flexibility in multihop routing approaches and makes then better suited for user-centric environments. In this paper we address and propose energy-aware routing metrics that can provide a level of stability in terms of network lifetime to shortest-path based routing, without incurring strong penalties in terms of operational changes and maintenance.

We have evaluated both heuristics under realistic settings for a specific case of on-demand routing, AODV. Albeit this work provides initial validation results concerning the proposed heuristics, results obtained are promising in the sense that the heuristics seem to overall improve the network operation without making the network incur a heavy cost.

We are carrying on this work both by fine-tuning not only scenarios but also the proposed metrics and also by evaluating their potential contribution for other forms of multihop routing, e.g. OLSR.

Acknowledgments. This work is supported by Fundação para a Ciência e Tecnologia (FCT) PhD scholarship number SFRH/BD/44005/2008 and sponsored by national fundings via FCT, in the context of the UCR project PTDC/EEA-TEL/103637/2008.

References

1. Chang, J.H., Tassiulas, L.: Energy conserving routing in wireless ad-hoc networks. In: Proceedings of IEEE INFOCOM 2000, vol. 1, pp. 22–31 (2000)
2. Ewing, G.C., Pawlikowski, K., Mcnickle, D.: Akaroa2: Exploiting network computing by distributing stochastic simulation. In: International Society for Computer Simulation, pp. 175–181 (1999)

3. Feeney, L., Nilsson, M.: Investigating the energy consumption of a wireless network interface in an ad hoc networking environment. In: Proceedings of IEEE INFOCOM 2001, vol. 3, pp. 1548–1557 (2001)
4. Guo, Z., Malakooti, S., Sheikh, S., Al-Najjar, C., Malakooti, B.: Multi-objective olsr for proactive routing in manet with delay, energy, and link lifetime predictions. Applied Mathematical Modelling 35(3), 1413–1426 (2011)
5. Junior, A., Sofia, R.: Energy-efficient routing in user-centric environments. In: 2010 IEEE/ACM Green Computing and Communications (GreenCom), pp. 436–441 (December 2010)
6. Junior, A., Sofia, R., Costa, A.: Energy-efficient routing. In: 19th IEEE International Conference on Network Protocols (ICNP), pp. 295–297 (October 2011)
7. Kim, D., Garcia-Luna-Aceves, J.J., Obraczka, K., Cano, J.C., Manzoni, P.: Routing mechanisms for mobile ad hoc networks based on the energy drain rate. IEEE Transactions on Mobile Computing 2(2), 161–173 (2003)
8. Mahfoudh, S., Minet, P.: Survey of energy efficient strategies in wireless ad hoc and sensor networks. In: Seventh International Conference on Networking, ICN 2008, pp. 1–7 (April 2008)
9. Misra, A., Banerjee, S.: MRPC: maximizing network lifetime for reliable routing in wireless environments. In: Wireless Communications and Networking Conference, IEEE WCNC, pp. 800–806 (March 2002)
10. Perkins, C.E., Belding-Royer, E.M., Das, S.R.: Ad hoc on-demand distance vector (aodv) routing. RFC 3561, Internet Engineering Task Force (July 2003)
11. Scott, K., Bambos, N.: Routing and channel assignment for low power transmission in pcs. In: 5th IEEE International Conference on Universal Personal Communications, vol. 2, pp. 498–502 (October 1996)
12. Singh, S., Woo, M., Raghavendra, C.S.: Power-aware routing in mobile ad hoc networks. In: MobiCom 1998: Proceedings of the 4th Annual ACM/IEEE International Conference on Mobile Computing and Networking, pp. 181–190. ACM, New York (1998)
13. Sofia, R., Mendes, P., Moreira, W., Ribeiro, A., Queiroz, S., Junior, A., Jamal, T., Chama, N., Carvalho, L.: UPNs: User-provided Networks, Technical Report: Living-examples, challenges, advantages. Tech. Rep. SITI-TR-11-03, Research Unit in Informatics Systems and Technologies (SITI), University Lusofona (March 2011)
14. Toh, C.K.: Maximum battery life routing to support ubiquitous mobile computing in wireless ad hoc networks. IEEE Communications Magazine 39(6), 138–147 (2001)
15. ULOOP: User-centric Wireless Local-Loop. EU IST FP7 Project (Grant 257418)
16. Vasseur, J., Kim, M., Pister, K., Dejean, N., Barthel, D.: Routing Metrics Used for Path Calculation in Low-Power and Lossy Networks. RFC 6551, Internet Engineering Task Force (March 2012)
17. Xie, Q., Lea, C.T., Golin, M., Fleischer, R.: Maximum residual energy routing with reverse energy cost. In: Global Telecommunications Conference, IEEE GLOBECOM 2003, vol. 1, pp. 564–569 (December 2003)
18. Yu, C., Lee, B., Youn, H.Y.: Energy efficient routing protocols for mobile ad hoc networks. In: Wireless Communications and Mobile Computing, vol. 3, pp. 959–973 (2003)

Towards *Wireless*HART Protocol Decentralization: A Proposal Overview

Ivan Müller[1], Jean Michel Winter[1], Edison Pignaton de Freitas[2,3], João Cesar Netto[2], and Carlos Eduardo Pereira[1,2]

[1] Electrical Engineering Department, Federal University of Rio Grande do Sul, Brazil
[2] Institute of Informatics, Federal University of Rio Grande do Sul, Brazil
[3] Electrical Engineering Department, University of Brasília, Brasília, Brazil
{ivan.muller,jean.winter}@ufrgs.br,
{epfreitas,netto}@inf.ufrgs.br, cpereira@ece.ufrgs.br

Abstract. Wireless industrial equipments for monitoring and process control are being widely adopted nowadays. Their main advantage lies in the ease of installation when compared with wired devices. However, link reliability and the strict real time communication requirements are frequently cited as real obstacles. To cope with this, adequate wireless protocols for industrial automation must present features such as deterministic temporal behavior, clock synchronization, and mesh topology. These characteristics lead to the adoption of centralized architectures such as the *Wireless*HART, the most relevant wireless communication protocol for industrial automation commercially available. By means of thorough analysis of this protocol and the current state of the art, several improvements can be proposed. This paper proposes a decentralized network manager, which should be able to coordinate with other network managers in order to perform a distributed message scheduling. This brings up several advantages such as the faster join and leave of field devices, faster and efficient scheduling schemes, increase of overall reliability and new possible mixed topologies. On the other hand, the proposal raises issues that must be solved in order to obtain practical results. The proposal is introduced in this paper as well as the first evaluations.

Keywords: Industrial wireless networks, Distributed systems, *Wireless*HART, Decentralized management.

1 Introduction

The usage of wireless industrial equipments for monitoring and process control is gradually increasing. The main advantage of this type of equipments lies in the ease of installation when compared with wired devices [1, 2]. The high wiring costs are one of the greatest incentives to consider the use of such systems, as they can permit 20% to 80% of cost reduction. However, link reliability and the strict real time communication requirements are frequently cited as real obstacles to the employment of industrial RF communication protocols [3]. In order to be considered adequate for

S. Andreev et al. (Eds.): NEW2AN/ruSMART 2012, LNCS 7469, pp. 248–259, 2012.

industrial automation applications, the communication protocols should present the following characteristics:

i) stringent real-time requirements, that are usually dictated by concurrent processes in the technical plant, so a deterministic temporal behavior is important;
ii) they have to integrate components that are geographically distributed over the technical plant area, so aspects such as clock synchronization and mesh topology should be supported;
iii) they must be secure, to prevent eavesdropping and denial attacks.
iv) they must cope with radio waves propagation phenomena, being resistant to jamming, coexistence, multipath effects, blockages and other effects.

Advanced manufacturing components are increasingly incorporating powerful embedded systems, which allow them to become autonomous entities, leading to a fully decentralized communication schema also from the automation decision making point of view. These characteristics tend to turn centralized architectures for automation systems, for instance centered on an industrial PLC, inadequate to face current requirements. The strategy chosen for actual industrial wireless protocols is centralized, because it simplifies control and management of wireless networks as well as hardware requirements.

The *Wireless*HART (WH) protocol was conceived to cope with these characteristics and, up to now, it is considered the only one capable to fulfill these requirements [3, 4]. By means of thorough analysis of this protocol and the current state of the art, several improvements can be proposed such as more agile and efficient scheduling and routing schemes by means of management decentralization. But the current version of the protocol, while allowing mesh networks formed by demand, has as a key element the centralized network manager (NM), which is responsible for the formation and maintenance of the network. This paper explores the proposal of a modified WH decentralized version by presenting and discussing one way to implement this.

The remaining text is structured as follows: Section 2 briefly presents current wireless industrial protocols. Section 3 presents a discussion about centralized versus decentralized management. The WH is presented, describing its characteristics and main drawbacks in Section 4. Section 5 describes the decentralization proposal and methodology. Finally, Section 6 summarizes the paper and provides directions for future works.

2   Current Wireless Industrial Protocols

There are few reliable wireless industrial protocols available in the market. It is possible to organize them in three groups: protocols available to use, under development, and adaptations. Some examples are WH and WIA-PA (available), SP-100 (under development) and several adaptations by means of WLAN, Bluetooth, ZigBee, applications over IEEE 802.15.4, among others. The last ones are considered adaptations because industrial applications were not the primary objective. Some organizations are making efforts to promote the use of their technology in industrial applications. However, only field devices (FD) that were specifically developed for industry can be truly reliable. Wireless technologies that work well at home and in office settings are not necessarily usable in stringent industrial environments. Protocols that are not able to change channels dynamically, such as Zigbee and WLAN are less able to cope with coexistence and jamming. Bluetooth, on the other hand, can employ adaptive frequency hopping but it fails in spatial diversity. Only full mesh topologies are adequate for industrial applications that intend to present high reliability, comparable with cable connections. In this case, all the nodes of the network must behave as routers. Neither WLAN nor Bluetooth are able to do this: their topologies are limited to star or multiple trees. Several task groups are currently making efforts to provide such features. Although Zigbee can form mesh networks, not all of the field devices are routers. This is not necessary for office and home applications. Other constrains are related to power consumption. The use of beacons to synchronize the active time need long periods of listening to the network and result in higher power consumption. WLAN cannot be considered a low power protocol especially because of its large bandwidth, which is unnecessary for most FD communications. Ethernet oriented wireless protocols are very inefficient when small payloads are transmitted and most of the industrial processes data fit in small packets. Also, they are connection oriented protocols, which leads to higher power consumption and difficulties in implementing radio cycling and mesh topologies.

One market tendency that is being noticed is the use of plain IEEE 802.15.4, by means of its original, non-modified MAC and LLC sub layers and an overall application layer. This is due to transceivers' low cost as well as good coexistence performance. But none of the desired diversities for industrial applications are present in these equipments: spatial (only star or tree networks are possible), frequency (only one channel is used at time), and time (the CSMA mechanism to arbitrate the channel is prone to collisions). Because of this, it is impossible to implement deterministic networks with 802.15.4 only.

ISA SP-100 can be considered a truly industrial grade wireless protocol. It is a wireless mesh protocol developed to be reliable and secure. The standard covers a wider range of specifications when compared with WH. It specifies the design of security manager with several policies, three different channel hopping techniques and several protocols tunneling, including HART. On the other hand, WH allows HART commands only, one channel hopping scheme and a simpler security policy. On the MAC sides both employ TDMA mechanism for temporal diversity but SP-100 presents variable size time slots. All devices are routers, allowing full mesh topologies, and the links are done through channel hopping, for better coexistence. The network layer permits several types of communications, including request-response and burst messages. The negative of SP-100 is not being widely adopted by the manufacturers, due to release lags occurred along the time. This lags probably occurred due to the wide coverage range of the standard that demanded long time to be released. A complete comparison between WH and ISA SP-100 can be found in [5].

The WIA-PA protocol (Wireless Networks for Industrial Automation and Process Control) is the Chinese standard for wireless communication architectures for industrial applications and process automation [6]. This protocol was accepted by the IEC in 2008 and became the second standard of wireless communication for industrial purposes, after the WH. The WIA-PA provides services to wireless communication devices used in field measurements and control loops in industrial processes.

The protocol adopts a mixed network topology (star and mesh), supporting field devices, handhelds, routers and gateways. The star networks are formed by the field devices and the mesh is done by the routers. The physical layer is based on the IEEE 802.15.4 protocol in two radio bands: 868/915 MHz and 2.4 GHz. The MAC sub layer is the native CSMA of 802.15.4. The active part of the frame is used for communication between field devices and the inactive part, for inter and intra cluster communication. In the active period, the contention part of MAC is used to join field devices, intra-cluster managements and retransmissions. The contention free period is used for communication between field devices and cluster heads. The network layer includes the management of networking, finding and maintaining routes and packet routing in multi hop. The employed routing technique is static one. The communication services include client/server, publisher/subscriber and report/sink modes.

It is noteworthy, that all the presented protocols here are centralized. For instance, Zigbee needs a PAN coordinator and WH, a central NM. Efforts are being made to decentralize these protocols [7, 8] but none of them are hardware approaches as this work proposes. With this proposal much more can be done, adding greater intelligence to the nodes of the network while maintaining compatibility with WH.

3   Centralized versus Decentralized Protocols

The matter regarding centralized versus decentralized network management has been discussed lately. The arguments vary a lot, but they can be resumed as follows. Concerning centralized wireless networks, the most relevant arguments are:

i) the use of centralized network managers are a better approach to solve real time requisites for a wireless industrial network because no interest conflicts will be generated by many distributed managers. For instance, WH and SP100 protocols adopt this strategy.
ii) with a centralized management, the field device hardware is reduced to the minimum necessary for the communication stack. This allows low cost and low power consumption and thus, battery powered devices.
iii) in a centralized network, the nodes are not able to schedule and route their own messages. This is programmed by the central network manager which is desirable for making important decisions such as better bandwidth occupation and controlled scheduling. The central manager is an AC powered device with much more computational power available.
iv) to cope with security issues, the messages must be ciphered and the keys must be generated by a centralized security manager.

On the other hand, there is the decentralized network manager approach. Several techniques can be employed, taking advantage of previously developed works related to wireless sensor networks [9, 10]. The most important arguments are:

i) with correct algorithms that are able to solve hierarchy conflicts, local network managers are much faster when solving real time and asynchronous demands.
ii) today microelectronics evolved to allow very small and powerful devices, such as system-on-a-chips composed by MCU, RF transceiver and Flash memory in a single

package. Technological barriers that today impede powerful computational devices to be battery powered will be surpassed with the passage of time.

iii) in a decentralized network, the routing decisions can be varied, according to different parameters such as maximum latency needs, spatial reliability, and network throughput. On the other hand, centralized managements usually adopts direct routing algorithm only. This leads to the need of a long time to plan all the routes and schedules.

iv) security mechanisms can be managed locally by means of dedicated hardware blocks to cipher messages. Eventually, the security reliability can be even greater than in a centralized system, because the keys are managed locally and can differ for each session as they do not need to be passed through the entire network, as in a centralized network.

Based on these ideas, this paper proposes a decentralized WH NM, to be able to coordinate with other distributed NMs in order to perform a decentralized message scheduling. This will brings up several advantages such as the faster join and leave of field devices, faster and efficient scheduling schemes, increase of overall reliability and new possible topologies. On the other hand, it raises several issues that must be solved in order to obtain practical results. Considering that such decentralized schema may impose some overhead to network nodes, the idea is to propose a hardware network coprocessor which is embedded into the field device. The coprocessor acts as a local NM to the WH protocol and works in collaboration with the so-called primary NM which contains the gateway to connect with the plant backbone.

4   *Wireless*HART Protocol

The WH is the advancement of HART technology for wireless links. It is the first wireless communication standard for industrial use approved by the IEC [11]. WH technology provides a secure network in the radio band of 2.4 GHz by means of IEEE 802.15.4, which enables the use of off-the-shelf RF transceivers. Nowadays, there are many WH device manufacturers such as Emerson Rosemount, Phoenix Contact, Nivis, Siemens and Awiatech. In this protocol, field devices are responsible for characterizing and controlling the industrial process, interacting directly with the plant. They produce and consume WH packets and are also able to route these packets to the other network devices. As all field devices are routers, there is no concept of full or reduced devices function. The adapters are responsible for connecting a wired HART device to a WH network. Handhelds are used by plant operators for commissioning and network diagnosis. The gateway is the element responsible for connecting the wireless network to the automation system of the plant, allowing data to transit between two or more separate networks by converting their protocols. The access point connects the gateway to the wireless field devices and can be redundant. The NM is responsible for set up the network by means of scheduling the communications between devices and managing the routes. This is done through the constant monitoring of the network and further rescaling. The communications between the devices are made on strictly time synchronized time slots. The TDMA scheme with fixed ten milliseconds time slots are defined to provide deterministic communication free of collisions. Transactions are carried out through multi-hop routes (graphs) created by the central NM. The network topologies can vary from simple stars to complete meshes, automatically programmed by the manager in accordance with the demands and characteristics of the network.

4.1   *Wireless*HART Drawbacks

Being a centralized protocol, each WH network has only one NM that is responsible for network formation and maintenance. Figure 1 shows a WH network with its main components. To understand how a WH network is formed, the join process of a FD is explained. The access point is connected to the NM through the gateway. The NM is responsible for the initial setup of the network and for this task it starts programming the gateway and the access point. After the first step, the access point starts generating broadcast advertisement packages. The advertisement package includes the absolute slot number (the number of the last time slot since the network formation), the join priority (based on the device capabilities), the channel map, the graph ID (in the case of Proxy devices), the superframes and links to provide join. The field device that wants to join the network, answers the message after some time listening the advertisement packages for synchronization.

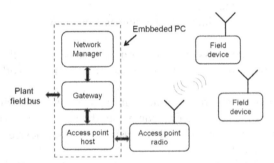

Fig. 1. WirelessHART network with its main components. The field devices that want to join the network must listen and synchronize with advertisement packages.

The field device sends the join reply message that is ciphered with the join key which is already programmed in the network manager and in the device itself. In the answer, commands 0, 20, and 787 are passed to the NM through the access point/gateway. These commands are the device's unique ID, long tag and neighbors signal levels. After this point of the join procedure, the NM must start scheduling algorithms for the network in order to allow the new field device to communicate with other nodes. The rescheduling is done, and the following commands (CMD) are exchanged between the new FD and the manager:

CMD 963: Write session, to permit communication between device and gateway or other devices.
CMD 961: Write network key, to authenticate messages by means of CCM* algorithm.

CMD 962: Write device nickname. Most of the commands are addressed in the network using a nickname chosen by the manager instead of long address.

CMD 965: Write superframe. This command is sent as many times as needed. There are normal superframes for the process variables and special superframes for other purposes such as maintenance and commissioning.

CMD 967: Write link, sent as many times as needed. Each link is designated to a super frame and has its own channel offset. The links can be normal, join, broadcast and discovery. The options are TX, RX and shared (with CSMA-CA).

CMD 971: Write neighbor flags, to determine if the new device can be a time source for the network.

CMD 777: Read device capabilities. The new device will answer its capabilities: power source, packets per second, peak packets before power drain, time to recover from power drain, threshold for good RSL, minimum keep-alive time.

CMD 64512: Read wireless transceiver module revision.

CMD 805: Enable / disable CCA mode, for the shared links to use CSMA scheme.

CMD 795: Write timers interval. This command sets the many timers used by the device along the time within the network. There the advertisement, discovery, broadcast reply, keep alive, and path failure timers.

CMD 793: Write UTC time and date.

CMD 808: Read time to live interval. Determines how many hops a packet can go in a network.

CMD 973: Write service. This command is sent for the device to allow service for it.

CMD 974: Write route, to establish paths for messages in the network.

Depending on the network strategy, routes are defined by graphs or directly, by the superframe. At any time after the first command, the new device may send a request service by means of command 779. This is usual for sensor devices for example, for temperature or pressure data publishing.

Some commands are grouped in one message, depending on their size. Others are repeated many times during network initialization (967 is an example). The above list of commands was obtained (sniffed) from commercial WH devices during the join process. Different brands of equipments behave slightly different, following some basic steps. In this example, the overall time for the first process variable publishing is less than a minute, counting from the advertisement response. Notice that this is a direct communication with the gateway. Things can go really bad with proxy devices, many hops and many scheduled times: five to ten minutes were verified. Also, if the device is battery operated, it takes much more time to synchronize with the advertisement message because of radio cycling.

With the previous information, one can notice that because of the centralized manager, WH networks are not able to cope with highly dynamical devices such as mobile stations. Also, it is not possible to make direct connections between two or three devices as in a sensor, controller, and actuator distribution. Although suggested in the standard, these direct connections are not possible in today's implementations. Figure 2 shows two possible scenarios for a joining event. The routes are programmed by the manager taking into account the reported data by the field devices. These data include RF signal levels, power supply condition, and number of packets that passed through the device. Consider a new device listening two advertisement messages propagated by two network devices nearby. The best route to be chosen is that followed by the arrow 1*a* of the Figure 2 and the worst, 1*b*.

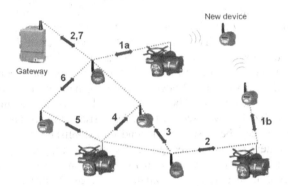

Fig. 2. Two possible scenarios for a join event, where the massage traffic goes through a short or a long path

Although the standard recommends at maximum four hops from any device to the gateway, one can imagine that for the join process all commands for rescheduling the network must be up and down linked through it. Because of this, the preferred topology for the same network in this example has been noticed and is shown in Figure 3. Actually, the preferable star topology is a common sense for this type of centralized network in order to guarantee lowest latency and high reliability. Unfortunately this topology leads to lower reliability to the protocol.

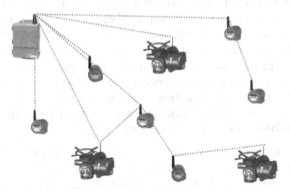

Fig. 3. Final preferred WH network topology observed by means of several experiments

5   *Wireless*HART Decentralization Proposal

In a totally decentralized system, the network nodes are autonomous and thus produce their own messages, which can lead to hierarchy problems. They also need to produce paths for messages between nodes and the network access points. If message scheduling is done locally on a network, the priorities may be chosen according to the "judgment" of the node itself. Even if a node has its dedicated time slots it may not differentiate the data from other nodes, because it ignores the requirements of others. It may also be difficult for each network node to form communications paths according to the communications between the neighbors. The selection of a route based only on signal strength, for example, leads to battery depletion of the most popular nodes in the network. These are the bad consequences of the employment of simple distribution algorithms. On the other hand, if the distributed management is properly done, the network will be reliable because it does not depend on a central coordinator. It will be easier to add new nodes to the network if it can negotiate its join directly with the nearest neighbor, if this neighbor is able to manage the join process. Still, centralized network controls are less efficient regarding the aggregation or loss of network nodes, because they depend on established paths involving other nodes to reach the new node or to recognize if someone stopped to communicate. In addition, many network resources are used for formation and maintenance: network information concentration requires that all the messages from the field devices are uploaded to the manager and vice versa. This is the weak point of a centralized network, which inevitably leads to throughput issues and overall latency. Distributed networks have the advantage of the faster discovery of network blockages, new devices and new routes. This leads to greater overall reliability.

There are several possible approaches to decentralize a network protocol. Some functions of the manager can be distributed along the network. For example, algorithms to decide routes can be deployed in the field device's firmware. The same can be done for scheduling the messages. Alternate approach is to deploy an entire manager in each node and this can be done by means of a network coprocessor. That is the proposal of this work: distribute network coprocessors through the field devices in a way that more than one can perform network scheduling at any needed time. Another possibility is to have more than one gateway to connect the network to external bus, increasing overall reliability, but this is not explored in this work. Figure 4 left depicts the proposal. A modified field device includes a network coprocessor that remains most of the time in idle mode. Periodically requested and by the communication stack, it starts to operate and reschedule the network.

Besides the field device modifications the main network manager which has the essential gateway to connect the wireless network with outside world, is maintained, (with its own modifications). These changes are:

(i) Field device: there are many commands which are restricted to be sent by the central manager, the WH stack is modified in order to permit local programming by the coprocessor. The case study is related to new device join process only. Figure 5 depicts the state machine implemented in the coprocessor that is implemented in a C++ PC software for evaluation purposes. Further efforts will be concentrated to port the code for a MCU. The modified FD sends the advertisement response to the coprocessor, which starts the rescheduling process. During the *Learn* state, the coprocessor receives the actual network schedule by means of commands 783 and 784, where current superframe and links are obtained. In the next state (*Sched*) the coprocessor reschedule a superframe provisioning link for the new device. A Bellman-Ford based algorithm is used to create the new links. After these procedures, the coprocessor orders the FD to send the new schedule for the main manager as well as for the other distributed managers. It ends the operation entering in a sleep state. Periodically, it wakes up and obtains the network status again.

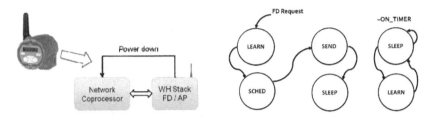

Fig. 4. Left: Block diagram of a modified field device for WH decentralization. Right: State machine for the local network coprocessor.

ii) Main network manager modifications: In order to permit distributed NM, the main NM is programmed to produce a special superframe and links to exchange rescheduling information. This superframe has links to all distributed NM and it is used every time the network is rescheduled, allowing the managers to be synchronized. Also, the main NM is modified in order to accept commands from the wireless network. These commands are related to rescheduling done by another NM through the special superframe. The remaining functions of the NM are kept and thus, keeping compatibility to the standard.

5.1   Evaluation System

To evaluate the system, several network scenarios are developed. The case study is limited to the network join of a new device. The network is divided in levels which are determined by the number of hops that are necessary to reach the gateway. Figure 5 depicts one of these scenarios. One can notice that FDs 1 and 2 are in the level zero, with one hop to the gateway and FDs 3 and 4 are in the level one, with two hops to the gateway. Even if the new device to join the network (FD 5) listen the devices in level zero, it enters in the network through FD 3, which has the network manager coprocessor. This is done through the join priority variable that is higher for the FDs with embedded coprocessor.

Related to the join of a new device, the real network scenarios include several possible variations. However, these are limited to the most common ones:

a)   A new device listening to two advertisements, each from two NMs in the same network level. In this case, it may choose anyone.

b)   A new device listening to two advertisement, one but weak from level zero (the main access point) and other and strong from a greater level. It may choose level zero, if latency is constrain or greater than zero, if reliability is constrain.

Tests are done in a PC platform (emulated coprocessor) connected with a previously developed WH radio. Figure 6 depicts the test topology. To verify the coprocessor correct behavior, a WH sniffer is used (Wi-Analysis tool) in conjunction with software debugging. Also, the FDs are debugged through serial port.

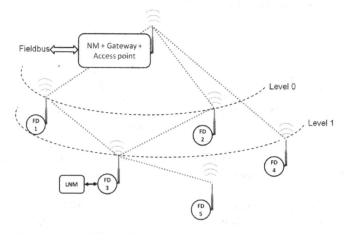

Fig. 5. Test scenario for case study

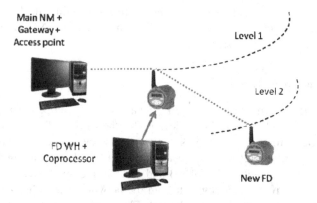

Fig. 6. Implemented setup for case study

6   Conclusion and Future Work

This paper presents a discussion about centralized versus decentralized management of industrial wireless protocols. Some pros and cons of each one are presented to provide comprehension of the implications of one or the other choice. We believe that the main obstacles of decentralization can be surpassed by means of local hardware coprocessor, to be deployed in some strategically selected points of the network. These coprocessors are initially evaluated by means of large computer systems, such as a PC but can be substituted by a single board computer or even dedicated logic devices. In future, they can be part of a system-on-a-chip, as subject of microelectronics evolves.

We propose and present the first steps of hardware based WH decentralization scheme. The radio field device was modified in order to accept commands from the coprocessor for network reschedule. After the observation of the correct functionality of the basic scenarios, future works include the development of a local network manager and a gateway, to provide redundant access to plant backbone.

References

1. Willig, A., Matheus, K., Wolisz, A.: Wireless Technology In Industrial Networks. Proceedings of the IEEE 93(6) (June 2005)
2. Khakpour, K., Shenassa, M.H.: Industrial Control Using Wireless Sensor Networks. In: Information and Communication Technologies: From Theory to Applications, ICTTA (April 2008)
3. Song, J., Song, H., Mok, A., Chen, D., Lucas, M., Nixon, M.: WirelessHART: Applying Wireless Technology in Real-Time Industrial Process Control. In: Real-Time and Embedded Technology and Applications Symposium, RTAS 2008, April 22-24, pp. 377–386. IEEE (2008)
4. Chen, D., Nixon, M., Mok, A.: WirelessHART: Real-time Mesh Network for Industrial Automation. Springer (2010)
5. Wang, G.: Comparison and Evaluation of Industrial Wireless Sensor Network Standards ISA100.11a and WirelessHART. Master of Science Thesis, Communication Engineering (2011)
6. Industrial wireless networks WIA-PA. System Architecture and Communication Specification for Process Automation (2008)
7. Di Benedetto, M.D., D'Innocenzo, A., Serra, E.: Dynamical Power Optimization by Decentralized Routing Control in Multi-Hop Wireless Control Networks. In: 18th IFAC World Congress, September 2 (2011)
8. Tinka, A., Watteyne, T., Pister, K.: A Decentralized Scheduling Algorithm for Time Synchronized Channel Hopping. In: Zheng, J., Simplot-Ryl, D., Leung, V.C.M. (eds.) ADHOCNETS 2010. LNICST, vol. 49, pp. 201–216. Springer, Heidelberg (2010)
9. Kumar, S., Chauhan, S.: A Survey on Scheduling Algorithms for Wireless Sensor Networks. International Journal of Computer Applications (5) (April 2011)
10. Zhuo, X., Hua, S., Miao, L., Dai, Y.: Direction Matters: A Decentralized Direction-based TDMA Scheduling Strategy for VANET. In: IEEE 13th International Conference on Communication Technology, pp. 566–571 (2011)
11. ISA. Wireless Systems for Industrial Automation: Process Control and Related Applications. ISA-100.11a-2009 (2009)

Process Mining Approach for Traffic Analysis in Wireless Mesh Networks

Kirill Krinkin[1], Eugene Kalishenko[2], and S.P. Shiva Prakash[3]

[1] Open Source and Linux Lab., Saint Petersburg, Russia
kirill.krinkin@fruct.org
[2] Computer Science Dept. Saint Petersburg Electrotechnical University "LETI",
Saint Petersburg, Russia
[3] Mobile Innovation Lab., Sri Jayachamarajendra College of Engineering,
Mysore, India
{ydginster,shivasp26}@gmailcom

Abstract. Short-time traffic flow prediction in particular systems will expedite discovering of an optimal path for packet transmitting in dynamic wireless networks. The main goal is to predict traffic overload while changing a network topology. Machine learning techniques and process mining can help analyze traffic produced by several moving nodes. Several related approaches are observed. Research framework structure is presented. The idea of process mining approach is proposed.

Keywords: Wireless mesh networks, adaptive routing, process mining, traffic hotspots, power aware metric.

1 Introduction

Nowadays, a representative class of program systems for intellectual data processing allows extracting of data structures and analyzing of regularities in them. Usually such systems consist of a raw application environment model, experimental information sources, data mining methods and instruments. The timely task now is to develop such methods for the process analysis system that allows accumulation and systematization of system's behavior templates and to use this information in order to correct the application environment model and to predict system state in the near future.

Traffic flow prediction is aimed to escape situations presented on Fig. 1. For instance, intensive nodes moving resulted in traffic flow overload on several nodes. Theoretically, such situations in mesh networks can be predicted and traffic redistribution can be made in advance.

Network traffic prediction plays an important role in guaranteeing network QoS. It is important to note that algorithms should be constructed to operate in real time and should be based on the minimal amount of historical information. Let us observe several main traffic prediction methods.

S. Andreev et al. (Eds.): NEW2AN/ruSMART 2012, LNCS 7469, pp. 260–269, 2012.

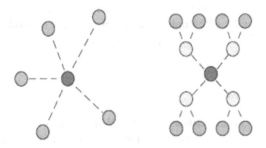

Fig. 1. Overload situations. Center - stressed node.

2   Relative Works

2.1   Prediction Using Wavelet Neural Networks

The traditional prediction methods such as time series analysis, regression method etc. are difficult to put in practice as WMN's traffic is complicated and lacks effective mathematical model. Moreover, most of traditional prediction algorithms are re-source-intensive. Thus the paper [1] presents a multipath routing algorithm based on wavelet neural networks.

Wavelet neural network prediction model adopts three-layer structure. There are P input for input layer, namely one input containing P values of the time series. There are N neurons on the Hidden layer. And there is one neuron on the Output

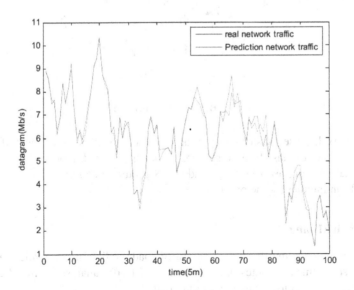

Fig. 2. Five-step prediction of wavelet-neural network

layer. The offered equation shows that the function can be approximated by orthogonal function sets and thus can be implemented using a neural network.

As shown in Fig. 2 in some cases such an approach leads to perfect results and can be used for traffic forecasting.

2.2   Clustering Approach

Apart from communication between nodes in WMNs there is a problem of load balancing between base stations which provide access to external services like Internet or telephony. A given geographical area consists of hexagonal cells each served by a base station (BS), as on Fig. 3. A base station is a part of the wireless infrastructure that controls one or multiple cell sites and radio signals [2].

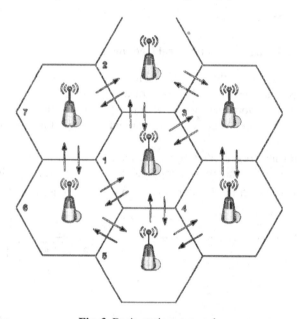

Fig. 3. Basic stations network

The authors [2] use clustering approach to form clusters based on traffic load. Clustering is used to classify base stations as heavily loaded, moderately loaded and idle base stations. In case of heavy loaded BS group detection, radial stations with lesser load can intercept some part of the traffic.

2.3   Graph Mining

One more approach comes from the road traffic load prediction. As wireless network's load forecast is aimed at routes optimization, road traffic analysis is aimed at choosing the optimal way considering the arterial load.

The basic model of [3] algorithm is a graph where vertexes are cross-roads and edges are streets. The main metric of each edge is speed. It is needed to mention that in case of transport modeling it is impossible to determine the end point of a particular unit. But in networks a destination node is always known. The main idea of presented in [3] algorithm is to determine edges' weights at some moment in the near future.

2.4   Time Series Analysis

Time series analysis is widely used in wired networks. But it needed to be modified to use in client wireless networks to decrease amount of information being analyzed. The solution proposed in [4] considers internal traffic only.

The input is a sequence of network load values (in bytes per hour, for instance). After several filter steps and prediction the authors achieved results that can be used in traffic prediction (Fig. 4)

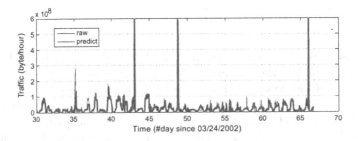

Fig. 4. Time series analysis

3   Process Mining Approach

Each algorithm briefly described above has its limitations:

1. *Wavelet neural networks:* the algorithm can be less efficient if lacking of representative traffic statistics
2. *Clustering approach:* the algorithm is designed for static basic stations and can't be effectively applied to the dynamic mesh topology
3. *Graph Mining:* this algorithm is resource-intensive and is mainly dedicated for static networks
4. *Time series analysis:* the algorithm takes into account internal traffic with periodic-type traffic only

The other proposed approach concerns with the extraction of template topologies using process mining algorithms [5]. The main idea is to look for topology changes and to redistribute traffic in case of overloaded nodes prediction (Fig. 5).

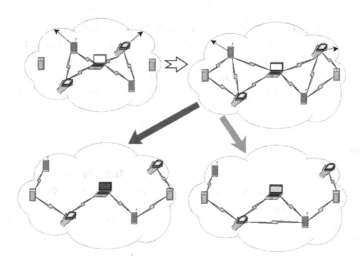

Fig. 5. Traffic redistribution

3.1   NS-3

NS-3 is a discrete-event network simulator used to model different kinds of networks both wired and wireless. It has a flexible object-oriented plug-in architecture for creating particular network topologies and it can be extended with new traces output formats [6]. NS-3 is used to build and model a set of dynamic mesh networks (like Fig. 5) in order to get traces for analysis using process mining techniques.

To model dynamic mesh network in NS-3 it is necessary to do the following (in C++):

1. Create a set of network nodes
2. Install WiFi devices on created nodes: NS-3 provides a set of "helpers" for some kinds of networks. Using such a helper for mesh network is enough to make a topology with a given set of nodes
3. Set node mobility properties: an initial position in a grid and a vector of velocity
4. Install IP addresses
5. Install network applications on each node
6. Start modeling: NS-3 gives a trace as a result of modelling, for instance in tcp dump format, which can be analyzed in, for example, Wireshark (an application for network traffic visualization)

3.2   ProM

ProM is a generic framework for implementing process mining tools in a standard environment. The ProM framework inputs logs in the XES or MXML (Mining XML) format. Currently, this framework has plug-ins for process mining, analysis, monitoring and conversion [8].

NS-3 architecture provides a plug-in mechanism for extending a set of formats for trace generation. Thus it is possible to implement a MXML plug-in for NS-3 and to pass modeling results directly to ProM for analysis.

The main goal of using ProM is to find a set of effective process mining algorithms to determine topology changes that can lead to traffic overload in some nodes of a network.

3.3   Algorithm

The basic research approach consists of several steps:

1. *Dynamic network modeling:* use NS-3 to model dynamic mesh network with several nodes witch form spatial configurations with potentially hotspot nodes.
2. *Trace converting:* convert NS-3 trace into MXML format for ProM.
3. *Process extraction:* process mining techniques are aimed to extract, analyze, extend or optimize different kinds of processes. Process extraction algorithms are used to extract information about network topology and traffic distribution changes.
4. *Template search:* if current extracted process is similar to a some one in the database of dangerous process templates, a potential hotspot node has to be informed about traffic overload in the near future. If overload has already occurred then a new template is added to the database to prevent such routes distribution in future.
5. *Recommendations for redistribution:* some of the metrics of a potentially overloaded node are changed to prevent new routes through it. If process is still similar to the one of dangerous templates, active routes redistribution may be done between the overloaded node neighbors.

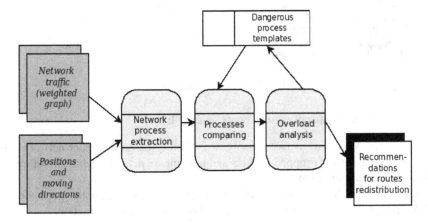

Fig. 6. Basic scheme

Main advantage of this approach is minimal information about traffic history. Taking into account nodes position leads to the ability of extraction different kinds of topologies and traffic load portraits.

Now the research is in process of defining the best process mining algorithms for topology analysis.

ProM presents the results of its work as a Petri Net. This Net is supposed to aggregate common and anomalous network behavior. It is important to note that the same network should be suitable for separation of common and anomalous process instances.

4   Research Framework

Current basic framework scheme is shown below on Fig. 7.

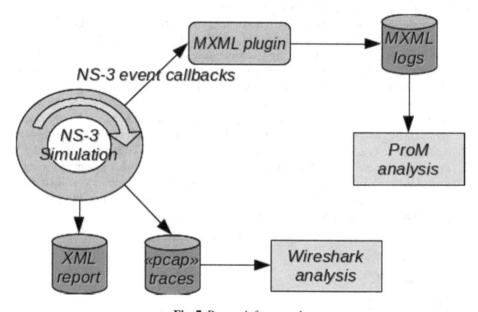

Fig. 7. Research framework

There are 3 main data collections being analyzed:

1. *pcap traces:* Wireshark network analyzer is used to monitor current network state to collect statistics of dropped packets and to compare network behavior before and after routes redistribution
2. *XML report:* NS-3 report scheme allows to see the complete statistics per node
3. *MXML logs:* this is the main data source for process network extraction

Now the basic version of the plug-in that converts NS-3 trace into MXML format for ProM is implemented as a library that can be used with an arbitrary NS-3 model. In order to use NS-3 logging framework, plug-in should be able to implement a set of callbacks and to register them on devices, applications etc... Callback receives an event with parameters depending on its type. For example, common signature of the network type callback is: *PacketSentCallback (string context, Ptr<Packet const> packet)*. Where "context" is an identifier of an originator (for instance, *"/NodeList/0/DeviceList/1/$ns3::WifiNetDevice/PhyRxBegin"*) and "packet" is a specific for the callback type parameter. The library is extensible in terms of new event and callback types. In order to use ProM several event types were chosen to register callbacks:

1. packet has begun transmitting (or being received) over the channel medium
2. packet has been completely transmitted (or being received) over the channel
3. packet has been dropped by the device during transmission
4. packet has been received
5. node moved...

5   Current Work

Current work is concerned with active development of the research framework based on NS-3 and ProM in order to define methods and algorithms suitable for network analysis. Thus, the next main research steps are:

1. *Extend a set of overloaded mesh networks:* extend a set of networks being analyzed. Now several simple dynamic networks with up to ten moving nodes are implemented using NS-3, for instance "dispersion" model, where a set of nodes run away from the central one (Fig. 8). In order to check process mining algorithms stability it is needed to model several networks witch are closer to reality.

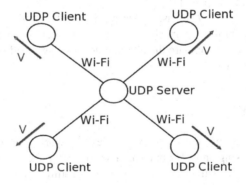

Fig. 8. Simple model

2. *Choose the best algorithms:* after using ProM framework for network traces analysis we need to choose a set of algorithms which can bring out topology templates and a potentially dangerous network that can be overloaded in a short-time period. As MXML-plug-in is already implemented, this stage is in focus of our research now. An extracted network process using genetic miner algorithm is shown below on Fig. 9.

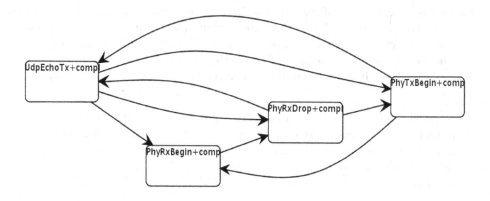

Fig. 9. Sample of an extracted process

3. *Add network events concerned with power consumption:* in WMN the lifetime of a node plays an important role. Regular recharging of battery is difficult, so network nodes operate with finite battery capacity which is managed efficiently to increase reliability of individual nodes and lifetime of the whole network [9]. Wireless network lifetime can be increased by distributing of routing equally over all nodes. This can be achieved by balancing individual node battery level. Calculating energy consumption speed gives information about the energy draining in data transmission. Once got estimated energy consumption speed and residual energy it is possible to compute the expected node lifetime assuming a node continues to consume energy with that speed [10].
4. *Implement correction metric:* use process mining algorithms to implement a new routing metric in NS-3 for mesh networks and integrate it into existing mesh routing protocol. Then compare the average network overload using the developed metric and without it.

6   Conclusion

Suggested process mining approach for wireless networks can be used for on-line analysis in WMN. Using short history and a set of data patterns allows us to identify particular processes and situation and therefore to estimate the network state in the nearest future.

References

1. Li, Z., Wang, A.R.: Multipath routing algorithm based on traffic prediction in wireless mesh networks. In: Proc. 5th IEEE Conf. on Natural Computation, Tianjin (2009)
2. Usha, J., Kumar, A., Shaligram, A.D.: Clustering approach for congestion in mobile networks. International Journal of Computer Science and Network Security 2 (2010)

3. Kriegel, H.-P., Renz, M., Schubert, M., Zuefle, A.: Statistical density prediction in traffic networks. In: Proc. 8th SIAM International Conference on Data Mining (2008)
4. Dai, L., Xue, Y., Chang, B., Cao, Y., Cui, Y.: Optimal routing for wireless mesh networks with dynamic traffic demand. Mobile Networks and Applications 1 (2008)
5. Elizarov, S., Bargesyan, A., Tess, M., Kupriyanov, M., Holod, I.: Data and process analysis, BHV. Saint-Petersburg (2009)
6. NS-3 description, http://www.nsnam.org/wiki
7. Zhang, Y., Luo, J., Hu, H.: Wireless Mesh Networking: Architectures, Protocols and Standards. Auerbach Publications (2006)
8. ProM decsription, http://www.processmining.org/
9. Akram, A., Shafqat, M.: Battery and Frequency Optimized AODV for Wireless Mesh Networks. Canadian Journal on Multimedia and Wireless Networks (April 2010)
10. Romdhani, L., Bonnete, C.: Energy Consumption Speed-Based Routing for Mobile Ad-Hoc Networks. Eurecom Institute

A Risk-Reward Competitive Analysis for Online Routing Algorithms in Delay Tolerant Networks

Maziar Mirzazad Barijough[1,2], Nasser Yazdani[1], Djamshid Tavangarian[2],
Robil Daher[2], and Hadi Khani[1]

[1] Department of Electrical and Computer Engineering, College of Engineering,
University of Tehran, Tehran, Iran
{mirzazad,yazdani,Hkhani}@ut.ac.ir
[2] Institute of Computer Science, University of Rostock, Rostock, Germany
{maziar.mirzazad,djamshid.tavangarian,
robil.daher}@uni-rostock.de

Abstract. Delay Tolerant Networks (DTNs) suffer from frequent disruptions and partitioning. Because of dynamic environment and lack of knowledge about the whole network topology, routing is quiet challenging in DTNs. Prediction based algorithms in DTNs are able to predict nodes future contacts in the case where each node has a movement model. In this paper we suppose that a new node is added to the network which is about to send messages to a destination node and has no knowledge about the upcoming contacts. Deciding about the next node to forward the messages to, is an online problem. We are going to analyze a greedy forwarding strategy using competitive analysis and also propose a risk-reward analysis framework for the risky version of the algorithm and obtain risk and reward values for different forecasts. Risk-reward analysis can be used to develop better forwarding strategies which decide about the next hops more wisely.

Keywords: Delay Tolerant Networks, Online Algorithms, Competitive Analysis, Risk Reward Analysis.

1 Introduction

Delay Tolerant Networks (DTNs) usually suffer from frequent disruptions and network partitioning and so may never have an end-to-end connected path. Unstable paths can be the result of several challenges at the link layer, for example high node mobility, low node density, and short radio range [13]. Sparse Vehicular Ad Hoc networks (VANETs) usually face intermittent connectivity and network partitioning. In this type of VANETs, traditional routing protocols such as DSR [17] or AODV [18] would not work well. Instead using store-carry-forward mechanism used in DTNs would increase network performance as in other challenged networks [15], [16].

Routing in DTNs is quiet challenging because of the dynamic topology of such networks [14]. In the forwarding based routing algorithms, nodes always have to

S. Andreev et al. (Eds.): NEW2AN/ruSMART 2012, LNCS 7469, pp. 270–279, 2012.

decide about the next hop to forward the bundles, among nodes which may be connected to in the future [6].

Researchers have proposed lots of methods in dealing with these issues. Some of these methods predict the future contacts between nodes using history of past contacts and also geographic information [5]. These methods are becoming fairly advanced such that connection times can be predicted with a high confidence level for special DTNs where nodes movements are not completely random and usually each node has a mobility model such as VANETs [1].

In this paper we are going to consider a certain case of DTN routing: when a node "S" is added to the network and has some messages to be sent through network. Having no knowledge about upcoming contacts for the new node and deciding about waiting time for better contacts leads us to an online routing problem. The simplest algorithm which is already used in many papers is greedy algorithm without waiting for future better contacts. At the first stage we are going to derive competitive ratio of greedy algorithm against offline optimal strategy. In addition we are going to generalize the competitive analysis and propose a risk-reward strategy framework. Risks and rewards obtained for different forecasts will help routing algorithm in making decisions about best next node to forward the messages to.

The reminder of this paper is organized as follows. In section 2, some related works for DTN routing, Competitive Analysis and Risk-Reward model is reviewed. In section 3 problem statement and formulations are presented. In section 4 the greedy algorithm is proposed and competitive analysis is made for this algorithm. In section 5 competitive analysis for the risk-reward model of this problem is reviewed. Finally the paper is concluded in section 6.

2   Related Works

Many different methods proposed in literature for routing in delay tolerant networks and VANETs. Some of these methods are flooding based and some are forwarding based. Flooding based algorithms such as Epidemic routing [4] all nodes are relays and a node copies it messages to all nodes that come in contact with it until one copy gets to destination node. Some other methods are prediction based approaches which are usually unicast algorithms. These algorithms use the history contacts information and also geographical information to predict future contact probabilities and also contact time of nodes. A. Lindgren et al. in [6] proposed PROPHET routing algorithm which uses contact history information to derive probability of node in delivering messages to a destination. Weak state paradigm was introduced in [5] to collect partial information about the region of the destination node. Weak state paradigm used in [1] to collect geographical information about destination node and make prediction about the cost of message transferring in such networks so that messages will be forwarded more wisely.

Competitive analysis suggested by Sleator et al. in [7] is a method used to evaluate online algorithms that compares the online algorithm to its optimal offline version which knows entire sequence of inputs in advanced and so is optimal. Competitive analysis can be thought of as a game between an online player and an adversary [9]. Adversary generates input sequence and tries to maximize competitive ratio knowing the strategy of online algorithm. Y. Xu on [8] presents two strategies (a greedy strategy and a comparison strategy) to solve Canadian Traveller Problem with failed edges in general networks and evaluates these methods using competitive analysis. They proved competitive ratios of $2^{k+1} - 1$ and $2k + 1$, respectively, for each strategy where k is the number of failed edges. Sven O. Krumke in [9] analyzes online-Dial-a-Ride and Online-TSP problems and discusses the competitive ratios for different algorithms for these problems.

Risk-Reward framework was first introduced for financial games by S. al-Binali in [3] which is an extension of competitive analysis and provides a framework in which investors can develop optimal trading strategies based on their risk tolerance and forecast. In [10] a frame work developed for managing reward and risk of the newsboy problem with range information. The proposed approach helps the newsboy flexibly choose the optimal reward strategies according to his own risk tolerance levels and different forecasts. In [11], B. Su and Y. Xu introduce a risk-reward competitive analysis for the Recoverable Canadian Traveller Problem (RCTP). RCTP is an extension of CTP in which blocked edges will be recovered in some unknown time. They also present an optimal online travel strategy and prove its competitive ratio for RCTP.

3   Problem Statement and Formulation

As told previously, lots of researches have been done on prediction based routing in Delay Tolerant Networks and also Vehicular Ad-hoc Networks recently. These algorithms use historical information of contacts between nodes and also geographical information to predict the future contact times [1], [2]. In this type of algorithms, each node has a matrix of contact probability and contact time between each pairs of nodes and forwarding decisions are made based on this matrix. Prediction based algorithms are becoming fairly advanced for special DTNs where nodes movements are not completely random and mostly use a mobility model such as VANETs.

In this paper, we are going to analyze behavior of a newly added node S to the network. At the beginning this node does not have any information about the network. By getting in contact with the first node, it updates its information about the network, connection time between each pairs of nodes. But node "S" does not have any information about its direct contact times. Consider S wants to send some data to destination node D. Node S gets a matrix of contact times T from other nearby nodes where $T_{i,j}$ is the message passing time from node i to node j. In this case for all j, $T_{s,j}$ is not known. In the first contact, S can get information about all nodes which are in the same region as S.

The Goal of the problem is to minimize the message transfer time from node S to node D. In the first step of message transferring process, S needs to pass the message to a nearby node which is more probable to meet Node D directly or indirectly. So the problem is to choose the best node among all nodes in the same region (which are more probably get in contact with S) to forward the messages toward destination D, by which message transfer time would be minimum. Suppose the problem as in Figure 1. Node S is currently in contact with node i. S knows that transfer time of node i is $T_{i,D}$. Also node S knows that it probably will be in contact with node j which has a transfer time $T_{j,D}<T_{i,D}$. Node S does not have any information about the contact time with node j ($T_{S,j}$). So it has to decide whether to send the message through node i or wait to node j.

As it is clear, lack of information about node S's mobility and contact times, makes the routing algorithm an Online Routing problem. Below we are going to analyze the performance of the greedy strategy using competitive analysis and also we are going to discuss the competitive ratio of a risk-reward algorithm.

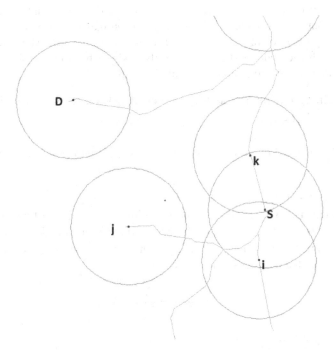

Fig. 1. Smaple DTN topology. Node S has some data to send to node D. Node S is currently in contact with node i.

4   Competitive Analysis of Greedy Strategy without Waiting

Let C_A be the cost of the algorithm A to send messages from source S to destination D. The online algorithm A is called α-competitive if $C_A \leq \alpha_A. C_{OPT} + b$ holds [19]. α_A and b are constants where α_A is called the competitive ratio of algorithm A. C_{OPT} is the cost of the optimal offline algorithm sending messages from S to D which knows the entire topology and future contacts in advance. In order to compute the competitive ratio of the algorithm, we make the following assumptions:

- Greedy routing algorithm knows the travel times of $T_{i,j}$ where $i, j \in N, i \neq S$, S is the newly added node to the network and N is the set of nodes in the network.
- Algorithm does not find the $T_{S,j}$ $S, j \in N$ until S gets in contact with any j.
- Algorithm finds the nodes which are in the same region of the S and more probably will get in contact with S.
- In all theorems assume node i is the first node that gets into contact with node S and node j is the best node in region which has the least travel time to D.
- In all cases $T_{S,D} > T_{i,D}, T_{j,D}$.

Greedy Strategy: In Greedy Strategy (GS), each node will forward the message to the first contact that has less transfer time. Consider node S wants to send message to node D as in Fig 1. Suppose that $T_{i,D}=C_1$ and $T_{j,D}=C_2$ and $C_1 \geq C_2$. In this case if node S gets in to contact with node i, it forwards the messages to this node regardless of node j that may get into contact with node S in future. In the case where node S gets into contact with two nodes at the same time, the node with lesser transfer time to D is known as node i and the other node is simply ignored.

Theorem 1: In Greedy routing strategy for newly added node S, competitive ratio is β, where β is $\frac{T_{i,D}}{T_{j,D}}$. $\beta \geq 1$.

Proof: Let $T_{S,j}$ be the time in which node S and j will get into contact after connection of S and i.

Case 1: $T_{S,j} > T_{i,D} - T_{j,D}$

Cost of sending messages to destination through node i is $T_{i,D}$ and cost of sending through node j is $T_{S,j} + T_{j,D}$. As it is obvious sending through node i has lesser cost than node j. In this case offline optimal algorithm is the same as online Greedy strategy and Forwards the messages to the node i. So:

$$\alpha_{GS} = \frac{C_{GS}}{C_{OPT}} = \frac{T_{i,D}}{T_{i,D}} = 1$$

Case 2: $T_{S,j} \leq T_{i,D} - T_{j,D}$

In this case greedy strategy forwards the packets to the node i and cost of this algorithm is $T_{i,D}$. Offline optimal algorithm waits $T_{S,j}$ and then forwards the messages to the node j. so the cost of this algorithm is $T_{S,j} + T_{j,D}$ which is lesser than cost of Greedy Strategy $T_{i,D}$. So:

$$\alpha_{GS} = \frac{C_{GS}}{C_{OPT}} = \frac{T_{i,D}}{T_{S,j} + T_{j,D}} \leq \frac{T_{i,D}}{T_{j,D}} = \beta$$

From (1) and (2) we conclude that competitive ratio of GS is β.

□

5   Competitive Analysis of the Risk-Reward Model

Concept of risk-reward analysis was first defined by Al-Binali in [3] for financial games. In this paper Risk, Reward and Forecasts are defined as follows: Let $cost_A(\sigma)$ be the cost of the online algorithm and $cost_{OPT}(\sigma)$ be the cost of the offline optimal algorithm where σ is an instance of the problem $\sum$. The competitive ratio of the algorithm A is:

$$\alpha_A = \sup_{\sigma \in \Sigma} \frac{cost_A(\sigma)}{cost_{opt}(\sigma)} \ .$$

Optimal competitive ratio of the algorithm A is defined as:

$$\alpha^* = \inf_A \alpha_A$$

Risk of the algorithm A is defined as:

$$\tau = \frac{\alpha_A}{\alpha^*}$$

A forecast F is assumed to be subset of problem instances.

The restricted competitive ratio which is restricted to the case where forecast becomes true is defined as:

$$\hat{\alpha}_A = \sup_{\sigma \in F} \frac{cost_A(\sigma)}{cost_{opt}(\sigma)}$$

Optimal restricted ratio is defined as

$$\hat{\alpha}^* = \inf_A \hat{\alpha}_A$$

When the forecast becomes true, reward of the algorithm A is defined as:

$$f_A = \frac{\alpha^*}{\hat{\alpha}_A}$$

Theorem 2: Competitive ratio for any deterministic algorithm is not better than β. So the optimal competitive ratio is β.

Proof: Suppose that we have two deterministic algorithms:

The first algorithm is the same as greedy algorithm and forwards the messages to the first node that gets into contact. The worst case in this algorithm happens when $T_{S,j} \leq T_{i,D} - T_{j,D}$. Algorithm sends the packets to node i and adversary sets $T_{S,j} = \varepsilon$ near zero which means node S and j get into contact just after sending messages to node i. Optimal offline algorithm waits for $T_{S,j} = \varepsilon$ and sends messages to node j. Competitive ratio in this case is :

$$\alpha_{ALG} = \frac{C_{ALG}}{C_{OPT}} = \frac{T_{i,D}}{T_{S,j} + T_{j,D}} \approx \frac{T_{i,D}}{T_{j,D}} = \beta$$

In the second algorithm node S always waits until it reaches the node with least message transfer time which we can suppose is node j. The worst case in this algorithm happens when $T_{S,j} > T_{i,D} - T_{j,D}$. Adversary will set $T_{S,j}$ to very large number. In this case competitive ratio is:

$$\alpha_{ALG2} = \frac{C_{online}}{C_{OPT}} = \frac{T_{S,j} + T_{j,D}}{T_{i,D}}$$

Competitive ratio based on value of $T_{S,j}$ can be very large. From the above analysis we conclude that optimal competitive ratio is:

$$\alpha^* = \beta$$

□

Two forecast can be made: $T_{S,j} \le T_{i,D} - T_{j,D}$ and $T_{S,j} > T_{i,D} - T_{j,D}$.

Forecast 1: $T_{S,j} \le T_{i,D} - T_{j,D}$.
Based on this forecast online algorithm decides to wait until connection between node S and j is established. So the transfer time from S to D would be minimum. Cost of online algorithm is $C_{online} = T_{S,j} + T_{j,D}$.

Theorem 3: If the forecast $T_{S,j} \le T_{i,D} - T_{j,D}$ becomes true, then the restricted competitive ratio is 1.

Proof: If forecast comes true, then offline algorithm would act the same as online algorithm. So:

$$\hat{\alpha}_A = \frac{T_{S,j} + T_{j,D}}{T_{S,j} + T_{j,D}} = 1.$$

□

From theorem 2 and theorem 3, reward of this forecast is

$$f_A = \frac{\alpha^*}{\hat{\alpha}_A} = \frac{\beta}{1} = \beta.$$

Theorem 4: If the forecast $T_{S,j} \le T_{i,D} - T_{j,D}$ becomes false, then competitive ratio is $\frac{T_{S,j}}{T_{i,D}} + \frac{1}{\beta}$.

Proof: If the forecast is not true, which means $T_{S,j} > T_{i,D} - T_{j,D}$, offline optimal algorithm forwards the packets to node i in which the cost of offline algorithm would be $C_{offline} = T_{i,D}$. So:

$$\alpha_A = \frac{C_{online}}{C_{OPT}} = \frac{T_{S,j} + T_{j,D}}{T_{i,D}} = \frac{T_{S,j}}{T_{i,D}} + \frac{1}{\beta}.$$

□

From theorem 2 and theorem 4 the risk is:

$$\tau = \frac{\alpha_A}{\alpha^*} = \frac{T_{S,j}}{\beta \, T_{i,D}} + \frac{1}{\beta^2}.$$

Forecast 2: $T_{S,j} > T_{i,D} - T_{j,D}$
Based on this forecast the online algorithm forwards the messages to the node I without waiting to node j. So cost of online algorithm is $C_{online} = T_{i,D}$.

Theorem 5: If the forecast $T_{S,j} > T_{i,D} - T_{j,D}$ becomes true, then the restricted competitive ratio is 1.

Proof: If forecast comes true, then offline algorithm would act the same as online algorithm. So:

$$\hat{\alpha}_A = \frac{T_{S,i}}{T_{S,i}} = 1.$$

□

From theorem 2 and theorem 5, reward of this forecast is

$$f_A = \frac{\alpha^*}{\hat{\alpha}_A} = \frac{\beta}{1} = \beta.$$

Theorem 6: If the forecast $T_{S,j} > T_{i,D} - T_{j,D}$ becomes false, then competitive ratio is β.

Proof: If the forecast is not true, which means $T_{S,j} \leq T_{i,D} - T_{j,D}$, offline optimal algorithm waits until connection between nodes S and j is established and forwards the packets to node j. Cost of offline algorithm would be $C_{offline} = T_{S,j} + D_{j,D}$. So:

$$\alpha_A = \frac{C_{online}}{C_{OPT}} = \frac{T_{i,D}}{T_{S,j} + T_{j,D}}.$$

□

From theorem 2 and theorem 6 the risk is:

$$\tau = \frac{\alpha_A}{\alpha^*} = \frac{1}{\frac{T_{S,j}}{T_{j,D}} + 1}.$$

Using the above analysis forwarding decisions can be made more wisely based on different parameters. In each forwarding step, the message carrying node calculates values of risk and reward for different forecasts and based on these values can decide about the next node to forward the messages to. Also these decisions can be different based on risk tolerance of the forwarding algorithm.

6   Some Empirical Results

To clarify the excellence of risk-reward strategy, we show some numerical results for two cases of the problem and compare risk-reward strategy with deterministic algorithms discussed above using rewards derived for different forecasts.

Table 1. Emperical results for two cases of the problem and rewards obtained using risk-reward strategy

CASE	$T_{i,D}$	$T_{j,D}$	$T_{S,j}$	Reward of True Forecast	Risk of False Forecast
$T_{i,D} \geq T_{S,j} + T_{j,D}$	150	100	20	1.5	0.83
$T_{i,D} < T_{S,j} + T_{j,D}$	150	100	80	1.5	0.79

We also compare the risk-reward strategy with Greedy Strategy Without Waiting using numerical samples for two main cases of the problem.

Table 2. Emperical results for two cases of the problem and comparison of Greedy Strategy Without Waiting (GS) and Risk-Reward Strategy (RR). Suppose $T_{i,D} = 150$, $T_{j,D} = 100$, $T_{S,j}$ for the first case is 20 and for the second case is 80.

CASE	Competitive ratio of GS	Restricted Competitive Ratio of RR	Competitive Ratio of RR
$T_{i,D} \geq T_{S,j} + T_{j,D}$	$\beta = 1.5$	1	$\beta = 1.5$
$T_{i,D} < T_{S,j} + T_{j,D}$	1	1	$\frac{T_{S,j}}{T_{i,D}} + \frac{1}{\beta} = 1.2$

7   Conclusion

In this paper a special case of DTN routing is discussed in which a new node which has no knowledge of future contacts is added to a network and is about to send some messages to a destination node. This node gets partial information about other nodes and their contact times with its first contact. We analyze two strategies for deciding about the next node to forward the messages to using competitive analysis: A greedy strategy without waiting and a risk-reward strategy. The proposed risk-reward framework can help us in developing better routing algorithms.

Acknowledgements. Authors would like to acknowledge the support of DAAD (German Academic Exchange Service) and ITRC (Iran Telecommunication Research Center) for research grant.

References

1. Acer, U.G., Kalyanaraman, S., et al.: DTN routing using explicit and probabilistic routing table states. Wireless Networks 17(5), 1305–1321 (2011)
2. Yuan, Q., Cardei, I., et al.: Predict and relay: an efficient routing in disruption-tolerant networks. ACM (2009)
3. Al-Binali, S.: A Risk-Reward Framework for the Competitive Analysis of Financial Games. Algorithmica 25(1), 99–115 (1999)
4. Vahdat, A., Becker, D.: Epidemic routing for partially connected ad hoc networks, Technical Report CS-200006, Duke University (2000)
5. Acer, U.G., Kalyanaraman, S., et al.: Weak state routing for large-scale dynamic networks. IEEE/ACM Transactions on Networking (TON) 18(5), 1450–1463 (2010)
6. Lindgren, A., Doria, A., et al.: Probabilistic routing in intermittently connected networks. ACM SIGMOBILE Mobile Computing and Communications Review 7(3), 19–20 (2003)
7. Sleator, D.D., Tarjan, R.E.: Amortized efficiency of list update and paging rules. Communications of the ACM 28(2), 202–208 (1985)
8. Xu, Y., Hu, M., et al.: The canadian traveller problem and its competitive analysis. Journal of Combinatorial Optimization 18(2), 195–205 (2009)
9. Krumke, S.O.: Online optimization: Competitive analysis and beyond. Habilitation Thesis, Technical University of Berlin (2001)
10. Zhang, G., Xu, Y.: A Risk-Reward Competitive Analysis for the Newsboy Problem with Range Information. In: Du, D.-Z., Hu, X., Pardalos, P.M. (eds.) COCOA 2009. LNCS, vol. 5573, pp. 334–345. Springer, Heidelberg (2009)
11. Su, B., Xu, Y., Xiao, P., Tian, L.: A Risk-Reward Competitive Analysis for the Recoverable Canadian Traveller Problem. In: Yang, B., Du, D.-Z., Wang, C.A. (eds.) COCOA 2008. LNCS, vol. 5165, pp. 417–426. Springer, Heidelberg (2008)
12. Ding, L., Xin, C., Chen, J.: A Risk-Reward Competitive Analysis of the Bahncard Problem. In: Megiddo, N., Xu, Y., Zhu, B. (eds.) AAIM 2005. LNCS, vol. 3521, pp. 37–45. Springer, Heidelberg (2005)
13. Burgess, J., Gallagher, B., et al.: Maxprop: Routing for vehicle-based disruption-tolerant networks, Barcelona, Spain (2006)
14. Zhang, Z.: Routing in intermittently connected mobile ad hoc networks and delay tolerant networks: overview and challenges. IEEE Communications Surveys Tutorials 8(1), 24–37 (2006)
15. Kurhinen, J., Janatuinen, J.: Delay Tolerant Routing In Sparse Vehicular Ad Hoc Networks. Acta Electrotechnica et Informatica 8(3), 7–13 (2008)
16. Wisitpongphan, N., Bai, F., et al.: Routing in sparse vehicular ad hoc wireless networks. IEEE Journal on Selected Areas in Communications 25(8), 1538–1556 (2007)
17. Johnson, D.B., Maltz, D.A.: Dynamic source routing in ad hoc wireless networks. Mobile Computing, 153–181 (1996)
18. Perkins, C.E., Royer, E.M.: Ad-hoc on-demand distance vector routing. IEEE (1999)
19. Borodin, A., El-Yaniv, R.: Online computation and competitive analysis. Cambridge University Press (1998)

Scalable MapReduce Framework on FPGA Accelerated Commodity Hardware

Dong Yin, Ge Li, and Ke-di Huang

School of Electromechanical Engineering and Automation
National University of Defense Technology
Changsha, Hunan, 410073, China
yindong1982@gmail.com

Abstract. Running MapReduce framework for massive data processing on a cluster of commodity hardware requires enormous resource, especially high CPU and memory occupation. To enhance the commodity hardware performance without physical update and topology change, the highly parallel and dynamically configurable FPGA can be dedicated to provide feasible supplements in computation running as coprocessor to CPU. This paper presents a MapReduce Framework on FPGA accelerated commodity hardware. In our framework, a cluster of worker nodes is designed for MapReduce framework, and each worker node consists of commodity hardware and special hardware. CPU base worker runs the major communications with other worker node and tasks, while FPGA base worker operates extended mapreduce tasks process to speed up the computation process. Due to internal pipeline in computing operations, FPGA base worker offers the high performance which enhances 10x faster task processing. Furthermore, CPU base worker can reconfigure FPGA chip immediately when it fails. In this period, data will be migrated and continuously processed in commodity hardware. Meanwhile a local memory in commodity hardware is implemented to recover the lost data. Moreover, most frequent computation modules are provided in FPGA module library which are convenient for user to configure operations in special hardware. Experimental results proves that our framework offers high performance and flexibility in applications.

Keywords: MapReduce framework, FPGA, Scalability.

1 Introduction

MapReduce is a parallel programming model proposed by Google for the ease of massive data processing and has been successfully implemented to many applications [1–3] in the past few years. It was first used in Google's MapReduce library to utilize large scale clusters for parallelized data processing applications [1]. For MapReduce framework, high-performance, flexibility and scalability are key architectural requirements. Commodity hardware based MapReduce frameworks have been proposed in [4, 5], which can achieve high performance by processing data in parallel. However, in those frameworks, each commodity hardware

S. Andreev et al. (Eds.): NEW2AN/ruSMART 2012, LNCS 7469, pp. 280–294, 2012.
© Springer-Verlag Berlin Heidelberg 2012

node generally suffers from low computing rate , such as double precision floating data multiplication and division operations. For better performance, one general solution is to upgrade hardware or add more commodity equipments. However, additional number of machines always causes the changes in topology and routing, which may degrade the flexibility in scalable applications. Therefore, to improve the single node performance in MapReduce framework without any unnecessary changes to user is more necessary in reality. For achieving high-speed operation in computing, special-purpose hardware is used to implement MapReduce task [6–8]. For example, GPU and FPGA equipped on commodity hardware are designed to operate the whole framework process in parallel at a high rate. However, current dedicated hardware base frameworks provide poor flexibility in configuration and low scalability while working with other commodity hardware platforms. Those disadvantages always bring out problems to users who absolutely have no developing experience on special hardware. As a result, previous frameworks can not meet all the requirements in reality.

In this paper, we build up a MapReduce framework based on both commodity hardware and special hardware coexisting system. It consists of a large cluster of worker nodes, each of which contains one CPU base worker and FPGA base workers. In this kind of worker node, CPU base worker, which is connected to other worker nodes, runs to send data to FPGA base workers for process and receive the processed results. Meanwhile, FPGA baes workers, which are plugged in commodity hardware by using Ethernet link, are dedicated to run MapReduce tasks in parallel. Because of the limit in resources of FPGA chips, our framework supports the additional tasks running in CPU base worker. Moreover, to tolerant FPGA's fault gracefully, data migration and recovery policy has been adopted in our framework. When FPGA fails, CPU base worker can reconfigure it immediately and the data supposed to be processed in special hardware should be migrated to CPU to process. Finally, a local backup memory is applied to store the most-current data for data recovery from FPGA failure. We have implemented our MapReduce framework using a Xilinx V2P50 Pro FPGA on the NetFPGA board [9], but our design can be implemented in any FPGA.

The remainder of the paper is organized as follows. Section 2 discusses related work on MapReduce framwork. Section 3 presents the design of our MapReduce framework. Section 4 details the implementation of our design. The experimental methodology and results used to evaluate the system performance are described in Section 5. Section 6 concludes this paper and projects future work.

2   Related Work

We survey related work on MapReduce framework in both commodity hardware and special programmable hardware.When MapReduce framework was design in [1] few years ago, it was first implemented in a cluster of machines. For high data processing performance, a modified MapReduce architecture in which intermediate data was pipelined between operators, while preserving the programming interfaces and fault tolerance models of previous MapReduce framework [10]. Meanwhile, multicore commodity hardware is used for parallelizaion. In reference [3], a mapreduce framework based on multicore computers is designed to parallelized some machine learning algorithms which fit the Statistical Query model. However, grouping intermediate <key,value> pairs is often a performance bottleneck on multicore processors. Therefore, one optimized MapReduce data structure is explored in [11] based on workload characteristics such as the number of keys and the degree of repetition of keys. Unfortunately, the previous frameworks based on commodity hardware have limits in scalability, which can not afford increasing volume of data without changes in the topology and the commodity hardware upgrade.

Compared with commodity hardware, special hardware has been widely explored in various high performance computing applications such as FPGA [12] and GPU. Mars, a MapReduce framework on graphics processors (GPUs) is implemented in [6]. GPUs have an order of magnitude higher computation power and memory bandwidth compared with CPUs, however, are harder for software users to program due to their special-purpose architecture design. Recently FPGA is implemented to support MapReduce framework. One MapReduce framework on FPGA is presented in [8] which provides a on-chip scheduled and reconfigurable architecture and also implements RankBoost and PageRank module. Although, FPGA can dramatically accelerate processing performance by using common data path and parallelization, it is still difficult to deal with massive distributed data in a cluster of FPGAs. Moreover, programming FPGA such as coding in Verilog HDL is another obstacle for software developer to implement applications on FPMR. Additionally, an implementation of a mapreduce library supporting both parallel FPGAs and graphic processing units (GPUs) is described in [7]. In that work, parallelization methodology based on the map-reduce higher order functions common in functional language such as APIs is given to support both FPGA and GPU.

However, in those programmable hardware such as GPU and FPGA based MapReduce frameworks, the translation from API to FPGA programming is not clearly defined, which prevents it from providing convenient application for users. Moreover, the scalability of the framework with both FPGAs and GPUs coexisting [7] is still not discussed.

Our MapReduce framework is using an FPGA based platform accelerating CPU performance. Ethernet is adopted to plug FPGA platform into commodity hardware to extend mapreduce tasks operated in FPGA. Meanwhile, map/reduce tasks running in extended processor are the same as tasks in CPU which can be configured by user program and dynamically managed by Master. Therefore, our design can be implemented in any existing MapReduce framework based on commodity hardware. Moreover, APIs for configure map or reduce task in FPGA are also provided. More importantly, a rough library based on Verilog HDL is designed for user configuration, which abstracts basic common operations in mapreduce task.

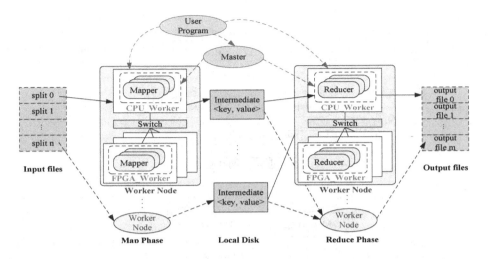

Fig. 1. Framework Overview

3   Framework Design

In this section, we present the top view of our MapReduce architecture and detail the design of worker node in our framework.

3.1   Framework Overview

A framework overview is illustrated in Figure 1. The architecture of our MapReduce framework is similar as the original framework which consists of a cluster of machines called workers. However, in our framework, one worker is represented by worker node which contains both a CPU-based worker and FPGA-based workers. As shown in Figure 1, CPU worker has two connections. One is for input and output file stream and the other is linked to FPGA worker through Ethernet. For every work node, when remotely or locally reading the input files, CPU worker dispatch data into each task running in either FPGA or software. After completing computation, CPU worker collects all the results and send out. And User Program is used to initialize map/reduce functions in all nodes in the system including Master and Workers [1]. However, in our implementation, the configuration to FPGA worker is completely different from software. Master is designed as the manager for all workers, which is supposed to configure and dynamically monitor tasks running in both software program and special hardware.

In our framework design, we focus on the scalability and performance without changing the existing networking structure, such as topology, and current MapReduce implementation. For scalability, each worker node is scalable by adding a "plug and play" FPGA base worker to CPU base worker. Ethernet

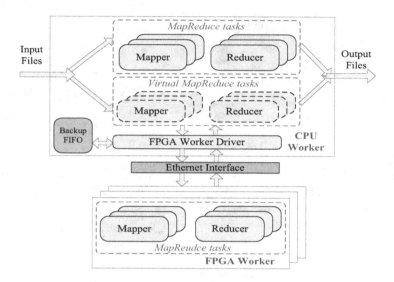

Fig. 2. Design in one Worker Node

is used to connect FPGA and CPU workers, and map/reduce tasks can be operated in both workers. User can add or remove additional FPGA workers according to different application. Meanwhile, due to the high speed process in FPGA and pipeline operation design in both workers, the process performance of worker node should be significantly improved. Therefore, our framework has higher throughput compared with other MapReduce frameworks. Moreover, in our design, we provide some APIs to simplify the programming FPGA for users. The details of design and implementation are presented in following subsections.

3.2   Design of Worker Node

As shown in Figure 2, MapReduce tasks are independently implemented in both CPU and FPGA worker. Two kinds of tasks are configured in CPU worker: real tasks and virtual tasks. Virtual task treated as the normal task running in CPU worker actually is an intermediate layer for FPGA worker to communicate with MapReduce program running in the Operation System. Each task configured in FPGA worker has one relative virtual task interface in CPU worker for transmission. For example, virtual map task remotely/locally read input files split by MapReduce library and dumped into FPGA baes map task using FPGA Worker Driver. All operations of tasks are processed in parallel. The final results generated in FPGA worker should be collected by virtual task running in CPU worker which then dispatch them out.

The most important component in worker node is FPGA Worker Driver, a bridge between two workers, which has two main functions: Communication and Configuration. **Communication.** The general transmission between CPU and FPGA worker is mapreduce files such as initial and intermediate <key,value> pairs carried by ethernet packets. Moreover, the status of each mapreduce task should be checked and reported to Master. In our design, Ethernet packets are used for data delivery. UDP packets are used for data delivery and every packet has a position tag which presents the sequence number during the transmission. When all packets are sent from CPU worker, FPGA worker starts to check all received tags. FPGA worker should ask CPU worker to resend the loss packets. The details of implementation of worker node is discussed in section 4.1. **Configuration.** Two kinds of configuration are applied to deploy FPGA. One is initial configuration. In this step, FPGA worker driver initially configures MapReduce tasks in hardware according to user's program. The other one is reconfiguration. When fault occurs that FPGA worker can not operate task or has no response in a period, reconfiguration is requested to reset the MapReduce task. Also while the task operation is changed, the hardware should be reconfigured.

4   Implementation

In our implementation, the open-source Hadoop project is modified to operate the MapReduce module, while NetFPGA is designed to run our FPGA worker functions. In this section, we first detail the worker node implementation and then focus on how to set up and configure worker node. Finally we discuss fault tolerance of the work node.

4.1   Worker Node

Each worker node has the scalability to extend additional workers by connecting multiple FPGA boards through ethernet. CPU worker runs the modified Hadoop MapReduce program which processes file system, transmit data to FPGA base workers and operates computational functions. Meanwhile, extended MapReduce functions can be implemented in programmable FPGA board which has virtual mappings in CPU worker.

CPU Worker in Linux System. In Figure 2, CPU worker is designed to process mapreduce functions in software and communicate with locally additional worker. Map/Reduce tasks running in CPU read the data from input files, generate the results and send out to output files. Every mapreduce task implemented in FPGA worker has a mapping task, virtual mapreduce task in CPU worker which dumps the input files into mapreduce task in FPGA worker and transmits the result to the output port.

Due to open-source Hadoop project, we modify the map and reduce class and add some new modules for data transmission and status monitor. Meanwhile, socket tunnel is applied to implement intermediate transmission between virtual and real map/reduce tasks. As shown in Figure 3, one virtual map task running in CPU worker is the interface for real map task running in FPGA worker to communicate with MapReduce implementation in linux system. Sockets are used in FPGA Worker Driver for data delivery. Two types of sockets are implemented to transfer <key,value> pairs and instructions from Master.

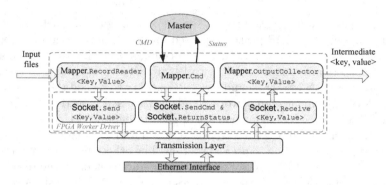

Fig. 3. Virtual Task in Linux OS

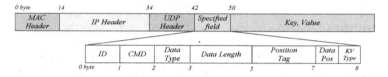

Fig. 4. Packet Structure

During map process, virtual map task first calls *RecordReader* function to read input files split files from file system and translate into initial <key,value> pairs which then are transferred to networking transmission layer by using *Socket.Send* process. After that, initial <key,value> pairs which are encapsulated into IP packet with ethernet header can be sent out to FPGA worker through ethernet. While dispatching the data, virtual map task adopt *Socket.Receive* function to listen and receive the results processed by FPGA worker. Function *Output-Collector* is used to roughly merge the results with the same key and send out intermediate <key,value> pairs to local disk. Meanwhile, for monitoring status and commanding task , *Mapper.Status* process sends the commands from Master to map/reduce task running in FPGA worker using *Socket.Cmd* and reports the dynamic status obtained from *Socket.ReturnStatus* which works to get status packets returned from FPGA worker. Note that, blocking sockets strategy is adopted in virtual task. For example, after sending a bunch of packet for one <key,value> pair, virtual task has to wait until receives the result.

In Hadoop MapReduce application, the structure of map and reduce task are similar. As a result, we implement the virtual reduce task in the same way as virtual map task.

In our implementation, IP packets with ethernet header packet is used to transmit data and instructions between virtual task in CPU worker and real task in FPGA worker. The input initial file is carried by data stream, that is the total length of stream is various. That is, the size of initial <key,value> pairs generated from input files is unpredictable. As a result, when the length of one pair is beyond the capacity of one packet, we have to divide one <key,value> pair into several blocks which can be encapsulated in Ethernet packets. To label every packet sending to FPGA base task, a special field in the packet is set to

Table 1. Specified Field in Packet

Field	Description
ID	Mapreduce task ID
CMD	Commands from Master translated by virtual task
Data Type	Data Type of KV: 32bits integer or 64bits double
Data Length	Length of data field
Sequence Tag	Packet sequence tag
Data Pos	Data position label. $'0'$ means final packet of KV pair(s)
KV Type	Type of KV structure. $'0'$ means one key with multiple values; $'1'$ means one key with one values

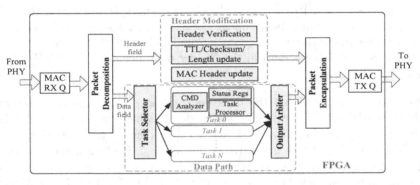

Fig. 5. FPGA Worker Node Implementation

indicate information from master and software program. As shown in Figure 4, the specified field contains several areas which are defined in Table 1. We will discuss the application of this specification in following subsection.

FPGA Worker on NetFPGA Board. Our MapReduce framework can be implemented in any FPGA with ethernet interface and on-board memory. Due to the facility of packets processing, we build up our FPGA worker in NetFPGA. As shown in Figure 5, incoming packets received from ethernet interface are first classified and decomposed by Packet Decomposition into two parts header field and data field. Header of the packet should be updated in Header Modification while data of the packet can be processed in Data Path. Due to only one connection between two platform, the source address of incoming packet is the same as the destination address of outgoing packet. All the header of the returning packets to commodity hardware is the same. Therefore, we separate the header modification from data process.

Header Modification. Header field of the packet is separated by Packet Decomposition, and verified and updated in parallel. First, the header is verified whether the packet is valid. If not, the packet should be discarded. Second, TTL and Checksum value needs to be updated to indicate that the packet is processed in

Table 2. Command definition

Cmd Code	Description
0x01	Check task status
0x11	Run map task and store the data
0x12	Run map computation
0x04	Run reduce computation
0x08	Reset task
0x10	Stop task

FPGA. According to the specified data block size, length of the packet should be modified, such as 84(bytes) when block size is 64(bytes). At last, the MAC address in the header may be exchange between original source and destination address for returning the processed data from FPGA to CPU. In our implementation, there are 32 bytes data in the Header field including 20-byte IP header and 12-byte MAC header.

Data Path. When receiving the data field of the packet, Task Selector first analyzes the ID label in specified field to decide which task module this data packet should be sent to. Then, task selector dispatchs the packet to destination task. The structure of task is illustrated in Figure 5. Each task presents either map task or reduce task. Command(CMD) Analyzer is designed to recognize the commands from master described in Table 2. The functional operations of the task are running Task Processor, while the status of task is recorded in Status Registers. When task completes computation, result will be taken away by Output Arbiter which adopts Round-Robin way to read task output port.

After processed in Header Modification and Data Path, these two field results should be encapsulated in Packet Encapsulation module and then sent out through MAC Output Queue.

In our implementation of FPGA worker, on-the-fly design of map/reduce task is adopted for limited resource in programmable hardware, which allows data to be pipelined between operators. This extends the map/reduce task beyond batch processing, and can reduce completion times and improve hardware resource utilization for batch functions as well. In this case, small capacity memory with low access latency should be needed to store the temporary data and final result can be sent out immediately when it is generated.

4.2   Worker Node Configuration

In the implementation of NetFPGA board, JTAG interface is used to configure FPGA. There are two types of configurations in our framework, initial configuration and map/reduce task configuration. In initial step, Linux host download the bitfile to FPGA for "Hello" test. After downloading the initial configuration file, commodity hardware will send out a Hello packet and wait for a reply from FPGA. While receiving a reply that show FPGA is ready to work, both hardware can be configured to run mapreduce program. Since that, User Program initializes every worker node and set one Linux machine as Master. From that moment, the MapReduce framework is built up and prepares for process files.

4.3 Fault Tolerance

Since the MapReduce library is designed to help process very large amounts of data using hundreds or thousands of nodes, the library must tolerate nodes including Master and Workers failures gracefully [1].

FPGA Worker Failure. In our MapReduce framework, a data migration method is implemented to help worker tolerate the FPGA fault gracefully. Local master send out Hello packet periodically to make sure that FPGA is normally running in good condition. If no response is received from NetFPGA board in a certain amount of time, the local master marks the FPGA as failed. In that case, local master will immediately reconfigure the FPGA to initial status by JTAG through the NetFPGA cube. It costs a short time to compete the configuration, such as approximately 12 seconds proved in [13]. During this period, the data originally allocated to the FPGA should be migrated to host workstation by Data Controller. While configuration is completed, the data will be transmitted to special hardware again. Due to mapreduce online applied in FPGA, the volume of the data concurrently being processed is considerable small compared with normal mapreduce library in [1]. As a result, only few data blocks will be lost when FPGA is failed. To solve the problem, we implement a backup FIFO (as shown in Figure 1) in workstation to store most-current data transferred to NetFPGA. when failure happens to FPGA, local master asks Data Controller first pick up the original data in Backup FIFO which is currently running in FPGA, then perform migration to direct all data to be processed in CPU. The performance of data migration and backup will be presented in the section5.

CPU Worker Failure. When Linux machine fails, the whole worker node needs to be reset. For simplicity, we follow the policy illustrated in [10] to deal with worker failure.

5   Experimental Results

This section introduces the experimental setup and presents the performance of two applications in our MapReduce framework.

5.1   Experimental Setup

To evaluate our framework in performance, flexibility and scalability, we build up a testbed for the worker node in our implementation. As shown in Figure 6,

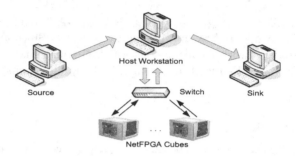

Fig. 6. The experimental testbed

Source node splits the original input data into blocks, then dispatches to worker nodes. Sink node derives final results from nodes in reduce phase. In the worker node, NetFPGA Cubes are connected to Host Workstation by ethernet link through switch. Packets encapsulated with MAC header are used to transmit data between Workstation and NetFPGA Cubes. FPGAs are remotely configured by Host Workstation through networking and also can be reconfigured momentarily when necessary. In following experiments, we adopt a host workstation with Due Core Intel CPU(2.66GHz), 4GB RAMS and Xilinx Virtex II Pro FPGA base NetFPGA cubes running at 125MHz.

Mapreduce framework is the most popular programming architecture which runs computation for web searching, geographic data process and so on. Matrix Multiplication is the most common operation among algorithms implemented in Mapreduce framework. As a result, we first implement matrix multiplication in our framework for experiment. Then, the most typical algorithm Pagerank is adopted in our implementation. We compare the performance of our framework with commodity hardware based Mapreduce framework and show the scalability and flexibility of our design.

5.2 Matrix Multiplication

For simplification, matrix multiplication, which is the most typical operation during algorithms such as pagerank, machine learning and geographic data process, is first implemented to evaluate the performance of our framework. In our experiment, UDP packet is adopted to transmit <key,value> paris. When the length of the pair is beyond the maximal length of one UDP packet, the data has to be divided into blocks and delivered by several packets. Meanwhile, we implement internal Block-RAM in FPGA to store parts of the matrix temporarily. Both map and reduce operation are running in the same worker node. Each task runs one row and one column multiplication, then sums up to generate the result.

Performance with Single FPGA Board. In MapReduce framework, map task always operates high-time cost computation while reduce task just runs simple merge process. Therefore, we design FPGA to run map tasks comparing the performance of the same tasks running in CPU. Hadoop file system is

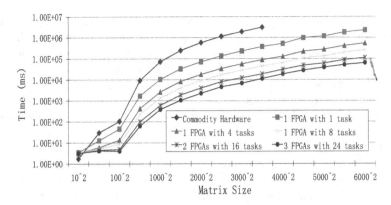

Fig. 7. Matrix Multiplication in CPU and FPGA

implemented to a cluster of commodity machines and Matrix Multiplication is normally implemented in data processing. We also choose square Matrix Multiplication AxB to test our framework performance. Task for map phase is running in both CPU worker and FPGA workers. while commodity machine reads the data stream from file system and sends the initial <key,value> pairs to both CPU base map tasks and FPGA base map tasks. When receiving the byte stream results from all tasks, commodity hardware sends out the intermediate <key,value> pairs. In CPU base map task, one matrix B is already locally stored and the other matrix is read from remotely. However, due to limited resource in FPGA, matrix with large size can not be stored in on-chip memory. In our implementation, we only store several columns of the local matrix B which are transmitted from commodity hardware. The whole matrix A is dumped to FPGA for a number of times which depends on the size of matrix B and the number of columns stored in FPGA. Each map task runs one row and one column multiplication to generate intermediate entries of final matrix.

As shown in Figure 7, when the size of matrixes is small, it is worst to run the computation in FPGA because interval in transmission costs too much. With matrix size increasing, time consumption of data processing in FPGA becomes less than CPU. When only one column is stored in FPGA, the performance of data processing is almost the time as CPU due to frequent data transmission from commodity hardware. Then we extend the number of tasks stored in FPGA from 1 to 16, the time cost in multiplication decreases dramatically. The FPGA can achieve almost 15x processing performance comparing to CPU, since 16-column of local matrix is temporarily stored in FPGA. Since sockets running in JAVA environment is adopted in virtual map task to transmit the data to FPGA, the sending speed is much slower than other programs operating in Linux kernel mode.

Performance with Multiple FPGA Boards. First single FPGA base worker is configured to run one task for computation, then the number of tasks is increased, finally more tasks are extend to more FPGA platforms. As shown in Figure 7, our framework with single FPGA base worker runs at least 10x faster than commodity hardware. However, when the size of matrix block is small, the consumption of socket transmission is rather more than computation in Linux system. In this case, our performance may be decreased. However, when more tasks are configured in one single FPGA worker, the matrix block size become larger and intervals in communication cost less. As a result, our framework can provide better performance. As described in Figure 7, while the number of on-chip tasks grows from 1 to 8, the performance of our platform achieve approximately linear increase.

Due to the limit in on-chip memory, we extend more tasks to more FPGA boards in which all tasks are running in parallel. The performance is presented in Figure 7. As allocating two FPGA base workers instead of one, our framework can double its processing speed. Unfortunately, increasing number of workers from two to three does not show proportional operation speed enhancement because of the bandwidth limit of Gbit Ethernet link.

5.3   PageRank

Implementation in FPGA. In this experiment, we implement Google PageRank in our mapreduce framework. To decrease the time cost in communication between CPU and FPGA, we run both map and reduce tasks in FPGA modules. For simplification, we set the number of iteration instead of checking the difference between the current and previous pagerank vector value. We implement the algorithm in parallel, while adopt pipeline operation among process modules to enhance the performance.

Performance. The dataset for Pagerank from Google webgraph is adopted to test our framework. We compare the average time consumption in each iteration of general MapReduce framework and our scalable platform. As shown in Table 3, on a 4-node cluster, for Google webgraph 916,417 nodes and 6,078,254 edges, it generally cost 39 seconds for each iteration on commodity hardware. While for our scalable framework with three FPGA boards, it takes less than 10 seconds to operation one iteration to update one node pagerank value, which is approximately 4x times faster than processing in JAVA program on commodity hardware. Due to the advantages in scalability and parallelization, our framework shows better performance than pure commodity hardware based implementation.

Table 3. Average cost of each iteration in Pagerank

Number of Nodes/Edges	Commodity hardware	Scalable framework with FPGA
102448/2162410	5.356s	2.056s
916417/6078254	38.211s	9.702s
1211756/9076553	92.145s	24.235s

6   Conclusion and Future Work

We have described a flexible and scalable MapReduce framework that integrates several software workers on commodity hardware with partially-reconfigurable workers on specially programmable hardware. Experimental results with multiple Virtex II Pro based FPGA boards indicate that our framework can offer remarkable improvement over previous commodity hardware base approach, and also provide good scalability and flexibility.

In the future, we plan to investigate the use of partial reconfiguration for better performance in larger Virtex 6 devices. Our future work will also focus on developing user-friendly interfaces to generate and download configuration bitstreams for framework operators with little hardware knowledge. More robust workload balancing algorithms will also be explored.

References

1. Dean, J., Ghemawat, S.: MapReduce: Simplified Data Processing on Large Clusters. In: OSDI 2004: Sixth Symposium on Operating System Design and Implementation, pp. 10–23 (2004)
2. Apache Group, Apache Hadoop (2008), http://hadoop.apache.org
3. Chu, C.-T., Kim, S.K., Lin, Y.-A., Yu, Y., Bradski, G., Ng, A.Y., Olukotun, K.: Map-Reduce for Machine Learning on Multicore, MapReduce: Simplified Data Processing on Large Clusters. In: Advances in Neural Information Processing Systems, pp. 281–288. MIT Press (2007)
4. Ranger, C., Raghuraman, R., Penmetsa, A., Bradski, G., Kozyrakis, C.: Evaluating MapReduce for multi-core and multiprocessor systems. In: HPCA 2007: IEEE 13th International Symposium on High-Performance Computer Architecture, pp. 13–24 (2007)
5. Yoo, R.M., Romano, A., Kozyrakis, C.: Phoenix rebirth: Scalable MapReduce on a large-scale shared-memory system. In: IISWC 2009: 2009 IEEE International Symposium on Workload Characterization, pp. 198–207 (2009)
6. He, B., Fang, W., Luo, Q., Govindaraju, N.K., Wang, T.: Mars: A MapReduce framework on graphics processors. In: PACT 2008: 17th International Conference on Parallel Architectures and Compilation Techniques, pp. 260–269 (2008)
7. Yeung, J.H.C., Tsang, C.C., Tsoi, K.H., Kwan, B.S.H., Cheung, C.C.C., Chan, A.P.C., Leong, P.H.W.: Map-reduce as a Programming Model for Custom Computing Machines. In: FCCM 2008: 16th IEEE Symposium on Field-Programmable Custom Computing Machines, pp. 149–159 (2008)
8. Yi, S., Wang, B., Yan, J., Wang, Y., Xu, N., Yang, H.: FPMR: MapReduce framework on FPGA. In: FPGA 2010: Proceedings of the 18th Annual ACM/SIGDA International Symposium on Field Programmable Gate Arrays, pp. 93–102 (2010)
9. NetFPGA Group (2010), http://netfpga.org
10. Condie, T., Conway, N., Alvaro, P., Hellerstein, J.M., Elmeleegy, K., Sears, R.: MapReduce Online, EECS Department, University of California, Berkeley, Tech-Rep. UCB/EECS-2009-136 (2009), http://www.eecs.berkeley.edu/Pubs/TechRpts/2009/EECS-2009-136.html

11. Mao, Y., Morris, R., Frans Kaashoek, M.: Optimizing MapReduce for Multicore Architectures, Computer Science and Artificial Intelligence Lab (CSAIL), Massachusetts Institute of Technology, Tech-Rep. MIT-CSAIL-TR-2010-020 (2010)
12. Herbordt, M.C., Van Court, T., Gu, Y., Sukhwani, B., Conti, A., Model, J., Di Sabello, D.: Achieving High Performance with FPGA-Based Computing. Computer 40(3), 50–57 (2007)
13. Unnikrishnan, D., Vadlamani, R., Liao, Y., Dwaraki, A., Crenne, J., Gao, L., Tessier, R.: Scalable network virtualization using FPGAs. In: FPGA 2010: Proceedings of the ACM/SIGDA International Symposium on Field Programmable Gate Arrays, pp. 219–228 (2010)

A Self-organizing P2P Architecture for Indexing and Searching Distributed XML Documents

Carmela Comito[1], Agostino Forestiero[2], and Carlo Mastroianni[2]

[1] University of Calabria, Rende (CS), Italy
ccomito@deis.unical.it
[2] ICAR-CNR, Rende (CS), Italy
{forestiero,mastroianni}@icar.cnr.it

Abstract. This paper presents *X-Pastry*, a peer-to-peer system that, maintaining the tree-based overlay of Pastry, exhibits enhanced functionalities thanks to the activity of ant-inspired mobile agents. The agents move the resource keys across the network, and sort them in a self-organizing fashion. The sorting ensures that discovery operations are executed in logarithmic time, and helps to improve load balancing, adaptivity and the efficient execution of range queries. X-Pastry is particularly efficient when resources are represented with hierarchical names, as in the case of XML objects, because different levels of the tree overlay are used to manage different name components. The paper presents the architecture and the ant-based algorithm of X-Pastry, and evaluates its performance when it is used to index and search a large and distributed collection of XML documents.

1 Introduction

In recent years, peer-to-peer (P2P) systems have overstepped their mere role of middleware solution for traditional file-sharing applications and are now adopted in all kinds of large-scale distributed computing systems, thanks to their advantages over centralized and hierarchical solutions, among which their scalability and robustness.

New fields of application for P2P systems are Grid and Cloud infrastructures and Content-Centric Networking. In Grids, different types of computing resources are shared among the nodes of a community, ranging from data to programs and software facilities. Resources are not typically discovered through their names, as in file sharing systems, but through a set of characteristics described by metadata documents. P2P solutions can also be exploited to manage the information systems of big companies, like Google and Amazon, which offer pay-as-you-go services through their Cloud platforms. For example, Amazon Dynamo [4], a key-value storage system adopted by several Amazon's core services, is a variant of the Chord P2P system [13]. Recently, Jacobson et al. [7] presented a bold vision of a content-centric network (CCN), in which content is searched by its name instead of its location. CCN nodes are collectively responsible for finding named content and returning it to the requestor. CCN adopts hierarchical names and longest prefix matching to scalably route requests to named content.

Most modern P2P architectures are *structured*, meaning that peers are organized in a predefined structure, for example a ring (as in Chord [13]), a multi-dimensional grid (adopted by CAN), or a tree (as in Pastry [12]). A hash function is used to give each

S. Andreev et al. (Eds.): NEW2AN/ruSMART 2012, LNCS 7469, pp. 295–306, 2012.

resource a key, and the key is assigned to a node whose code, also computed with a hash function, is equal or as close as possible to the key. Structured P2P systems are preferable because they use *informed* algorithms to drive user queries towards the desired keys in a short and bounded time [1]. Unfortunately, they have a major drawback: discovery algorithms are exclusively driven by the key value, and cannot take advantage from the knowledge of specific characteristics of target resources. As a consequence, there is no simple way of efficiently serving *range* queries that need to discover a set of resources sharing common features, for example a set of machines with CPU speed and RAM memory comprised in a given range, or a set of mathematical tools with given characteristics. The difficulty derives from the fact that the keys of similar resources are spread by the hash function over the P2P overlay [11]. Other notable drawbacks of structured P2P systems concern the load balance (the nodes that are assigned the most popular keys may be significantly more loaded than the others) and the dynamic behavior (an immediate reassignment of keys is necessary every time a node joins or leaves the system).

In recent years, there have been attempts to reinforce the adaptive and fault-tolerance characteristics of P2P networks by imitating the self-organizing behavior of biological systems, such as insect swarms and ant colonies [8]. These systems are sometimes referred to as "self-structured" [5] [2], because their structure is constructed with self-organizing techniques.

In this paper we present X-Pastry, a P2P system that uses the same tree overlay as Pastry [12], but adopts an ant-inspired algorithm to map resource keys to the different peers of the overlay. X-Pastry is specifically tailored to the cases where resources are represented with hierarchical names. This naming strategy, encouraged by the Content-Centric Network mentioned above, is widely adopted for the management of XML data. Indeed, as an XML document is represented as a labeled tree, each resource in the document can be denoted with a hierarchical name where each name component is associated to a specific level of the tree, from the root to the leaf node containing the resource. Some hash-based P2P systems have been proposed to index and retrieve distributed collections of XML data [3] [10] but these systems, even if optimized for the discovery of XML resources, still present the drawbacks mentioned before, specifically concerning the execution of range queries.

In X-Pastry each path in an XML document is associated with a key, obtained as the concatenation of the components of the path. The keys are sorted over the tree overlay by ant-inspired mobile agents, in such a way that the different levels of the tree overlay are associated with the different components of XML paths. X-Pastry avoids the use of hash functions, which inevitably disperse similar resources to distant nodes of the structure, and allows the keys to maintain the semantic characteristics of resources. This strategy improves the efficiency of range queries, because similar resources are assigned to peers that are close to each other: this means that a discovery message can find the target resources in a single peer or in a small number of close-by peers. Other advantages of X-Pastry are its capacity of balancing the load among the peers and of rapidly reacting to environmental modifications. The performance of X-Pastry is analyzed in the paper with reference to the management of a large collection of XML documents, the DBLP bibliography of scientific papers.

The paper is organized as follows: Section 2 describes the coding of XML documents; Section 3 presents the X-Pastry overlay and the ant-inspired algorithm; system performance is analyzed in Section 4, and Section 5 concludes the paper.

2   Hierarchical Naming of XML Resources

X-Pastry is able to manage any kind of resource, provided that an ordering is defined in the space of key values. However, the tree overlay is particularly suited for the management of keys that have a hierarchical representation. An interesting example is the construction of indexes of distributed XML databases. In the rest of the section we describe the adopted data model and then introduce the encoding scheme of XML documents.

2.1   Data Model

XML is a markup language for documents containing structured information. An XML document is represented as an ordered, labeled tree according to the DOM standard (*http://www.w3.org/DOM/*). Each node in the tree corresponds to an element, an attribute, or text data; edges between nodes represent element/subelement or element/attribute relationships. Leaf nodes correspond to textual data values and internal nodes correspond to XML elements that form the structure of the document. An XML *path* identifies a route from the root node to a leaf node, where the intermediate nodes are XML elements and the edges are the parent/child relationships between such elements. Thus, a path includes a *structure* – the sequence of element labels together with their parent-child relationships – and the *text* contained in the leaf node. All the paths sharing the same structure form a *path class*.

We defined an encoding scheme that allows XML paths to be represented with a concise notation, with the objective of generating semantic meaningful resource keys. Each XML path is represented as a numerical key, built by encoding and concatenating the intermediate elements forming the path and the text contained in the leaf. Details of such encoding are given in the following.

2.2   Encoding of XML Documents

We devise an encoding strategy that associates each path with a key – or "code" – composed of two parts: the *structure code*, representing the sequence of intermediate nodes, and the *content code*, representing the textual data value of the leaf node. The XML instances belonging to the same path class have the same structure code and differ only for the content code.

The structure code is obtained with a prefix-based node labeling scheme, similar to those used in [9,6], in which the parent of a node is encoded as a prefix of the node label. The XML tree is explored with a depth-first visit strategy in such a way that nodes that are consecutive following the XML document order are traversed and the labels are generated accordingly. Based on this scheme, a code of a given path is composed of as many digits (separated by dots) as the number of levels in the path, where each digit represents a specific tree level. For each element, the corresponding digit represents the order of this element among its siblings, i.e., among the other elements having the same ancestor.

The content part of a code consists of a single digit. For each class path, the instances are numbered respecting an order defined in the space of admissible values. This encoding ensures that similar XML instances (for example, instances that share the same structure code) are associated with similar codes. As codes are sorted over the X-Pastry overlay, similar codes can be found in a restricted region of the structure, which enables the efficient execution of range queries (e.g., all the papers published between 2002 and 2006 by a given author or about a specified subject).

We can represent a path code through the following notation $< d_1.d_2.,...,.d_L.d_C >$, where digit d_i is the numerical representation of the element at level i, L is the depth of the path class to which the considered element belongs, and d_C is the digit associated to the instance. A number of bits b is assigned to each level. The code can also be expressed converting the positional notation into a sum of powers:

$$\sum_{i=1}^{L}(d_i \cdot (2^b)^{L-i})$$ (1)

This paper focuses on the use of X-Pastry to index distributed XML documents, and the adopted example is the well known DBLP bibliography of scientific papers (http://www.informatik.uni-trier.de/ley/db/). The DBLP database, dblp.xml, has a hierarchical structure (see Figure 1), where the root node is dblp and branches and sub-branches correspond to the different types of papers included in the bibliography. At the first level there are 8 XML elements corresponding to the different types of papers: inproceedings, proceedings, mastersthesis, article, phdthesis, www, book and incollection. In turn, each of these nodes has a specific number of child elements. For example, inproceedings has 11 children, some of which are shown in the figure. Starting from the third level, nodes can be terminal elements, i.e., leaves of the tree, in which case the nodes contain a text. For example, they can contain the paper title or the author name.

The procedure used to code each level of a path is explained with an example, in the case that each level is represented with a number of bits b equal to 4. Let us assume that we need to build the code of a conference paper entitled "PaperTitle", whose path is dblp/inproceedings/title="PaperTitle". The root node is coded with the number 0. The first node of the path is inproceedings, the first below the root. Since 4 bits are available at each level, the node must be encoded with a number between 0 and 15. In this case there are 8 possible values for the first level (8 different categories of paper), which are encoded with values 0, 2, 4 ... 14. For the given example, the code for the first level is 0. For the second level, the possible children of inproceedings are 11: title, year, month, author, etc. In general, if the number of children is N_c, the i-th child is encoded with the value $d = \lfloor (i-1) \cdot 2^b/N_c \rfloor$, so as to obtain a value between 0 and $2^b - 1$. In our case, the element title is the second among its siblings, so the integer code for the second level is $d = \lfloor 1 \cdot 2^4/11 \rfloor = 1$. Thus, the structure part of the code is $< 0.0.1 >$.

A slightly different method is used to code the text value of the leaf element, for example the title of the paper, in this case "PaperTitle". Of course, it is not possible to

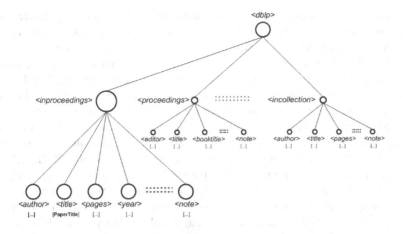

Fig. 1. A fragment of the XML tree that represents DBLP papers

determine a specific number of possible values, so the approach consists of converting the first characters of the text string into integer numbers. If the first value, say "AAAA", is converted to the integer 0, and the last possible value, "ZZZZ", is converted to $2^b - 1$, the value of the string is converted into a number comprised between 0 and $2^b - 1$, using a proportion. In our case, the paper title is converted into the number 11. The complete code of the element is thus computed, using expression (1), as: $0 \cdot 16^2 + 1 \cdot 16 + 11 = 27$ (the code of the root node is omitted).

It should be noted that each entry of the DBLP database can be indexed on several elements. For example, the sample paper considered so far can be indexed on its first author, say "Smith", its publication date, etc. This allows queries, and in particular range queries, to be executed on each indexed field. For example, a query may discover all the conference papers written by "Smith" or the conference papers written in March 2011.

3 The X-Pastry System

In the previous section, we defined a mechanism to associate each XML path to a code composed of a sequence of digits, each corresponding to a hierarchy level. This type of encoding can be efficiently mapped to the hierarchical P2P structure of Pastry. In Pastry, the peers are logically organized in a tree structure, and each resource key is consigned to the peer having the same code as the key, or a very similar one. Discovery messages are routed through two sets of pointers: i) the *routing table* contains pointers to peers whose codes share a common prefix, of any desired length, with the code of the local peer; ii) the *leaf set* contains pointers to peers that are adjacent or very close to the local peer. The details about Pastry can be found in [12]. Thanks to these pointers, Pastry routes messages towards the target objects in a number of steps that is logarithmic with respect to the number of peers. Pastry, however, shares with the other structured P2P systems some significant drawbacks, mainly deriving from the use of a hash function to assign keys to resources. In this way, the semantic aspect of a resource is lost, and similar resources are inevitably associated with completely different keys and then mapped to distant peers, which hinders the efficient execution of range queries.

The X-Pastry framework presented here uses the same overlay as Pastry to connect the peers with each other, but a set of ant-inspired agents move the keys across the network and sort them in a self-organizing fashion. This sorting ensures that discovery operations can be still executed in logarithmic time, exploiting the tree-based overlay. In addition, X-Pastry features several important benefits with respect to Pastry: (i) peer codes and resource keys can be defined over different spaces and there is no obligation to assign a resource key to a specific peer. This allows the semantic meaning of resources to be preserved, so that similar resources (for example, similar XML elements) are assigned similar keys, which are then mapped to X-Pastry peers that are close to each other, in this way improving the efficiency of range queries; (ii) in X-Pastry, the keys are fairly distributed over the peers, irrespective of the popularity distribution of key values. Conversely, load balancing is a major issue in Pastry, as in other structured systems, because the nodes storing the most popular keys must sustain a higher load; (iii) in X-Pastry, the mobile agents are always active and continuously reorganize the keys. Therefore, it is not necessary to perform any operation when a node joins the network or when new resources are published. This feature improves scalability (keys are continuously reordered as the network grows) and robustness with respect to environmental changes.

The rest of this section focuses on the X-Pastry overlay and then on the ant algorithm performed by agents to map the keys over the overlay.

3.1   The X-Pastry Overlay

In X-Pastry, as in Pastry, peers are assigned codes with B bits, through pair-wise hash functions. A code is a sequence of digits having base 2^b: for example, with $B=128$ and $b=4$, the code is a sequence of $B/b=32$ digits. The Pastry architecture is inherently hierarchical. This can be clarified with the help of Figure 2. Let us consider an example in which $B=6$ and $b=2$: this means that each peer is identified with a 3-digits code, and each digit has values between 0 and 3. For example, the peer with the marked border is assigned the code 112. The leaf set of the peer contains pointers to some neighbor peers, for example the peers embraced by the curly bracket in the figure. On the other hand, the routing table contains pointers to remote peers that share a common prefix with the current peer. Specifically, the first row of the routing table contains pointers to one leaf peer for each of the big shaded rectangles, i.e., to three peers whose codes begin with digits 0, 2 and 3, respectively. Three peers of this kind – specifically, peers 021, 203 and 331 – are indicated by the continuous-lined arrows. Analogously, the second row of the routing table contains pointers to peers included in the small shaded rectangles, whose codes start with digits 10, 12 and 13. Example of such peers – 101, 122 and 133 – are indicated by the dashed-lined arrows.

X-Pastry adopts a similar architecture. However, while Pastry uses the same space for peer codes and resource keys, X-Pastry is free to adopt any coding technique for the resources. There is no predefined association among keys and peers, differently from classical structured systems. However, keys must be ordered to let discovery operations be executed in logarithmic time. It is convenient to code resources in a way that

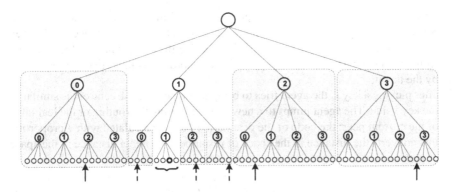

Fig. 2. Pastry/X-Pastry tree architecture. The peer with marked border has code 112. The leaf set of this peer has pointers to the peers included in the curly bracket. The pointers of the routing table are indicated with arrows.

preserves their semantic characteristics, as explained in Section 2, since this helps to perform complex and range queries. In fact, since the X-Pastry algorithm places similar keys in neighbor peers, range and class queries can be executed efficiently, as target keys can be found in the same or in close-by peers.

3.2 The X-Pastry Ant Algorithm

This section illustrates how ant-inspired agents order the resource keys, computed with the technique illustrated in Section 2, over the X-Pastry overlay.

An agent can move a key towards a close-by peer through the leaf table, or towards a remote peer through the routing table, depending on the value of the *centroid* of the current peer. The centroid of a peer represents, with a single value, the keys stored locally. It is defined as the value between 0 and N_r (N_r is the number of admissible values of resource keys) that minimizes the average distance between itself and the keys stored in the local region, i.e., in this peer and in the two adjacent leaf peers. Each agent cyclically performs a set of operations to sort the keys over the X-Pastry tree, as described in the following:

1. while it is not carrying any key, an agent hops towards any of the peers pointed by the leaf table, chosen randomly;
2. at any new peer, the agent decides whether or not to *pick* a key k through a Bernoulli trial, whose success probability P_{pick} is:

$$P_{pick} = \frac{\alpha_p}{\alpha_p + f(k,c)} \quad \text{with } 0 \le \alpha_p \le 1 \tag{2}$$

where c is the centroid of the current peer, and $f(k,c)$ is the similarity between k and c. The similarity, which can assume values between 0 and 1, is defined as:

$$f(k,c) = 1 - \frac{d(k,c)}{N_r/2} \tag{3}$$

where $d(k,c)$ is the distance between the key and the centroid of the local peer. Since the distance is normalized with respect to the maximum distance $N_r/2$, it can also assume values between 0 and 1^1. Notice that the a key whose value is distant from the peer centroid – i.e., a key that is not in order – is more likely to be picked by the agent;

3. after picking a key k, the agent tries to bring it to a peer whose centroid is similar to the key value. The agent jumps to a new peer exploiting the routing or the leaf table of the current peer; the code of the destination peer is calculated with a proportion that converts the distance in the space of resource keys into a distance in the space of peer codes:

$$P_d = P_s + \frac{N_p}{N_r}(k-c) \qquad (4)$$

where P_s and P_d are the codes of the current and the destination peer, respectively, N_p is the number of admissible values of peer codes, and c is the centroid of the current peer. The agent jumps to the leaf node of the tree whose code is the closest to P_d, in other words to the peer that shares the largest number of digits with P_d.

4. at the new peer, the agent decides whether or not to *drop* the carried key through a Bernoulli trial whose probability P_{drop} is:

$$P_{drop} = \frac{f(k,c)}{\alpha_d + f(k,c)} \qquad \text{with } 0 \le \alpha_d \le 1 \qquad (5)$$

In this case the key is deposited with high probability if its value is similar to the centroid c of the local peer, i.e., if the key is in order with the keys stored locally.

Operations (2) and (4) are repeated until the agent picks or drops a key, respectively. Pick and drop operations, executed by a multitude of agents, assure that both centroids and resource keys are sorted over the leaf nodes of the X-Pastry tree. Figure 3 shows an example of how peer centroids are ordered by agents. In this example, the systems is composed of 16 connected peers (those with marked border), and the admissible key values of XML documents are between 0 and 63. When peer centroids are ordered, the average distance between the centroids of two consecutive peers is equal to 4, that is, the ratio between the number of admissible values N_r (64 in the example) and the number of connected peers (16).

4   Experimental Evaluation

To evaluate the performance of X-Pastry and its scalability features, a set of experiments were performed using a Java event-based simulator. Firstly, we assessed the capacity of X-Pastry of sorting the keys over the network. The sorting allows both punctual and range queries to be executed efficiently, as will be shown in Section 4.1.

A method to assess the X-Pastry sorting process is to consider the distances (in the space of resource keys) between the centroids of every two consecutive peers, and compute the average of these values. As mentioned in the previous section, when the keys

---

[1] The space of keys is circular, therefore the maximum distance corresponds to the length of a semi-circle in this space, $N_r/2$.

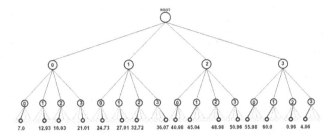

Fig. 3. Ordering of peer centroids in an X-Pastry system

are correctly sorted the average distance between two consecutive centroids should be comparable to the ratio between the number of possible values of keys, N_r, and the number of peers N_p. It is assumed that the average number of resources published by a peer is equal to 10 and the actual number of resources of each single peer is extracted with a Gamma probability function. The value of each key is generated with a uniform distribution, therefore at the beginning keys are not ordered; afterwards, the keys are sorted through the operations of X-Pastry agents. It is also assumed that every peer generates one agent, and that the lifetime of the agent is set to the average connection time of the generating peer. This ensures that the average number of agents is comparable to the number of connected peers, because dying agents are substituted with other agents generated by connecting peers. Furthermore, the average time between two consecutive movements of an agent is set to 3 seconds. As the value of this parameter modulates the velocity of the process, it will be used as a time unit in the experiments.

Figure 4 shows the trend of the average distance between consecutive centroids. Figure 4(a) reports the values of this index in a scenario with N_r set to 1024 and N_p ranging from 256 to 4096, whereas Figure 4(b) is related to a scenario in which N_p is equal to 1024 and N_r ranges from 4096 to 1048576. Starting at time 0 from a state with maximum disorder, and owing to agent operations, the average of the centroid distance decreases from very large values to the expected value N_r/N_p.

The keys are sorted in about 2000 time units, corresponding to 6000 seconds. It should be noticed that this transient phase is only necessary when starting from a disordered network: once keys and centroids are ordered, agent operations maintain the stability of this ordering by rapidly picking and relocating any outlier key.

4.1 Resource Discovery

The sorting of keys over the peers is exploited by the discovery procedure: when the user needs to find a key with a given target value, a search message is issued and directed towards the peer whose centroid is as similar to the key value as possible, through the repeated execution of the step 3 of the ant algorithm described in Section 3.2. The destination peer is determined as described there, then, if the centroid of that peer is closer to the target key than the local centroid, the search message is forwarded to the destination peer and the discovery procedure continues. Whenever this condition is not satisfied, the search terminates because with very high probability the current peer is the

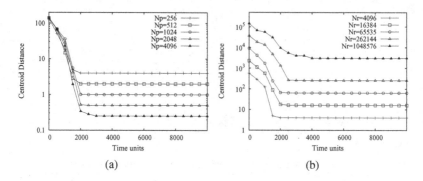

Fig. 4. Average distance between two consecutive centroids with: a) fixed number of admissible keys, N_r=1024, and variable number of peers N_p; b) fixed number of peers, N_p=1024, and variable value of N_r

one that stores the largest number of keys having the desired value. If a range query is issued to find a range of contiguous key values, the described procedure is used to find a key value comprised in the range. Then, by exploiting the fact that keys are sorted, the search proceeds along connections among adjacent peers, until both the lower and the upper bound of the target range have been found.

Figure 5 reports the distribution of the distance between every key and the local centroid, in a network with 1024 peers and $N_r = 1024$. The figure shows that the value of the large majority of the keys is very close to the peer centroid, which confirms that the search for a target key can be converted into the search for a peer centroid.

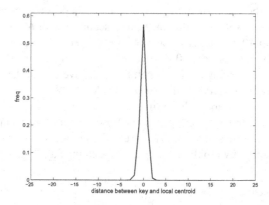

Fig. 5. Distribution of the distance between a key and the local centroid

To assess the discovery procedure we issued 200 queries. The queries were constructed by randomly selecting two XML paths in the DBLP tree: one specifies the target attribute and the other specifies the condition that must be satisfied by the results. For the latter attribute, the lower and upper bounds of the condition are selected randomly. For example, if the path /dblp/proceedings/title is selected as the target attribute, the path /dblp/proceedings/year is selected as the attribute for the condition, and the extreme values for the condition are the years 2006 and 2008, the issued query – expressed in *XPath* notation – is /dblp/proceedings[year>2006 and year<=2008]/title.

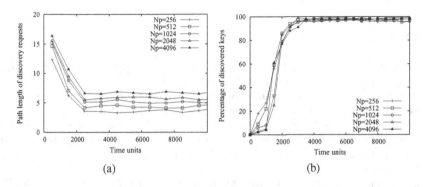

(a)   (b)

Fig. 6. Performance of range queries in a scenario with 1024 admissible key values and varying numbers of peers: a) average length of the path needed to find the keys of the target range; b) percentage of keys of the target range that are actually discovered by the search message

Figure 6(a) reports the average path length, defined as the number of jumps performed by a search message to reach a peer that contains some keys matching the query, i.e., comprised in the target range. After reaching this peer, the query explores the adjacent peers to find more keys included in the range. Figure 6(b) shows the percentage of target keys discovered by the query. The experiments were performed in a scenario with N_r equal to 1024 and N_p varying from 256 to 4096. The values are reported versus time, starting from a state in which keys are not ordered. At the steady state, the path length is nearly logarithmic with respect to the number of peers, while the percentage of discovered keys is close to 100%, which means that the discovery procedure successfully discovers nearly all the XML resources having the desired key values.

5 Conclusions

In this paper we presented *X-Pastry*, a P2P system that, maintaining the tree-based overlay of Pastry, exhibits enhanced functionalities thanks to the activity of ant-inspired mobile agents, which sort the resource keys over the network in a self-organizing fashion. The system inherits the main advantages of structured P2P systems – e.g., discovery operations are executed in logarithmic time – and offers further profitable properties inherited by biological systems, such as adaptivity and fast recovery from external perturbations. The paper illustrates how X-Pastry is used to index and search XML documents. An encoding scheme is adopted to associate each XML resource with a key composed of a sequence of digits, each corresponding to a hierarchy level of the XML tree.

This type of encoding allows resource keys be mapped to the hierarchical structure of the X-Pastry overlay, and helps the efficient execution of range queries, as confirmed by simulation results.

Acknowledgments. This research work has been partially funded by the MIUR project FRAME, PON01_02477.

References

1. Androutsellis-Theotokis, S., Spinellis, D.: A survey of peer-to-peer content distribution technologies. ACM Computing Surveys 36(4), 335–371 (2004)
2. Brocco, A., Malatras, A., Hirsbrunner, B.: Enabling efficient information discovery in a self-structured grid. Future Generation Computer Systems 26, 838–846 (2010)
3. Comito, C., Talia, D., Trunfio, P.: Selectivity-based xml query processing in structured peer-to-peer networks. In: Proc. of the 14th Int. Database Engineering and Applications Symposium (IDEAS 2010), Montreal, Canada, August 16-18. ACM Int. Conf. Proc. Series, pp. 236–244 (2010)
4. De Candia, G., Hastorun, D., Jampani, M., Kakulapati, G., Lakshman, A., Pilchin, A., Sivasubramanian, S., Vosshall, P., Vogels, W.: Dynamo: Amazon highly available key-value store. Technical Report, Amazon (October 2007),
   http://www.allthingsdistributed.com/files/
   amazon-dynamo-sosp2007.pdf
5. Forestiero, A., Leonardi, E., Mastroianni, C., Meo, M.: Self-chord: a bio-inspired P2P framework for self-organizing distributed systems. IEEE/ACM Transactions on Networking 18(5), 1651–1664 (2010)
6. Harder, T., Haustein, M.P., Mathis, C., Wagner, M.: Node labeling schemes for dynamic xml documents reconsidered. Data Knowl. Eng. 60(1), 126–149 (2007)
7. Jacobson, V., Smetters, D.K., Thornton, J.D., Plass, M., Briggs, N., Braynard, R.: Networking named content. Commun. ACM
8. Ko, S.Y., Gupta, I., Jo, Y.: A new class of nature-inspired algorithms for self-adaptive peer-to-peer computing. ACM Transactions on Autonomous and Adaptive Systems 3(3), 1–34 (2008)
9. O'Neil, P., O'Neil, E., Pal, S., Cseri, I., Schaller, G., Westbury, N.: Ordpaths: insert-friendly xml node labels. In: Proceedings of SIGMOD 2004, pp. 903–908 (2004)
10. Rao, P.R., Moon, B.: Locating xml documents in a peer-to-peer network using distributed hash tables. IEEE Trans. on Knowl. and Data Eng. 21(12), 1737–1752 (2009)
11. Rodrigues, R., Druschel, P.: Peer-to-peer systems. Commun. ACM 53, 72–82 (2010)
12. Rowstron, A., Druschel, P.: Pastry: Scalable, Decentralized Object Location, and Routing for Large-Scale Peer-to-Peer Systems. In: Guerraoui, R. (ed.) Middleware 2001. LNCS, vol. 2218, pp. 329–350. Springer, Heidelberg (2001)
13. Stoica, I., Morris, R., Karger, D., Frans Kaashoek, M., Balakrishnan, H.: Chord: A scalable peer-to-peer lookup service for internet applications. In: Proc. of the Conference on Applications, Technologies, Architectures, and Protocols for Computer Communications, SIGCOMM 2001 (2001)

Context-Aware Mobile Applications for Communication in Intelligent Environment

Andrey L. Ronzhin, Anton I. Saveliev, and Victor Yu Budkov

SPIIRAS, 39, 14th line, St. Petersburg, Russia
{ronzhin,budkov}@iias.spb.su

Abstract. Peculiarities of communication between users and devices in intelligent environment are described in the paper. The main problems are the heterogeneity of hardware and software at integration of mobile devices as well as the variability of natural signals at the development of multimodal user interfaces. The modern context-aware applications take into account preferences and abilities of user and adapt ones to conditions of physical environments, computing and network resources. The means of Android-based mobile devices, which could be used to acquire attributes of context, are analyzed.

Keywords: Pervasive computing technologies, multimodal interfaces, context, heterogeneous mobile devices.

1   Introduction

The first time the conception of ambient intelligence was mentioned by members of Philips Company for the description of a prototype of intelligent living room with embedded devices, which provides the proactive non invasive personified services for inhabitants [1]. Then three main groups of technologies, which contribute into the development of intelligent environments, were marked in the report of scientific and technical result forecast by European Commission: ubiquitous computing, ubiquitous networks and user-friendly interfaces [2]. In Russia the notion of ubiquitous computing was firstly mentioned in 2008 in [3]. Nevertheless the study of user devices networks integration based on multi-agent approach, as well as the development of multimodal user interfaces provided natural interaction with users were initiated much early [4, 5]. The necessity of three referred above technologies group development is justified by the fact that all the devices of intelligent environments are equipped by computing means, powerful network resources are required for connection and distribution of data between the devices, and finally the interfaces, which process natural modalities proper to inter-human communication, are required for interactions with users.

Usually ambient intelligence implies global intelligent environments, but it could not be developed now. Therefore the particular less scaled prototypes of intelligent environments, for instance smart room, smart home, smart city, are investigated [6, 7]. Also the existent prototypes could be classified based on applied area, for instance:

S. Andreev et al. (Eds.): NEW2AN/ruSMART 2012, LNCS 7469, pp. 307–315, 2012.

smart meeting room, intelligent train, smart medical room. The classification of intelligent environment prototypes based on two criteria: the assignment and the scale of serviced space, is presented in Figure 1.

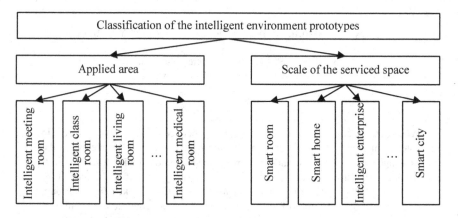

Fig. 1. The classification of intelligent environment prototypes

In spite of the popularity of smart technologies in most cases its intelligence is concerned only with automation of means of interaction between the devices, but the interaction between users and client's device is realized by standard input methods and graphical user interface. Although the one of the main task at the development of the intelligent environments is the creation of non-invasive natural and contactless interaction means with users.

The non-invasive proactive intelligent services are realized by analysis of the current context, prediction and satisfaction of the user's needs without explicit request. The natural ways of interaction between humans, for instance, speech, gestures, facial expression, are mainly used at the development of user interfaces. Most of the mentioned ways are contactless that allows user and machine to conduct the dialogue at a distance.

2   The Main Types of Communications in Intelligent Environment

The main types of communications in intelligent environment are shown in Figure 2: user-device, device-device, user-user. Now let us to consider the main tasks, which are connected with each of the mentioned types of communication. The connection between devices is based on standard protocols and transfer artificial signals with the preliminary known characteristics, and so, if the devices are located on the acceptable distance for connection and acceptable rate of signal/noise is provided, then connection and communication between the devices could be easily set. But, take into account wide variety of personal or stationary resources for telecommunication, life-support, security and entertainment, which are used everywhere, that become clear, that the integration few thousand devices with the variety soft and hardware architectures is difficult. At first the problem of devices heterogeneity was raised with development of wide spectrum of personal mobile devices, which have different operating systems, computing resources and ways for user interfaces of interaction. The development of cross-platform applications, which could work on most operation systems, not always helps, because of the differences in hardware and network resources lead to own constrains. Another problem, which arises at device communication, is the interoperability deal with joining usage and sharing of computing and informational resources. Decision of this problem depends on effectiveness of creation of the client-server or multi-agent architectures, access policy for shared resources and priority for client devices.

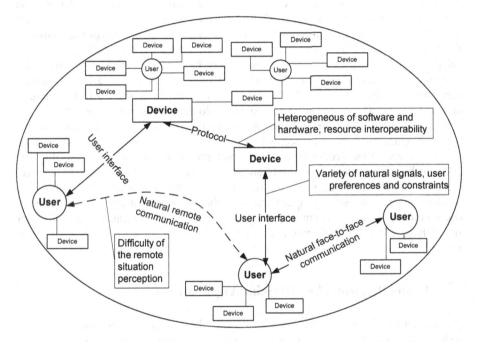

Fig. 2. Communications between users and devices in the intelligent environment

The human communication is distinguished by usage of all the natural means of human perception and data transfer via accessible communication channels. The quantity of participant, conditions and dialogue object influences on style of communication. Speech is the most often ways for people communication. The style and content of speech are selected by speaker according to the principle of communication reasonability, and to minimization of the time spent on the dialogue [8]. Nonverbal ways of communication are also important, and needed to show the emotions and psycho-and physiological state of a human. The pragmatic information, which includes the characteristics of speaker, listener and the current situation, influences on lingual structure, economy of verbal resources and provides the effective interaction in spite of background noises and unclear speaker's articulation [9].

The availability of general pragmatic knowledge and joint context accelerates interaction and improves the understanding of speech communication. Therefore the effectiveness of people dialogue is better at the face-to-face interaction. For the web meeting the capabilities of means of capture transfer and presentation limits the quantity of informational channel, volume and quality of the transferred information. Now the wide-screen, smart-board and multichannel means for audiovisual processing for transfer data with maximal resolution and quality are used at the communication between participants located inside the intelligent meeting rooms and some remote users.

Now let us to consider the specifics of interaction means between user and intelligent environment devices. The standard graphic user interface based on display, keyboard and mouse not always could be used in mobile devices. This depends on small size and constructive features of client devices, for example, computer keyboard in the domestic devices is unreasonably. Besides of input/output devices the quality of interaction scenario is important at the development user interfaces. Many scientific and technical advancements are not completely used owing to some functions are accessible only for limited number of specialists. Today the creation of natural interaction is important as well as device functionality. But the realization natural interaction leads to complication of the technological resources needed to process natural modalities. The variety of natural signals of users is the major reason, which is deters implementation of technologies of speech, gestures and mimic syntheses/recognition processing. Also it should be taken into account that the physical limitations and user preferences influence on interaction way choice. Therefore informational channels between user and device should be coordinated at the development of user interfaces. At last, the physical features of environment, where communication is occurred, influence on ways of data transfer, and the user should have the choice of interaction ways, which are most stable in the current situation.

3   Context-Aware Multimodal Applications

The capability of an application to analyze the current conditions during the exploitation including the current user state, physical environment, computing resources, and to dynamically adapt the scenario of interaction with a user is one of the major requirement at the development of pervasive applications. In order to provide this capability it is necessary that user and application acquire and process the same context information anytime and anywhere [10]. The fundamental problem of pervasive application development is connected with ambiguity of context data received from thousands various applications, which are implemented for serving an user. To solve the problem the powerful methods of context data fusion and distribution between sets of applications and users are required.

Moreover in most of the cases the acquired context information has ambiguity and heterogeneity. For instance, user position could be detected by video cameras, pressure sensors embedded in the floor, means of radio frequency identification, range finder and other ways. Each of the listed ways uses own set of software and hardware and possesses by different characteristics of speed and accuracy of object detection.

Also the ambiguity could be increased during the acquiring the higher level context, for example, the determination of the current activity of a user rather than his/her position [11].

Therefore the context ambiguity problem should be solved on two levels. First, it is necessary to minimize ambiguity of data received from numerous sensors, which are used by one application in the same physical and computing environment. Second, the dynamics of changing physical environment and the heterogeneity of available computing and network resources implemented for the intelligent services should be taken into account during the fusion and aggregation of high-level context information used by many applications and users.

Thus, the development of context-aware computer application depends on solving the problem of ambiguous data fusion received from heterogeneous sensors and other applications served a user. The creation of multimodal user interfaces to an application implemented in mobile devices requires to solve the following issues: (1) data received from various sensors, which have different degree of accuracy and deviation of parameters, contain information transferred by different natural modalities; (2) the data should be processed in real time in order to provide interactivity and user-friendless of the interface; (3) the situation is immediately changed with data received from different sensors.

In a smart meeting environment, context-aware systems analyze user' behavior based on multimodal sensor data and provide proactive services for meeting support, including active control of video cameras, microphone arrays, context dependent automatic archiving and web-transmission of meeting data at the interaction. However, there is a lack of universal approaches to the problem of context prediction, especially for acting on predicted context [12]. Two ways to act on predicted context can be marked: (1) rule-based engines containing action rules for every particular prediction result; (2) machine learning techniques based on neural networks, dynamical Bayesian networks, Markov models, etc.

Problems of context representation, sensor uncertainty and unreliability are considered in numerous recent papers; however, there is no any accepted opinion on types and number of context spaces and their attributes. For example, user's location, environment, identity and time were analyzed at the context definition by Ryan et al. [13]. In [14], Dey described context as the user's emotional state, focus of attention, location and orientation, date and time, objects and people in a user's environment. Three different categories of contexts were proposed in [15]: (1) real-time (location, orientation, temperature, noise level, phone profile, battery level, proximity, etc.); (2) historical (for instance, previous location, and previous humidity and device settings); (3) reasoned (movement, destination, weather, time, user activity, content format, relationship, etc.).

In [16], the context information used for service personalization and designing of multimedia applications for heterogeneous mobile devices were divided into the five categories: spatio-temporal (place, time), environment, personal, task, social. A personalization service based on user profile retrieves user context and context history information from context management services. It helps the user to get relevant content and services in the current situation.

Three types of contexts are proposed to use at fusion of multimodal information [17]: (1) a context of a subject domain, which contains some a priori knowledge, user's preferences, situation models, descriptions of objects and subjects, their possible activities and their locations relatively to each other; (2) a conversational context describing possible dialogues with the system and current e conditions; (3) a visual context including an analysis of gaze direction, gestures, actions of the user in the course of the observable situation.

The human beings, the physical and informational environments were considered by Dai et al. in the framework of two types of contexts [18]: interaction context representing interactive situations among people and environment context describing meeting room settings. They use propositions that the interaction context of a meeting has a hierarchical structure and expresses the context as a tree. User's standing-sitting states, changing user's location, face orientation, head gestures, hand actions, speaker turns and other events are analyzed for the context prediction. A Finite State Machine framework was introduced in order to classify these meaningful participants' actions. However, before the classification an event should be detected, so particular issues of signal capturing and feature extraction are appeared. The appropriate audio and video processing techniques are used for tracking participants in the smart environment [19]. The means of modern mobile devices provided data input and output, connection with external devices and registration of other parameters, which could be used for acquiring context information, will be considered in the next section.

4   Means of Context Attribute Registration in Mobile Devices

Peculiarity of context-aware applications for mobile devices is the high dynamics of environment change. In comparison to applications executed on personal computer or stationary systems the physical parameters of environment are changed during interaction session. User could walk or move in a transport vehicle that influences on changing the available connection means to other devices and physical parameters of environment.

In the given research the study of context analysis methods and the creation of context-aware application are conducted with the aim of development of the system, which supports distributed events in the intelligent environment [20]. For system development and testing the series of modern smartphones and tablet personal computers, including: Samsung Galaxy Tab 10.1 P7500 32Gb, Apple iPad 2 32Gb Wi-Fi + 3G, Apple iPhone 4S 16Gb, Apple iPhone 4 16Gb, Nokia N9 16Gb, Sony Ericsson Xperia S, Samsung GT-N7000 Galaxy Note. The applied devices have different operation systems and embedded hardware components. The list of main means, which are used at the interaction with users and devices, as well as could register various context attributes, is presented in Figure 3. In total five groups of the means oriented on data input and output, connection with external devices, determination of physical parameters of the environment and device location are selected. Each means has own parameters, which values could be used as context attributes. For instance, the context attributes of a display will be values of screen resolution and screen orientation type; for speakers the attributes will be volume, channel number, sampling frequency and other parameters.

For example, the smartphone Nokia N9 has the following technical means by the classification represented in Figure 3: (1) means of data output: screen (size: 3.9", resolution: 16:9 FWVGA (854 x 480 pixels)), speakers (supported formats: MPEG-4 AAC, eAAC, eAAC+, WMA Professional 9 and 10, Dolby Digital Plus, bit rate up to 320 kbps); (2) means of data input: touchscreen (on-screen alphanumeric keypad and keyboard), video cameras (HD quality video capture in 720p resolution at 30 fps, true 16:9 video recording), microphone (sample rate 48 kHz, format MPEG-4 AAC-LC, bit rate 160 kbps variable bit rate); (3) connection means: Bluetooth (2.1 +EDR), GPRS/EDGE (class B, multislot class 33), HSDPA (cat 10 maximum speed up to 14.4 Mbit/s), HSUPA (cat 6 up to 5.7 Mbit/s), WLAN (IEEE802.11 a/b/g/n), IrDA; (4) sensors: orientation sensor (accelerometer), proximity sensor, ambient light detector; (5) navigation means: GPS, A-GPS, Wi-Fi network positioning, compass (magnetometer).

At the development of universal context-aware application the extraction of context attribute values from the embedded means of mobile devices referred in Figure 3 is the non-trivial task owing to heterogeneity of software and hardware. Each operation system has own software developer kit. Moreover the used smartphones are equipped by not all the mentioned sensors and means of connections, data input and output. At this moment the facilities of Android API implemented for the development of Android-based user application were considered. In particular, the classes Display, Touch, MotionEvent, Camera, AudioRecord, Net, Bluetooth, Location and Os contain the functions required for work with means presented in Figure 3. In total around 50 functions used for reading more than 200 parameters, which should be used as context attributes, were accumulated. The formalization of context attributes and the assessment of the fact that the current context state belongs to a situation space and therefore the presence of the concrete situation will be calculated based on methods described in [12, 21].

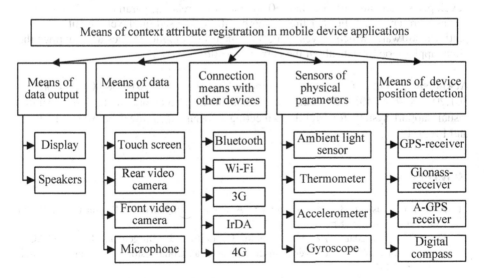

Fig. 3. The means of context attribute registration

Also it is should be marked that the reading of the parameter values of some embedded means is unavailable by standard functions of Android API and own special functions should be developed for them determination. For instance, in order to read or edit screen brightness it is necessary to handle the configuration file located in: /sys/class/leds/lcd-backlight. Software developer kits for other operation systems possess own abilities and constraints, so on the next step of the research the software modules for several main mobile operation systems, which could be implemented for access to embedded means delivered data for the context analysis, will be developed for providing the cross-platform compatibility and taking into account of hardware heterogeneity.

5   Conclusion

The realization of one of the main principle of intelligent environment – non-invasive user service is connected with decision of communication problems between users and devices. The heterogeneity of software and hardware of embedded and mobile devices as well as variety of natural signals are the major issues at the development of communication ways between objects in intelligent environment. The development of pervasive computer technologies, which take part at the intellectualization of ambient environment, is focused on concentration of user attention on the task performance and reduction of the time required for study of interaction means with applied services. Therefore the pervasive applications should be aware of the current situation, in which they are executed. At this moment the approaches to context formalization and analysis, as well as means for context attribute extraction in mobile devices are considered. Five groups of the means responsible for registration of context information are selected. The main classes of Android-based software developer kit contained more than 50 functions for reading parameters of embedded means were analyzed. Further the research will be focused on development of cross-platform software, which serves embedded means of mobile devices and supplies the client application by data about the current situation.

Acknowledgments. This work is supported by the grants #10-08-00199-a, #MD-501.2011.8, #14.740.11.0357 and performed in the framework of the joint "Audio-visual support system for Mobile E-meeting Participant" project between SPIIRAS and FRUCT.

References

1. Zelkha, E., Epstein, B.: From Devices to Ambient Intelligence. In: Digital Living Room Conference (June 1998)
2. Ducatel, K., Bogdanowicz, M., Scapolo, F., Leijten, J., Burgelman, J.-C.: ISTAG - Scenarios of Ambient Intelligence in 2010. European Commission Community Research (February 2001)
3. Gorodetsky, V., Karsaev, O., Samoilov, V., Serebryakov, V.V.: Agent platform for ubiquitous computing. Information Technologies and Computational Systems 4, 51–60 (2008) (in Rus.)

4. Gorodetsky, V., Karsaev, O., Samoilov, V.: Multi-agent technology for distributed data mining and classification. In: Proc. of IEEE/WIC International Conference on Intelligent Agent Technology, pp. 438–441 (2003)
5. Karpov, A.A., Ronzhin, A.L., Lee, I.V., Shalin, A.Y.: Speech technologies in multimodal interfaces. SPIIRAS Proceedings 1(2), 183–193 (2004)
6. Aldrich, F.: Smart Homes: Past, Present and Future. In: Harper, R. (ed.) Inside the Smart Home, pp. 17–39. Springer, London (2003)
7. Yusupov, R.M., Ronzhin, A.L.: From Smart Devices to Smart Space. Herald of the Russian Academy of Sciences, MAIK Nauka 80(1), 63–68 (2010)
8. Miller, G.R.: An introduction to speech communication (1972)
9. Karpov, A., Kipyatkova, I., Ronzhin, A.: Very Large Vocabulary ASR for Spoken Russian with Syntactic and Morphemic Analysis. In: Proc. INTERSPEECH 2011 International Conference, pp. 3161–3164. ISCA Association, Florence (2011)
10. Roy, N., Roy, A., Das, S.: Context-aware resource management in multi-inhabitant smart homes: A nash H-learning based approach. Pervasive and Mobile Computing Journal 2(4), 372–404 (2006)
11. TalebiFard, P., Leunga, V.: A Data Fusion Approach to Context-Aware Service Delivery in Heterogeneous Network Environments. Procedia Computer Science 5, 312–319 (2011)
12. Boytsov, A., Zaslavsky, A.: Extending Context Spaces Theory by Proactive Adaptation. In: Balandin, S., Dunaytsev, R., Koucheryavy, Y. (eds.) ruSMART/NEW2AN 2010. LNCS, vol. 6294, pp. 1–12. Springer, Heidelberg (2010)
13. Morse, D.R., Ryan, N.S., Pascoe, J.: Enhanced reality fieldwork using hand-held computers in the field. Life Sciences Educational Computing 9(1), 18–20 (1998)
14. Dey, A.K., Salber, D., Abowd, G.D.: A Conceptual Framework and a Toolkit for Supporting the Rapid Prototyping of Context-Aware Applications. The Human-Computer Interaction 16(2-4), 97–166 (2001)
15. Moltchanov, B., Mannweiler, C., Simoes, J.: Context-Awareness Enabling New Business Models in Smart Spaces. In: Balandin, S., Dunaytsev, R., Koucheryavy, Y. (eds.) ruSMART/NEW2AN 2010. LNCS, vol. 6294, pp. 13–25. Springer, Heidelberg (2010)
16. Goh, K.H., Tham, J.Y., Zhang, T., Laakko, T.: Context-Aware Scalable Multimedia Content Delivery Platform for Heterogeneous Mobile Devices. In: Proceedings of MMEDIA 2011, Budapest, Hungary, pp. 1–6 (2011)
17. Chai, J., Pan, S., Zhou, M.: MIND: A Context-based Multimodal Interpretation Framework. Kluwer Academic Publishers (2005)
18. Dai, P., Tao, L., Xu, G.: Audio-Visual Fused Online Context Analysis Toward Smart Meeting Room. In: Indulska, J., Ma, J., Yang, L.T., Ungerer, T., Cao, J. (eds.) UIC 2007. LNCS, vol. 4611, pp. 868–877. Springer, Heidelberg (2007)
19. Ronzhin, A.L., Prischepa, M., Karpov, A.: A Video Monitoring Model with a Distributed Camera System for the Smart Space. In: Balandin, S., Dunaytsev, R., Koucheryavy, Y. (eds.) ruSMART/NEW2AN 2010. LNCS, vol. 6294, pp. 102–110. Springer, Heidelberg (2010)
20. Budkov, V.Y., Ronzhin, A.L., Glazkov, S.V., Ronzhin, A.L.: Event-Driven Content Management System for Smart Meeting Room. In: Balandin, S., Koucheryavy, Y., Hu, H. (eds.) NEW2AN/ruSMART 2011. LNCS, vol. 6869, pp. 550–560. Springer, Heidelberg (2011)
21. Glazkov, S.V., Ronzhin, A.L.: Methods of context analysis of applications for mobile heterogeneous devices. Reports of TUSUR, №1 3(1), 236–240 (2012) (in Rus.)

Power Allocation in Cognitive Radio Networks by the Reinforcement Learning Scheme with the Help of Shapley Value of Games

Jerzy Martyna

Institute of Computer Science, Faculty of Mathematics and Computer Science
Jagiellonian University, ul. Prof. S. Lojasiewicza 6, 30-348 Cracow

Abstract. In this paper, we present an algorithm based on the reinforcement learning scheme with the help of Shapley value of game for power allocation in the cognitive radio networks. The goal is to optimize the achievable transmission rates for secondary users and simultaneously to maximize their usefulness in the coalition. A performance measure is formed as a weighted linear function of the probability of the idle channel amongst N cooperating secondary users. Then, the problem is formulated as a semi-Markov decision process with an average cost criterion and reinforcement learning algorithm is developed to an approximate optimal control policy. The proposed scheme is driven by an estimated dynamic model of cognitive radio network learning simultaneously with the use of the Shapley value of games, to form the best coalition. The simulations are provided to compare the effectiveness of the proposed method against other methods under a variety of traffic conditions with some well-known policies.

Keywords: cognitive radio networks, game theory, Shapley value method, reinforcement learning.

1 Introduction

Cognitive networking [1, 10, 16] deals with wireless communications systems that will have deeper awareness of their own operations and of the network environment, learn relationships among the parameters of the networks and tune to the current spectrum reuse. Cognitive radio (CR) technology lays the foundation for the development of smart flexible network of users that cooperatively increase the overall network performance. We recall that the terminology of cognitive radio, first formulated by Mitola [11], refers to smart radios which have the ability to sense the external environment, learn from the history and make intelligent decisions according to the current state of the environment.

Cognitive radios can be treated as autonomous agents learning their network parameters, whose interactions can be modeled with the help of game theory. The cognitive radios are players and their action influences the state of the system and their own performance, as well as the performance of all neighbouring players. We use Fig. 1 to illustrate the CR network, where multiple primary users (PUs) or wireless providers compete for a shared pool of secondary users. All the secondary users (SUs) are equipped with cognitive radio technologies. They are usually static or mobile. The primary users

S. Andreev et al. (Eds.): NEW2AN/ruSMART 2012, LNCS 7469, pp. 316–327, 2012.

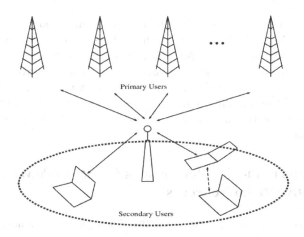

Fig. 1. Spectrum structure of the cognitive radio

are the infrastructure of wireless operators and then are responsible to throw unused frequency to the secondary users for a monetary payoff.

Reinforcement learning (RL) [15] is an unsupervised machine-learning technique to achieve context awareness and intelligence. The term *unsupervised* means that the machine-learning methods are used to learn knowledge about the environment ithout an external teacher or critic. The reinforcement learning for context awareness and intelligence in wireless networks in a variety of problems such as: routing, resource management, etc., was widely discussed by K.-L. A. Yau *et al.* in paper [18].

In this paper, we examine the problem of power allocation (PA) in the CR networks on a licensed and unlicensed band with the main emphasis on the concept of hierarchy of the existing between radios. This problem arises in the following situations: (a) when the primary and secondary systems share the spectrum, (b) when users have access to the medium in an asynchronous manner, (c) when operators deploy their networks at different times, (d) when some nodes have more power than others, such as the base station. One of the most popular model of the hierarchical spectrum of sharing is the Stackelberg equilibrium (SE) [14], [5]. This approach was motivated by the fact that the noncooperative Nash equilibrium (NE) is generally inefficient and nonoptimal. The Stackelberg equilibrium provides better outcomes as compared to the noncooperative approach. However, the mathematical framework of the Stackelberg equilibrium is not suitable for practical use. Therefore, we propose a new scheme of the power allocation (PA) problem in the CR networks which is based on reinforcement learning scheme with the help of Shapley's value of games.

The main goal of this paper is to show that power allocation for secondary users in the cognitive radio networks can be realized by means of model-based reinforcement learning with the help of Shapley value method. The proposed method lowers the hand-off failure probability while maintaining the traffic parameters and adepts the power allocation in a wireless environment with several limitations such as: interference, deterioration of received signal, etc. The proposed method does not presume a priori knowledge of the system model or traffic parameters. It can be used to direct the search for an optimal control policy of power allocation in the cognitive radio networks. A comparison with other methods, based on game theory policies, are also provided.

The remaining portion of this paper is as follows. The next section presents an application of the Shapley value of game to power allocation in cognitive radio networks. Section 3 describes the reinforcement learning scheme with the help of the Shapley value method for the power allocation in cognitive radio networks. Simulation and numerical results are provided in Section 4. Section 5 presents the conclusions and future work in this field.

2   Application of the Shapley Value of Games to Power Allocation in Cognitive Radio Networks

In this section we investigate applications of the Shapley value of games to power allocation in the CR network.

Our approach is based on two important multiuser channel models, namely: the multiple access channel (MAC) [3] and the interference channel (IFC).

In the first one, we assume use of the uplink channel in a single-cell multi-carrier cellular system in which each multiple access channel (MAC) consists of K transmitters aiming to communicate with a single receiver using a common channel. There exist N independent or parallel MACs. None of the transmitters in different MACs interferes with each other. The channel gain from transmitter i to the receiver over channel n is denoted by h_i^n. Let the channel realizations during the transmission of M consecutive symbols be constant. All the channel realizations i, $i \in \{1, \ldots, k\}$ and $n \in \{1, \ldots, N\}$ are drawn from a Gaussian distribution with a zero mean and a unit variance. Thus, the power allocated by transmitter i to channel n is denoted by p_i^n. We can formulate the following condition for the transmitter i, namely

$$\sum_{n=1}^{N} p_i^n \leq P_i^{max} \quad \forall i = \{1, \ldots, K\} \tag{1}$$

We assume that the noise at the receiver is described by w_i^n. It corresponds to the additive white Gaussian noise (AWGN) process with a zero mean and variance σ^2. The received signal can be written as

$$y^n = \sum_{i=1}^{K} h_i^n x_i^n + w_i^n \quad \forall n = \{1, \ldots, N\} \tag{2}$$

where x_i^n and h_i^n are the transmitted symbols and the channel realization of the transmitted symbols, respectively.

Assuming a single-user decoding (SUD) on channel n for transmitter i, the received signal to interference plus the noise ratio (SINR) is expressed as

$$SINR_i^n = \frac{p_i^n \mid h_i^n \mid^2}{\sum_{j \neq i}^{K} p_j^n \mid h_j^n \mid^2 + \sigma^2} \quad \forall i = \{1, \ldots, K\}, \ \forall n = \{1, \ldots, N\} \tag{3}$$

The interference channel model (IFC) described by T. Han *et al.* [6] and by Etkin *et al.* [4] consists of a set of K point-to-point links sufficient to produce mutual interference due to their co-existence on the same channel. Assuming that $N \geq 1$ channels are available, in the IFC model N independent or parallel channels exist, where transmitters in different IFCs do not interfere with each other. In essence, the IFC model corresponds to the transmission in pairs between nodes over a set of sub-carriers. We assume that the channel realization from transmitter i to receiver j on channel n is denoted by b_i^n, where $n = \{1, \ldots, N\}$, and $(i, j) \in \{1, \ldots, K\}^2$. Thus, the received signal at receiver i is given by

$$r_i^n = \sum_{j=1}^{K} h_{ji}^n x_i^n + w_i^n \tag{4}$$

where $\mathbf{w}_i = \{w_i^1, \ldots, w_i^n\}$ is the noise at receiver i over channel n.

Assuming a single-user decoding (SUD) on channel n for transmitter i, the received SINR can be expressed as

$$SINR_i^n = \frac{p_i^n \mid h_{ii}^n \mid^2}{\sum_{j \neq i}^{K} p_j^n \mid h_{ji}^n \mid^2 + \sigma^2} \quad \forall i = \{1, \ldots, K\}, \ \forall n = \{1, \ldots, N\} \tag{5}$$

The difference between Eqs. (3) and (5) is that each transmitter knows the channel realization h_i^n for all $\forall i = \{1, \ldots, K\}$ in the MAC model and h_{ij}^n for all $(i, j) \in \{1, \ldots, K\}^2$ in the IFC model.

In the PA game, the set of players includes transmitters, base stations, and mobile stations. In general, a game is presented in a normal form as follows:

Definition 1 (Normal form). *A game $\mathcal{G}$ in a normal form is given by $\{\mathcal{K}, \mathcal{S}, \{u_k\}_{k \in \mathcal{K}}\}$ and is composed of three elements:*

- *a set of players:$\mathcal{K} = \{1, \ldots, K\}$,*
- *a set of strategy profiles: $\mathcal{S} = S_1 \times \ldots \times S_k$, where S_k is the strategy set of player k,*
- *a set of utility functions: the k-th player's utility function is $u_k : S_k \to \mathcal{R}_+$ and is denoted by $u_k(s_k, s_{-k})$ where $s_k \in S_k$ and $s_{-k} = (s_1, \ldots, s_{k-1}, s_{k+1}, \ldots, s_k) \in S_1 \times \ldots \times S_{k-1} \times S_{k+1} \times \ldots S_K$.*

Considering that players are willing to cooperate to achieve a fair allocation of resources, we impose a condition that the utility function must account for both the interference perceived by the current players, and the interference that particular player is causing to neighboring players sharing the same channel.

The utility function is defined as follows:

Definition 2 (Utility function). *The utility function is given by*

$$u_k(s_k, s_{-k}) = - \sum_{j \neq k, j=1}^{N} p_j(s_j) G_{kj} f(s_j, s_k)$$

$$- \sum_{j \neq k, j=1}^{N} p_k G_{jk} f(s_k, s_j) \quad \forall k = 1, \ldots N \tag{6}$$

where G_{kj} is the link gain between transmitter j and receiver k, $f(s_k, s_j)$ is an interference function given by

$$f(s_k, s_j) = \begin{cases} 1, \text{ if } s_j = s_k, \text{ transmitter } j \text{ and } k \text{ choose} \\ \quad \text{the same strategy (same channel)} \\ 0, \text{ otherwise.} \end{cases} \tag{7}$$

The above utility function accounts for both the interference measured at the current user's receiver and the interference created by the user to others.

Definition 3 (Nash equilibrium (NE)). *A pure-strategy Nash equilibrium of a non-cooperative game* $\mathcal{G} = \{\mathcal{K}, \mathcal{S}, \{u_k\}_{\forall k \in \mathcal{K}}\}$ *is a strategy profile* $s_k^* \in S_k$ *if it satisfies*

$$\forall k \in \mathcal{K} \text{ and } \forall s_k \in S_K, \ u_k(s_k^*, s_{-k}^*) \geq u_k(s_k, s_{-k}^*) \tag{8}$$

In other words, a strategy is a pure-strategy Nash equilibrium if no player has an incentive to unilaterally deviate to another strategy, given that other players' strategies remain fixed.

The NE is generally a lower bound in the case of noncooperation between players.

Power allocation games in which the concept of hierarchy is taken taken into considerations are rerred to as Stackelberg games [14]. The Stackelberg game is a two-stage game where one player (leader) moves in the first stage and all the other (follower) players react simultaneously in the second stage. In such game, the Stackelberg equilibrium (SE) is defined as follows [9]:

Definition 4 (Stackelberg equilibrium (SE)). *A vector of power allocation* $\overline{p}^{SE} = (p_i^{SE}, p_{-i}^{SE})$ *is called a Stackelberg equilibrium (SE) if* $p_{-i}^{SE} \in \mathcal{U}^*(p_i)$ *where* $\mathcal{U}^*(p_i)$ *is the set of NE for the group of followers when the leader plays strategy* p_i *and the power* p_i^{SE} *maximizes the utility function of user i (game leader).*

As a solution of the coalition for the game a method introduced by L. S. Shapley is used [12], [13]. The main idea of Shapley's method lies in the defintion of player usefulness for the coalition and rewards assignment which is proportional to their potential contributions.

We introduce the Shapley value:

Definition 5 (The Shapley value of a game in a normal form). *Let v be a game given by* $\{\mathcal{K}, S, \{u_k\}_{\forall k}\}$. *The Shapley value of v,* $\Phi(v) = (\phi_1(v), \ldots, \phi_k(v)) \in \mathcal{K}$ *is defined by*

$$\phi_i(v) = (K - 1)! \frac{(a - 1)!}{K!} \sigma(C, i) \tag{9}$$

for each player i, $1 \le i \le K$ attached to coalition C couting $(a - 1)$ players as the a-th player, and $\sigma(C, i)$ is the usefulness of player i for the coalition C and is given by

$$\sigma(C, i) = \mu(C) - \mu(C \backslash \{i\}) \tag{10}$$

where $\mu(C \backslash \{i\})$ is the reward for coalition C without the i-th player and $\mu(C)$ is the reward for coalition C.

Each coalition can be assigned the usefulness function of all players for the formed coalition. We assume that the usefulness function of a dummy player is equal to 0.

A formal definition of the influence of the outgoing player for the coalition is given as follows:

Definition 6 (Influence of the player's going out into the coalition). *For the sake of the best possible coalition we have observed that the reward of coalition C changes its value from 1 to 0 after player i is going out of the coalition.*

According to the Shapley method [12], [13] we can univocally assign to each game the imputation which is reasonable partitioning of winnings. The following definition gives the terms of the player's participation in the coalition.

Definition 7 (Participation in the coalition). *The participation of the player in the coalition is determined by the values of the Shapley vector.*

For a simple example, consider the three-player game with coalition.

Example 1 *Three players are building a coaltion to utilize an existing system. A participation of each player in the possible coalition is different. Player usefulness for all possible coalitions are given in Table 1.*

Table 1. Player usefulnesses for all possible coalitions

{X}	{Y}	{Z}	{X, Y}	{X, Z}	{Y, Z}	{X, Y, Z}
80	0	80	120	240	80	400

The participations of players in formed coalitions are as follows:

$\sigma(\{X\}) = 80, \ \sigma(\{Y\}) = 0, \ \sigma(\{Z\}) = 80$
$\sigma(\{X, Y\}, X) = 120 - 0 = 120$
$\sigma(\{X, Y\}, Y) = 120 - 80 = 40$
$\sigma(\{X, X\}, X) = 240 - 80 = 160$
$\sigma(\{X, Z\}, Z) = 240 - 80 = 160$
$\sigma(\{Y, Z\}, Y) = 80 - 80 = 0$
$\sigma(\{Y, Z\}, Z) = 80 - 0 = 80$
$\sigma(\{X, Y, Z\}, X) = 400 - 80 = 320$
$\sigma(\{X, Y, Z\}, Y) = 400 - 240 = 160$
$\sigma(\{X, Y, Z\}, Z) = 400 - 120 = 280$

Table 2. Participations of all players in different coalitions

	X	Y	Z
{X, Y, Z}	80	40	280
{X, Z, Y}	80	160	160
{Y, Z, X}	320	0	80
{Y, X, X}	120	0	280
{Z, X, Y}	160	160	80
{Z, Y, X}	320	0	80
Together:	1080	360	960

The participation of all players in different coalitions are given in Table 2. Thus, the average participation of each player is given by $\frac{1}{6}(1080, 360, 960)$. It is just the Shapley value of this game.

3 The Spectrum Sharing for Cooperative Secondary Systems with the Use of the Shapley Value

In this section, we investigate the dynamic spectrum sharing in the CR network in which primary systems lease the spectrum to secondary system in exchange for cooperation in the PA game.

We assume that a primary transmitter wishes to send information to its primary receiver either directly with a rate R_{dir} or by means of the cooperation from a subset $S \subseteq S_{tot}$ of $\mid S \mid = k \leq \mid S_{tot} \mid = k$ secondary nodes/transmitters. The primary system can divide its data into two parts $(1 - \alpha)/bit$ durations, and αL bit durations with $0 \leq \alpha \leq 1$. The first part is dedicated to a direction transmission from primary transmitter to the primary receiver whereas the second αL bit duration is again divided into two parts. One part, consisting of $\beta \alpha L$, with $0 \leq \beta \leq 1$, is dedicated to sending information from the primary transmitter to the primary receiver using the secondary nodes by means of the distributed space time coding [8]. The remaining $\alpha(1 - \beta)L$ bits are devoted to the secondary network for the sake of its own data transmission. The problem of power allocation in the secondary system of the CR network can be solved by maximization of its utility function while deciding about the portion of time-slots α, β and $S \subseteq S_{tot}$ subset of secondary transmitters.

Given the set S and cooperation parameters α, β, the PA optimization problem is given by

$$\max_{\alpha, \beta, S} \left(\sum_{i \in S, i=1}^{C} u_i(s_i, s_{-i}) \right) \tag{11}$$

subject to $S \subseteq S_{tot}, 0 \leq \alpha, \beta \leq 1$.

The secondary system maximizes its utility function of the formed coalition C by means of maximization of the achievable trasmission rate along with taking into consideration the cost of the transmitted energy E_c. The optimization problem for the secondary system can be expressed as

$$\max_{s_i} \left(\sum_{i=1, i \in S}^{C} u_i(s_i, s_{-i}) \right) = \max_{s_i} \{ \alpha(1 - \beta)$$

$$\log_2 \left(1 + \frac{\mid h_{S,ii} \mid^2 s_i}{\sigma^2 + \sum_{j=1, j \neq i}^{k} \mid h_{S,ji} \mid^2 s_j} \right) - E_c s_i \} \tag{12}$$

subject to $0 \leq s_i \leq S_{i,max}$.

Solving Eq. (11) and assuming that the strategy profile $s_i = p_i$, we can obtain the value of power for transmitter i, namely

$$p_i = \max \left(0, \frac{1 - \beta}{E_c} - \frac{\sigma^2}{\mid h_{S,ii} \mid^2} - \sum_{j=1, j \neq i} \frac{\mid h_{S,ji} \mid^2}{\mid h_{S,ii} \mid^2} p_j \right) \tag{13}$$

In our approach the interaction between the primary and secondary users is modelled as cooperative game. The coalition maximizes its own utility function. Using the Shapley value we obtain the participation of each player in the game with the maximal utility function of the coalition.

$$\hat{p}_i = arg \max_i (u_i(p_1, \ldots, p_C)) = f \cdot \Phi_i(v) \cdot p_i \quad 0 \leq p_i \leq P_i^{max} \tag{14}$$

where $\Phi_i(v)$ is the Shapley value for transmitter i and f is a normalizing parameter.

4   The Reinforcement Learning Algorithm for the Coalition Game

In order to address more formally the approach in the formulation of the learning algorithm for the coalition game, we present a definition of the optimal joint action.

Definition 8 (Optimal joint action). *All the joint actions belonging to the set are treated as the optimal Nash equilibrium in the coalition which gives to all the players from the coalition a reward equal to 1.*

The proposed model for the coalition game based on reinforcement learning system, as illustrated in Fig. 2, has three main components: controlled system, policy learner and model estimator. The controlled system could be a real-world or simulated system. The policy learner has two elements: value functions and control policy finder. The value functions are real-valued mapping of the state-action pairs. The control policy is mapping of the system state to the corresponding action. This control policy is based on the Q-learning algorithm, first introduced by Watkins [17] for discrete-time Markov decision process (MDP).

The Q-learning algorithm (see Fig. 3) that first forms a coalition among all the players using Shapley's value, and subsequently, on basis reinforcement learning, can

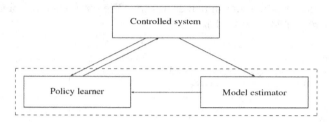

Fig. 2. Model-based system for the coalition game based on reinforcement learning

learn the game structure and optimal coordination. The complexity of presented learning algorithm is linear in the number of states, polynomial in the number of

1. Initialization step. $t = 0$, *best_coalition* := $false$,
  For all $s \in S$ and $a \in A$ **do**
  $n_t(s, a) = 1, T_t(s, a, a') = \frac{1}{|s|}$,
  $R_t(s, a) = 0, \epsilon_t = constant, A^{\epsilon_t}(s) = A$.

2. Formation of the best coalition.
  repeat
  while not *best_coalition* **do**
  begin
    compute the Shapley vector for coalition C;
    if *Shapley's value is greatest of possible* **then** *best_coalition* := *true*;
    for all $i \in C$ **do**
    compute p_i;
    find $\overline{p}_i = f \cdot \phi_i \cdot p_i$;
    otherwise *randomly selected agent left out the coalition;*
    end;
  until *coalition is empty;*

3. Learning of game structure for selected coalition.
  a) Compute reward r_t under the joint action a. **do**
    $n_t \leftarrow n_t(s, a) + 1$;
    $R_t(s, a) \leftarrow R_t(s, a) + \frac{1}{n_t(s,a)}(r_t(s, a) - R_t(s, a))$;
    $T_t(s, a, s') \leftarrow T_t(s, a, s') + \frac{1}{n_t(s,a)}(1 - T_t(s, a, s'))$;
    for all $s'' \in S$ **and** $s'' \neq s'$ **do**
    $T_t(s, a, s') \leftarrow (1 - \frac{1}{n_t(s,a)})T_t(s, a, s'')$;
    $r_t(s, a) \leftarrow r_t(s, a) + \frac{r_t(s,a)}{n_t(s,a)}$;
  b) $Q_{t+1}(s, a) \leftarrow R_t(s, a) + \gamma \sum_{s' \in S} T_t(s, a, s') \max_{a' \in A} Q_t(s', a')$;
  c) $t \leftarrow t + 1$; $N_t \leftarrow \min_{s \in S, a \in A} n_t(s, a)$;
  d) **for** all (s, a) update $Q_t(s, a)$;
    $A^{\epsilon_t}(s) \leftarrow \{a \mid Q_t(s, a) + \epsilon_t \geq \max_{a' \in A} Q_t(s, a')\}$;

Fig. 3. Reinforcement learning algorithm for the coalition game

actions, but exponential in the number of players. The given algorithm converges to optimal Nash equilibrium in any Markov team game. Convergence proof establishes the convergence of a general Q-learning process updated by a pseudo-contraction operator. Proof of convergence for Nash Q-learning general-sum stochastic games was presented by J. Hu [7].

In the reinforcement learning algorithm S means a finite state space, $n_t(s, a)$ is the number of times and action a has been played in state s by time t. The *constant* has any positive value, $R(S, a)$ is the expected reward for taking action a in a state s, $s \in S$, and a is transition function. $T_t(s, a, s')$ is the probability of ending in a state s, if the action a is undertaken in state s at the moment t.

5   Simulation Results

We have the following general settings for the simulation. We place the primary trans-miiter BS at coordinates and five secondary users which are uniformly located in the area 100 m × 100 m. The maximum power for a secondary user is $P_{max} = 100$ mW. The rest of the parameters are set as follows: antenna gain 5, threshold power $3 \cdot 10^{-7}$ W. The AWGN at all receivers has the same power $W = 5 \times 10^{-7}$ mW and the interference power threshold at all receivers is -50 dB.

By means of using the Nash and the Stackelberg equilibria [2] we find the optimal value of p_i^1 while keeping $p_1^2, \ldots, p_1^N$ fixed and then we find the optimal p_1^2 keeping the other p_1^n ($n \neq 2$) fixed and so on. Such a process guaranteeds to convergence because each iteration increases the objective function.

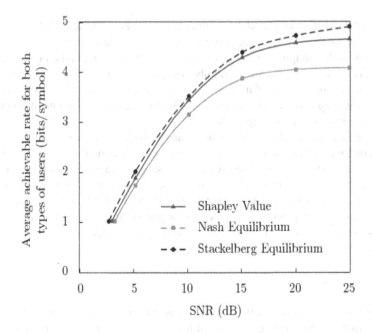

Fig. 4. Average achievable rate for both types of users versus the signal-to-noise (SNR) ratio for the Shapley value, Stackelberg and Nash equilibria approaches

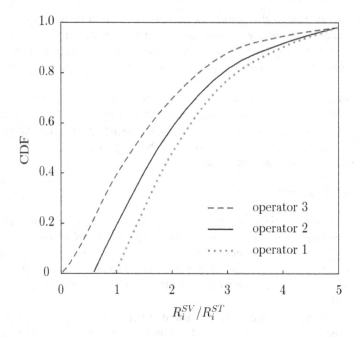

Fig. 5. Cumulative distribution function (CDF) versus the ratio of the rates determined by the Shapley and Stackelberg value (noncooperative game) approaches

In our implementation of the RL algorithm, the parameter γ is set to 0.5. The simulation results show that the learning approaches were capable of self-adjusting. Then, we evaluate how the Shapley vector of the formed coalition affects the power allocation in the CR networks with a varied number of secondary users. Figure 4 depicts the achievable rate for both types of users versus the signal-to-noise ratio for the NE, SE and Shapley value approaches. As can be seen, the average achievable rate of Shapley's vector is comparable to the SE approach.

Figure 5 shows the cumulative distribution function (CDF) of the ratio of the cooperative and noncooperative approach. In this scenario, we assume that $K = 3$ operators with one primary operator and three secondary users sharing the same spectrum. It is composed of $N = 5$ carriers. The entrance of each player to the coalition took place according to their index values. In order to achieve the CDF in a noncooperative approach we propose a repeated game in which the players will be added to the coalition in a strictly defined succession. This succession must guarantee the highest values of Shapley vector.

6 Conclusions

In this paper, we have addressed the power allocation game in the cognitive radio network. A reinforcement learning scheme with the help of Shapley value method has been developed. Simulation results show that for the considered traffic conditions, the proposed method can autonomously adjust the power allocation to the current traffic conditions. Also, the reinforcement learning algorithm which learns the game structure and with the help of Shapley value method finds the coalition of secondary users with the best usefulness for that coalition, allocates the power to each secondary user without undesirable power degradation.

The proposed method can be used to design a cognitive radio network for each traffic parameters and all possible situations. Here, we have focused on the formation of a coalition among secondary users that will have better or worse usefulness for the coalition; secondary users with conflicting interests can be captured in a similar manner. Future works could consider those situations, and also provide a more extensive simulation of reinforcement learning schemes.

References

1. Akyildiz, I.F., Lee, W.-Y., Vuran, M.C., Mohanty, S.: Next Generation/Dynamic Spectrum Access/Cognitive Radio Wireless Networks: a Survey. Computer Networks (2006)
2. Boyd, S., Vandenberghe, L.: Convex Optimization. Cambridge University Press (2004)
3. Cover, T.M., El Gamai, A., Salehi, M.: Multiple Access Channels with Arbitrarily Correlated Sources. IEEE Trans. on Information Theory 26(6), 648–657 (1980)
4. Etkin, R., Tse, D., Wang, H.: Gaussian Interference Channel Capacity to Within One Bit. IEEE Trans. Information Theory 54(12), 5534–5562 (2008)
5. Fudenberg, D., Tirole, J.: Game Theory. The MIT Press, Cambridge (1991)
6. Han, T., Kobayashi, K.: A New Achievable Rate Region for the Interference Channel. IEEE Trans. on Information Theory 27(1), 49–60 (1981)
7. Hu, J., Wellman, M.P.: Nash Q-Learning for General-Sum Stochastic Games. Journal of Machine Learning Research 4, 1039–1069 (2003)
8. Laneman, J.N., Wornell, G.W.: Distributed Space-Time Coded Protocols for Exploiting Cooperative Diversity in Wireless Networks. IEEE Trans. on Information Theory 49(10), 2415–2425 (2003)
9. Lasaulce, S., Debbah, M., Altman, E.: Methodologies for Analyzing Equilibria in Wireless Games. IEEE Signal Processing Magazine, Special Issue on Game Theory for Signal Processing 26(5), 41–52 (2009)
10. Martyna, J.: A Mathematical Framework for the Multidimensional QoS in Cognitive Radio Networks. In: Balandin, S., Koucheryavy, Y., Hu, H. (eds.) NEW2AN/ruSMART 2011. LNCS, vol. 6869, pp. 440–449. Springer, Heidelberg (2011)
11. Mitola III, J.: The Software Radio Architecture. IEEE Comm. Mag. 33(5), 26–38 (1995)
12. Shapley, L.S.: Rand Corporation Research Memorandum. Notes on the N-Person Game: Some Variants of the von Neumann-Morgenstern Definition of Solution, RM-812 (1952)
13. Shapley, L.S.: A Value for N-Person Games. In: Kuhn, H.W., Tucker, A.W. (eds.) Contribution to the Theory of Games. Princeton University Press, Princeton (1953)
14. Stackelberg, V.H.: Marketform und Gleichgewicht. Oxford University Press, Oxford (1934)
15. Sutton, R.S., Barto, A.G.: Reinforcement Learning: An Introduction. The MIT Press, Cambridge (1998)
16. Thomas, R.W., Friend, D.H., DaSilva, L.A., MacKenzie, A.B.: Cognitive Networks: Adaptation and Learning to Achieve End-to-End Performance Objectives. IEEE Communication Magazine 44(12), 51–57 (2006)
17. Watkins, C.J.C.H.: Learning from Delayed Rewards. Ph.D. Thesis, University of Cambridge, Cambridge, UK (1989)
18. Yau, K.-L.A., Komisarczuk, P., Teal, P.D.: Reinforcement Learning for Context Awareness and Intelligence in Wireless Networks: Review, New Features and Open Issues. Journal of Network and Computer Applications 35, 253–267 (2012)

The Internet Erlang Formula

Villy B. Iversen

Department of Photonic Engineering, Technical University of Denmark
Building 343, 2800 Kongens Lyngby, Denmark
vbiv@fotonik.dtu.dk

Abstract. This paper presents a robust and efficient algorithm for evaluating multi-service multi-rate queueing systems, including finite buffer systems and loss systems. Vint Cerf listed in 2007 seven research problems concerning the Internet. This paper responds to the second problem: an Internet Erlang Formula. The algorithm derived is based on reversible models and thus insensitive to service time distributions. For buffer-less systems we get the classical multi-rate teletraffic models. As the simplest special case we get the classical recursion formula for Erlang-B. The performance of the algorithm is $O\{N \cdot k\}$ where N number of services, and k is the total number of servers and buffers in basic bandwidth units. The memory requirement is $O\{N \cdot d\}$ where d is the maximum requested bandwidth in basic bandwidth units.

Keywords: multi-rate queueing model, generalized processor sharing, algorithm.

1 Introduction

The demands to a traffic model of a system is that it is simple, requires input parameters which are easy to obtain by measurements, and describes the performance of the system with few relevant parameters. The success of the classical teletraffic theory is based on fulfilling these properties. Erlang's B-formula describes the system by one parameter, number of channels n. The traffic is described by average offered traffic A which is the ratio between the arrival rate and the service rate. The performance is characterized by the call congestion B. The model is insensitive to the service time distribution and thus very robust. The model assumes Poisson arrival process which is fulfilled when the subscribers behave independently. By measuring we can estimate the average traffic, but in practice it is impossible to estimate parameters as variance. Erlang's model is generalized to the BPP (Binomial – Poisson – Pascal) model, which in addition to Poisson traffic also includes finite source Engset traffic (Binomial) and bursty traffic (Pascal). The classical models are generalized to include multi-rate traffic and restricted accessibility [4].

Models in classical queueing theory are very sensitive to the service time distribution. The first paper by Erlang in 1909 [2] deals with constant service times. Later Erlang realized that exponential service times are much simpler

S. Andreev et al. (Eds.): NEW2AN/ruSMART 2012, LNCS 7469, pp. 328–337, 2012.

to deal with. Due to this sensitivity, the literature on queueing theory is very extensive and applications are rather restricted. The only insensitive models in queueing theory are processor sharing models for single server systems, and generalized processor sharing for multiple server systems. They assume that all customers request the same bandwidth, one basic bandwidth unit (BBU).

The basis for insensitivity in both loss systems and delay systems is reversible state transition diagrams, where there is local balance. For Kolmogorov cycles the flow clockwise should equal the flow counterclockwise. When we have insensitivity, all performance measures based on state probabilities depend only upon the mean service time. The original idea behind the model presented below is to construct a reversible state transition diagram which combines the above two classes of insensitive models: multi-rate loss model for states below the capacity limit and processor sharing models above the capacity limit. By requiring a reversible diagram and usage of all resources when needed (no idle capacity when calls are delayed) we get a unique reversible diagram.

We find the detailed state probabilities by introducing reduction factors. The details of this are given in [5] [7]. In this paper we assume full accessibility and a single node. We are then able to find performance measures for each service without first deriving the detailed state probabilities. We get a simple, accurate and robust recursion scheme which yields performance measures for each individual service. For delay systems with unlimited buffer we get mean waiting time and mean queue length. When the buffer size is limited, we in addition also get the blocking probability. For loss systems the algorithm is easily generalized to include BPP-traffic and batch Poisson arrivals.

A similar model has been dealt with in [1] in a more mathematical way, and the algorithm derived is much more complex and less general than the one presented in this paper.

2   Traffic Model

We choose a basic bandwidth unit, [BBU], and split the available bandwidth into n [BBU]. Also the buffer size $k - n$ is given in [BBU]. The [BBU] is a generic term for channel, slot, server, etc. The smaller the basic bandwidth unit (i.e. the finer the granularity), the more accurate we may model the traffic of different services, but the bigger the state space becomes. We consider N different services offered to a system with n BBUs. Let $(j = 1, N)$. The arrival processes are Poisson processes with rates λ_j, and a call type j requests d_j [BBU]. If the bandwidth request is fulfilled, the mean service time is chosen to $s_j = 1/(d_j \mu_j)$. We choose this value for convenience, because when we later choose $\mu_j = \mu$, then the service rate in a state with x busy [BBU] becomes $x \mu$, independent of the call mix. In fact, the service requested is a traffic volume d_j/μ_j, for example a data file of a certain number of bytes. The offered traffic in connections becomes $\lambda_j/(d_j \mu_j)$ and thus the offered traffic in [BBU] becomes $A_j = \lambda_j/\mu_j$ [BBU].

In this paper, accepted customers in some way share the available capacity, and therefore they are being served all the time. But they may obtain less capacity than requested, which results in an increased sojourn time. The *sojourn time* is not split up into separate waiting time and service time as in classical queueing models. In this paper we use the definitions usually used in queueing networks. For a customer of type j we define:

$\overline{W}_j$ = mean waiting time, defined as total sojourn time, including service time.
$\overline{L}_j$ = mean queue length, defined as total number of customers in the system.

As an example we may think of the time required to transfer a file in the Internet. If the available bandwidth is at least equal to the bandwidth requested, then the mean service time s_j for a customer of type j is defined as the mean transfer (sojourn) time. If the available bandwidth is less than the bandwidth requested, then the mean transfer (sojourn) time $\overline{W}_j$ will be bigger than s_j, and the increase

$$W_j = \overline{W}_j - s_j, \tag{1}$$

is defined as the *mean virtual waiting time*. We shall introduce the virtual waiting time as the increase in service time due to limited capacity. In a similarly way we define the *mean virtual queue length* as

$$L_j = \overline{L}_j - Y_j, \tag{2}$$

where Y_j is the carried traffic of type j.

The systems considered in this paper are reversible, but they do not have product form. In service-integrated systems the bandwidth requested depends on the type of service. The approach is new and very simple. It allows for very general results, including all classical Markovian loss and delay models, and it is applicable to digital broadband systems, for example Internet traffic.

3   Reversible Multi–rate Multi-server Systems

We now consider a queueing system with n servers [BBU] and $k-n$ buffers [BBU]. This system is offered N multi-rate traffic streams. We assume traffic stream j has arrival rate λ_j, service rate $d_j\,\mu_j$, and requests d_j simultaneous channels for service. The state of the system can be specified either in [connections] or in [BBU]. Let $i_j \geq 0$ denote the number of connections of type j, then the number of basic bandwidth units occupied by service j is $x_j = d_j \cdot i_j$. The state of the system is defined by one of the following options:

$$\underline{i} = (i_1, i_2, \ldots, i_j, \ldots, i_N),$$
$$\underline{x} = (x_1, x_2, \ldots, x_j, \ldots, x_N),$$
$$= (i_1\, d_1, i_2\, d_2, \ldots, i_j\, d_j, \ldots, i_N\, d_N),$$
$$x = \sum_{j=1}^{N} x_j = \sum_{j=1}^{N} i_j\, d_j \leq k,$$

where x is the global state of the system in [BBU]. In the following we use number of basic bandwidth units (channels) because global number of connections yields no information about the number of busy basic bandwidth units.

3.1 Generalized Algorithm for State Probabilities

The relative state probabilities are denoted by $q(x)$, $x \geq 0$, and we usually choose $q(0) = 1$. The true state probabilities are denoted by $p(x)$ and they are obtained from $q(x)$ by normalization. The global relative state probability $q(x)$ is made up of contributions $q_j(x)$ from each traffic type:

$$q(x) = \begin{cases} 0 & x < 0, \\ 1 & x = 0, \\ \displaystyle\sum_{j=1}^{N} q_j(x) & x = 1, 2, \ldots, k, \end{cases} \tag{3}$$

where we find $q_j(x)$ by the following recursion formula which is derived below in (Sec. 3.2):

$$q_j(x) = \frac{1}{\min\{x, n\}} \left\{ \frac{d_j}{x} \lambda_j q(x - d_j) + \sum_{i=1}^{N} \left(\frac{x - d_i}{x} \lambda_i q_j(x - d_i) \right) \right\}, \quad x \geq d_j. \tag{4}$$

We let $q_j(x) = 0$ for $x < d_j$, see later (10). Remember that the requested service rate of a type j call is $d_j \mu_j$, but when buffers are used the number of servers obtained is less than d_j. The above recursion can be used to find global performance measures. To find performance measures for the individual services, we split the contribution to state x from service j up into two contributions:

$$q_j(x) = q_{j,y}(x) + q_{j,l}(x), \quad j = 1, 2, \ldots N. \tag{5}$$

1. The contribution $q_{j,y}(x)$ with index y relates to proportion of servers used by stream j in global state x, and from this we find the average number of servers allocated to a type j customer. Given that we are in global state x, then the mean number of channels serving type j calls (carried traffic) is:

$$n_{j,y}(x) = \frac{q_{j,y}(x)}{q_j(x)} \cdot x,$$

Of course, we have:

$$\sum_{j=1}^{N} n_{j,y}(x) = \begin{cases} x, & x \leq n, \\ n, & x \geq n. \end{cases}$$

For global state x we have the following total contributions to relative state probabilities:

$$q_y(x) = \sum_{j=1}^{N} q_{j,y}(x),$$

2. The contribution $q_{j,l}(x)$ with index l relates to the proportion of queueing positions used by stream j. Given that we are in global state x, then the mean queue length [BBU] of type j calls is:

$$n_{j,l}(x) = \frac{q_{j,l}(x)}{q_j(x)} \cdot x \,.$$

For the global state x we have the following total contributions to relative state probabilities:

$$q_l(x) = \sum_{j=1}^{N} q_{j,l}(x) \,,$$

Of course we have:

$$\sum_{j=1}^{N} \{n_{j,y}(x) + n_{j,l}(x)\} = x \,, \qquad 0 \le x \le k \,.$$

For $x \le n$ the above algorithm is identical with the generalized algorithm for a multi-rate loss system with Poisson arrival processes (19). For states $x \le n$ when all customers obtain the requested rate and there is no delay we have:

$$q_j(x) = q_{j,y}(x) \quad \text{and} \quad q_{j,l}(x) = 0 \,, \qquad x \le n \,. \tag{6}$$

For all traffic streams we have:

$$q(x) = q_y(x) + q_l(x) \,, \qquad 0 \le x \le k \,. \tag{7}$$

We find $q_{j,y}(x)$ in the following way. If we look at the local balance for type-j customers between global state $[x - d_j]$ and global state $[x]$, then under the assumption of statistical equilibrium we have:

$$\lambda_j \cdot q(x - d_j) = (x \cdot q_{j,y}(x)) \cdot \mu_j \,.$$

The requested bandwidth for type j is $d_j \, \mu_j$ per connection. The total bandwidth obtained by service j is $x \cdot q_{j,y}(x)$. Thus we find:

$$q_{j,y}(x) = \frac{\lambda_j}{x \, \mu_j} \cdot q(x - d_j) \,, \tag{8}$$

$$q_{j,l}(x) = q_j(x) - q_{j,y}(x) \,. \tag{9}$$

Initialization, Iteration, and Normalization

The initial values of the relative state probabilities are

$$q_j(x) = q_{j,y}(x) = g_{j,l}(x) = 0 \,, \qquad x < d_j \,. \tag{10}$$

Thus (4) is not valid for $x = 0$, as the traffic stream's contribution to state zero is undefined.

Knowing the state probability $q(0)$ for a system with $x = 0$ (3), we calculate the relative state probabilities $q_j(1)$ of the next state $x = 1$, using (4). From (3) we get the global state $q(1)$. At the same time we may calculate $q_{j,y}(x)$ and $q_{j,l}(x)$ and the contributions to all performance measures, given below. We notice that state probability $q_j(x)$ depends both upon the global term $q(x - d_j)$, the local term $q_j(x - d_j)$, and the traffic parameters of all traffic streams.

To obtain accurate numerical values for any size of parameters, we should normalize the state probabilities during each step of iteration so that they add to one. This is what is done in the accurate recursion formula for Erlang-B. Let us assume we have obtained the normalized state probabilities for all states from 0 to $(x - 1)$ for a system with $(x - 1)$ [BBU]. Thus all these states add to one. Then we increase the number of [BBU] by one to x and calculate the performance for this system. We first find the relative values $q_j(x)$ and $q(x) = \sum_j q_j(x)$ for state x. To obtain the new true state probabilities we have to divide all state probabilities from zero to x by the normalization constant $\{1 + q(x)\}$. Also performance measures already accumulated from individual state probabilities should be re-normalized during this step. In this way we only need to store probabilities for states $(x - d_j, x - d_j + 1, \ldots, x - 1)$ to calculate the following normalized state probabilities. As special cases we have for example the well-known accurate recursions for the Erlang-B and Engset formulæ. The recursion is very stable and eliminates errors because we always divide with a factor $1 + q(x)$ which is bigger than one. If we use the recursion in opposite direction, the numerical errors will accumulate as we then divide with a factor less than one $1 - q(x)$, and blow errors up.

Wrapping up the algorithm in this way, the computer memory requirement becomes:

$$2 (N + 1) \cdot \max_j \{d_j + 1\}.$$

The number of operations becomes of the order of size:

$$2 (N + 1) \cdot (k + 1).$$

If the number of states is unlimited, then we continue until, say, $p(x) < 10^{-c}$. Then experience shows that the number of corrects digits is at least $c - 2$. Due to reversibility, we may truncate the state space and limit the maximum state to $k \geq n$. Then we have a finite system with n [BBU] and $k - n$ buffers operating in reversible processor sharing mode (mixed delay and loss system). For $k = n$ we get the multi-rate loss system.

3.2 Derivation of Recursion Formula

The above recursion formula for $q_j(x)$ (4) is based on detailed local balance for each service (reversibility). For type-j calls, the total flow out os state x due to departing calls of type-j, i.e. the total service rate in state x, must be equal to the total flow into state x due to arriving type-j calls. We rewrite (4) to:

$$q_j(x) \cdot \min \{x, n\} = \left\{ \frac{d_j}{x} \cdot \lambda_j \cdot q(x - d_j) + \sum_{i=1}^{N} \left(\frac{x - d_i}{x} \cdot \lambda_i \cdot q_j(x - d_i) \right) \right\}$$

1. Left-hand side: Flow out of state x due to departure of type-j calls

   This is the flow down from state x due to termination of type-j calls. We choose all service rates $\mu_i = 1$ so that the requested service rate of a type-j call becomes d_j. The total service rate in state x is $\min\{x, n\}$, and $q_j(x)$ denotes the relative proportion of channels serving type-j calls.

2. Right-hand side: Flow into state x due to arrival of customers

   – Right hand side term one:
     This is new contribution to $q_j(x)$ because a new call type-j arrives. Arrival rate of type-j calls is λ_j. A new call type-j adds d_j channels to the new state x. Therefore the ratio of type-j channels in state x is increased by d_j/x. Channels occupied by type-j which already exist when a new call arrive is taken into account by the following term.
   – Right hand side term two:
     Already existing occupied channels of type-j in state $x - d_i$ are given by the proportion $q_j(x - d_i)$. If a type-i call arrives in state $x - d_i$, then the contribution of these occupied channels is shifted up to the new state x, so that the relative ratio of channels serving type-j calls in state x is increased by $(x - d_i)/x$. The additional contribution from a new call when it happens to be a type-j call is taken into account by term one above.

3.3   Performance Measures

For Poisson arrival processes we do not need to calculate the detailed state probabilities to obtain detailed performance measures for each service (4). We assume that the true state probabilities $p(x)$ have been obtained from the relative state probabilities $q(x)$ by normalization. In the following we specify the performance measures in [BBU]. To get the measures in [connections] for service j from measures based on [BBU], we have to divide with d_j. The number of servers is n and the number of buffers is $(k - n)$ so that the total number of states is $(k + 1)$. Contribution to the carried traffic of type j from state x is

$$y_j(x) = x \cdot p_{j,y}(x).$$

Total carried traffic of type j becomes:

$$Y_j = \sum_{i=0}^{k} y_j(i) = \sum_{i=0}^{k} x \cdot p_{j,y}(x). \tag{11}$$

If we have a finite buffer (k limited), then $Y_j < A_j$ and the traffic congestion C_j for type-j customers will be positive:

$$C_j = \frac{A_j - Y_j}{A_j}. \tag{12}$$

The time congestion can easily be obtained from the global state probabilities. Due to the *PASTA*-property, time, call, and traffic congestion are equal.

Contribution to the virtual queue length of traffic type j from state x is:

$$l_j(x) = x \cdot p_{j,l}(x).$$

The total virtual queue length expressed in [BBU] of traffic type j becomes:

$$L_j = \sum_{i=0}^{k} l_j(i) = \sum_{i=0}^{k} x \cdot p_{j,l}(x). \tag{13}$$

The mean number of [BBU] (servers and buffers) occupied by type j is (2):

$$\overline{L}_j = Y_j + L_j \tag{14}$$

From the mean virtual queue length, we obtain by using Little's theorem the virtual mean waiting time, expressed in mean service times. The average number of type-j calls in the system is $\overline{L}_j/d_j$, and the average number of customers in the virtual queue is L_j/d_j. We get the following mean sojourn time, respectively mean virtual waiting time:

$$\overline{W}_j = \frac{\overline{L}_j}{d_j \lambda_j (1 - C_j)}, \tag{15}$$

$$W_j = \frac{L_j}{d_j \lambda_j (1 - C_j)}. \tag{16}$$

For the total system we get by proper weighting:

$$\overline{W} = \sum_{j=1}^{N} \frac{\lambda_j (1 - C_j)}{\lambda (1 - C)} \cdot \overline{W}_j = \frac{1}{\lambda (1 - C)} \cdot \sum_{j=1}^{N} \frac{\overline{L}_j}{d_j}, \tag{17}$$

$$W = \sum_{j=1}^{N} \frac{\lambda_j (1 - C_j)}{\lambda (1 - C)} \cdot W_j = \frac{1}{\lambda (1 - C)} \cdot \sum_{j=1}^{N} \frac{L_j}{d_j}. \tag{18}$$

4   Generalizations

For multi-rate traffic during underload we get the multi-rate loss models where each user get the requested bandwidth, and during heavy overload we get the (Generalized) Processor Sharing model, where each user get the same bandwidth. In between, broadband users are allocated more bandwidth than narrowband users.

For a bufferless system ($k = 0$) we get a multi-rate loss system with Poisson arrivals and (4) becomes (remember service rate of a type-j call is $d_j \mu_j$, and that we choose $\mu_j = 1$):

$$q_j(x) = \frac{\lambda_j}{x} \cdot q(x - d_j) \tag{19}$$

This can be generalized to BPP multi-rate traffic [5]:

$$q_j(x) = \frac{\lambda_j}{x} \frac{1}{Z_j} \cdot q(x - d_j) - \frac{x - d_j}{x} \cdot \frac{1 - Z_j}{Z_j} \cdot q_j(x - d_j) \tag{20}$$

where Z_j is the peakedness (variance/mean–ratio of state probabilities) of service j in an infinite group. For Poisson arrival processes we have $Z = 1$, for Engset $Z < 1$, and for Pascal traffic $Z > 1$. We cannot use this generalization of loss systems to BPP traffic for systems with buffers, as erroneously done in [6]. We get a solution which is not exact, but a very good approximation. The problem is that the flow rate into state x due to an arrival cannot be described by a mean value of the state probability (for Engset average number of idle sources), as the service rate in a buffered system for a given service is a non-linear function of the states. For Poisson arrival processes we don't have this problem.

For buffered systems we deal with finite number of sources by considering closed queueing networks. As the above systems are reversible, they can be combined into networks, and thus we get queueing networks with multi-rate traffic. If we define a closed two-node network, where the capacity of one node always is sufficient, then the seconde node is a queueing system with finite number of sources. We first assume Poisson arrival processes to each node, then during convolution fix the number of customers in each chain and re-normalize the state probabilities [7]. For multi-rate queueing networks we have to calculate the individual state probabilities, and due to state space explosion we get into trouble.

We may put upper limits upon the number of channels occupied by all calls of a given type. We may also allocate each service a guaranteed minimum bandwidth. The state transition diagram is still reversible. But also here we have to calculate the detailed state probabilities.

5   Conclusions

Above we have presented a general loss-delay model with multi-rate traffic. The most important property for applications is the insensitivity to the service time (packet size) distribution. The model is evaluated by a simple, robust, and accurate algorithm which can deal with thousands of services and basic bandwidth channels. The algorithm is implemented in [3].

References

1. Bonald, T., Virtamo, J.: A recursive formula for multirate systems with elastic traffic. IEEE Communication Letters 9(8), 753–755 (2005)
2. Brockmeyer, E., Halstrøm, H.L., Jensen, A.: The Life and Works of A.K. Erlang. Transactions of the Danish Academy of Technical Sciences (2), 278 p. (1948)
3. Iliakis, E., Kardaras, G.: Resource allocation in next generation Internet. Master thesis, COM, Technical University of Denmark, 99 p. (2007)
4. Iversen, V.B.: Modelling Restricted Accessibility for Wireless Multi-service Systems. In: Cesana, M., Fratta, L. (eds.) Euro-NGI 2005. LNCS, vol. 3883, pp. 93–102. Springer, Heidelberg (2006)

5. Iversen, V.B.: Teletraffic Engineering and Network Planning. Photonic Engineering, Technical University of Denmark, 380 p. (2011), http://oldwww.com.dtu.dk/education/34340/material/telenook.pdf
6. Iversen, V.B.: Reversible Fair Scheduling: The Teletraffic Theory Revisited. In: Mason, L.G., Drwiega, T., Yan, J. (eds.) ITC 2007. LNCS, vol. 4516, pp. 1135–1148. Springer, Heidelberg (2007)
7. Iversen, V.B., Ko, K.-T.: Algorithm for queueing networks with multi-rate traffic. In: Conference Proceedings of First European Teletraffic Seminar, Poznań, 5 p. (February 2011) ISBN: 978-83-925375-5-7; Also published in the Journal: Advances in Electronics and Telecommunications 2(3), 3–7 (2011)

Ubiquitous Sensor Networks Traffic Models for Medical and Tracking Applications

Anastasia Vybornova and Andrey Koucheryavy

Saint-Petersburg State University of Telecommunications,
Pr. Bolshevikov, 22, Staint-Petersburg, Russia
a.vybornova@gmail.com, akouch@mail.ru

Abstract. The Internet of Things (IoT) is a new concept for telecommunication development. The Ubiquitous sensor Network (USN) is one of the general IoT components. The traffic models for such network should be studied well. The USN traffic models study results for medical and tracking applications are considered in this paper. The paper results show that the traffic flows for medical and tracking USN applications are self-similar with the middle level of self-similarity in both cases. The R/S and Higuchi methods were used for Hurst parameter estimation. The Hurst parameter dependence on the length of interval between packets for medical USN and the Hurst parameter dependence on the packet rate and OFF perion length for tracking USN are considered.

Keywords: IoT, USN traffic, medical application, tracking application, Hurst parameter.

1 Introduction

The 7 trillion wireless telecommunication units for 7 billion people could be form the network at 2017-2020 in according with forecast [1]. Moreover, the estimation of the total number of things that could be connected with the future networks is 50 trillion (up to 5000 things for every human) [2]. Therefore we could suppose that future networks will be self-organizing and the most important traffic sources will be things.

The Internet of Things (IoT) is the new ITU-T concept for the network development. The ITU-T determines IoT as a future global infrastructure: "In a broad perspective, the IoT can be perceived as a vision with technological and societal implications. From the perspective of technical standardization IoT can be viewed as a global infrastructure for the information society, enabling advanced services by interconnecting (physical and virtual) things based on, existing and evolving, interoperable information and communication technologies" [3,4]. Things are considered as things from the nature (physical and virtual things) or from the world of information [3] (non-physical [4] things).

The IoT will be based on the Ubiquitous Sensor Network (USN) [5]. So the USN traffic models should be studied well. Unfortunately, now there is not much research activity in this area. The Poisson arrival process was assumed as traffic

S. Andreev et al. (Eds.): NEW2AN/ruSMART 2012, LNCS 7469, pp. 338–346, 2012.

model for each individual sensor node in [6]. The ON/OFF method [7] for USN traffic models was analyzed in [8]. Authors proved that ON period distribution and OFF period distribution could be described by generalized Pareto distribution. The pseudo long range dependent (LRD) traffic model was proposed in [9] for mobile sensor networks. The USN traffic models for telemetry applications was studied in [10], where the Hurst parameter estimations were determined for telemetry USN with the stationary and mobility sensor nodes. But the main problem is that the different USN applications produce the types of traffic with the quite various characteristics.

There are many USN classifications. The appendix 1 to [11] determines the USN classification by such objects characteristics as size, mobility, power, connectivity, ability, people involvement, physical/logical objects, object with tag, IP/non-IP. The USN classification by application groups based on traffic characteristics was proposed in [10]. The following USN traffic classes are considered: voice, signaling, telemetry, pictures, and reconfiguration, local positioning.

We assumed that telemetry applications could be detailed by application functionality in some specific cases. The USN traffic models for medical and tracking applications will be searching in paper.

There are two important parameters of USN that affect traffic: type of sensor nodes (fixed, mobile, or both in the same network) and application scenario (periodically or event-given information reporting).

The first application considered in this paper is a medical USN intended for communication between hospital data center and medical devices for continuous health monitoring (e.g. pulsometers, sphygmomanometers, and modern glucometers). The network architecture in this case is hybrid: mobile health sensors periodically send data to the traffic sink through the network composed of stationary nodes. The application scenario is periodic; therefore source traffic could be modeled by Constant Bit Rate (CBR) [8].

The second application is target tracking. In this case network consists of only stationary sensors spatially distributed on some territory, but the application scenario is event-given: the sensing node sends information to the sink only when it has a "target" in its sensing area. According to [8], the source traffic in this case may be captured by ON-OFF model with the length of ON and OFF periods following the Pareto distribution.

The self-similarity of USN traffic in the applications described above was taken as a hypothesis. To test this assumption we used two methods for self-similarity level estimation: the R/S and the Higuchi methods.

2   Methods of Analysis

The rescaled range analysis (R/S analysis) defined by Hurst [12] is one of the most widely used methods for estimation of self-similarity level. The Hurst parameter according to this method can be calculated as follows.

For a given finite set of time series, $X = X_1, X_2, \ldots, X_n$:

1. Calculate mean value m:

$$m = \frac{1}{n} \sum_{i=1}^{n} X_i \tag{1}$$

2. Calculate mean adjusted series Y:

$$Y_t = X_t - m \text{ for } t = 1, 2, \ldots, n \tag{2}$$

3. Calculate cumulative deviate series Z:

$$Z_t = \sum_{i=1}^{t} Y_i \text{ for } t = 1, 2, \ldots, n \tag{3}$$

4. Calculate range series R:

$$R_t = \max(Z_1, Z_2, \ldots, Z_t) - \min(Z_1, Z_2, \ldots, Z_t) \text{ for } t = 1, 2, \ldots, n \tag{4}$$

5. Calculate standard deviation series S:

$$S_t = \sqrt{\frac{1}{t} \sum_{i=1}^{t} (X_i - u)^2} \text{ for } t = 1, 2, \ldots, n \tag{5}$$

Here u is the mean value from X_1 to X_t.

6. Calculate rescaled range series (R/S):

$$(R/S)_t = \frac{R_t}{S_t} \text{ for } t = 1, 2, \ldots, n \tag{6}$$

7. $(R/S)_t$ is averaged over the regions $[X_1, X_t], [X_{t+1}, X_{2t}]$ till $[X_{(m-1)t+1}, X_{mt}]$ where $m = \text{floor}(n/t)$. The values of t should be divisible by n.

8. For finding the Hurst parameter value, plot (R/S) versus t in log-log axes. The slope of the regression line approximates the Hurst parameter value.

Estimation of the Hurst parameter with regression analysis is shown on the Fig. 1.

The Higuchi method [13] considers the fractal dimension D of a time series such that:

$$H = 2 - D \tag{7}$$

The Hurst parameter according to Higuchi method can be calculated as follows. For given set of time series observations $X(1), X(2), X(3), \ldots, X(N)$:

1. Construct a new time series $X_k{}^m$:

$$X_k{}^m; \ X(m), X(m+k), X(m+2k), \ldots, X\left(m + \left[\frac{N-m}{k}\right] \cdot k\right) \tag{8}$$

where $m = 1, 2, \ldots, k$, $\left[\frac{N-m}{k}\right]$ denotes the Gauss' notation or the "floor" function of value $\frac{N-m}{k}$, and both k and m are integers. Parameters m and

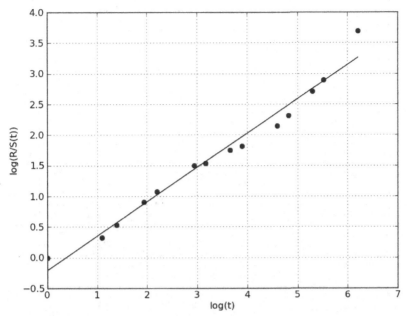

Fig. 1. Approximation of the R/S set with a line. Experimental data set is distributed by the Gaussian noise with the mean zero and a standard deviation of 1 (theoretical Hurst parameter value $H = 0.5$).

k represent the initial time and the interval time respectively. For the time equal to k, we get k sets of the new time series. For example, in the case of $k = 3$ and $N = 100$ three time series obtained by the above process are described as follows:

$X_1{}^3$; $X(1), X(4), X(7), \ldots, X(97), X(100)$
$X_2{}^3$; $X(2), X(5), X(8), \ldots, X(98)$
$X_3{}^3$; $X(3), X(6), X(9), \ldots, X(99)$

2. Calculate length of the curve for each $X_k{}^m$:

$$L_m(k) = \frac{1}{k} \cdot \left\{ \left(\sum_{i=1}^{\left[\frac{N-m}{k}\right]} |X(m+ik) - X(m+(i-1) \cdot k)| \right) \frac{N-1}{\left[\frac{N-m}{k}\right] \cdot k} \right\}$$
(9)

where $\frac{N-1}{\left[\frac{N-m}{k}\right] \cdot k}$ represents the normalization factor for the curve length of subject time series.

3. Find average curve length over k sets $L(k)$:

$$L(k) = \frac{1}{k} \sum_{m=1}^{k} L_m(k)$$
(10)

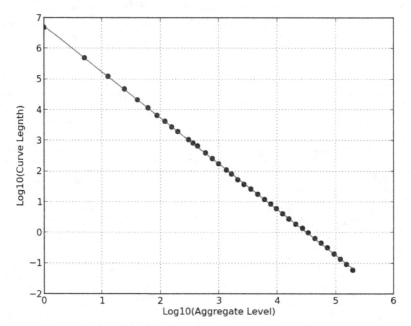

Fig. 2. Approximation of the data set with a line. Experimental data set is distributed by the Gaussian noise with the mean zero and a standard deviation of 1 (theoretical Hurst parameter value $H = 0.5$).

4. Plot values of function $L(k)$ versus k on a log-log scale; the data should fall on a straight line with a slope $-D$.
5. Calculate H according to (7).

Estimation of the parameter D value is shown on the Fig. 2.

Both methods were verified on the data sets of length 1000 distributed by the Gaussian noise with the mean zero and a standard deviation of 1 (theoretical Hurst parameter value $H = 0.5$). Twenty measurements were done and the results presented in Table 1.

The estimation of the Hurst parameter value of Gaussian noise less than 0.5 obtained with the Higuchi method agrees with the same result in [13]. The rather large standard deviation obtained with the both methods could be explained by the small size of test data sets.

Table 1. Estimation of Hurst parameter of Gaussian noise with R/S and Higuchi methods

	R/S method	Higuchi method
Mean	0.5877	0.4953
Standard deviation	0.0598	0.1448

3   Simulation Parameters

The USN applications described above were simulated with the Network Simulator-2. The data processing was made by the Python language libraries Numpy and Scipy. The data visual process was support by the library Matplotlib.

For both models the following protocol stack was used: IEEE 802.15.4 (physical and data lint OSI layers), IP (network layer) and UDP (transport layer). Ad-hoc On-demand Distance Vector protocol (AODV) was used for routing.

For the medical USN, 25 mobile sensor nodes and 25 stationary nodes were used. The stationary nodes were evenly distributed on the plane size 30*30 m as shown in Fig. 3. The mobile nodes were distributed randomly and moved with a fixed speed randomly changing direction every 100 seconds. The data sink were placed in the center of the test area. The mobile nodes send to the sink CBR traffic with different rates.

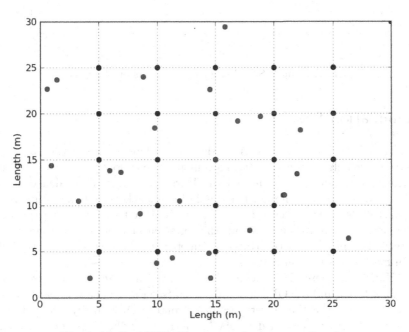

Fig. 3. Medical USN. Placement of sensors on the plane.

Target tracking USN consisted of 25 stationary nodes evenly distributed on the plane size 30*30 m as shown in Fig. 4. Source traffic for each node was described by the ON-OFF model with the length of the ON and OFF periods fitted to the Pareto distribution with the shape parameter equal to 1.5. Mean value of ON period length was 10 seconds; mean value of OFF period length and data rate during ON period was variable.

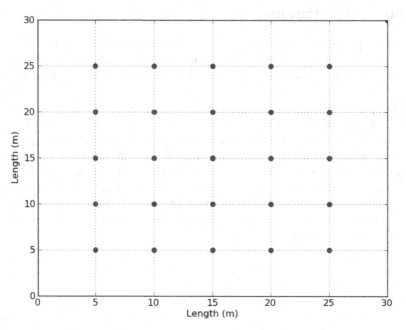

Fig. 4. Tracking USN. Placement of sensors on the plane.

4   Simulation Results

Simulation showed that there was no significant dependence of the Hurst parameter on such characteristics as a movement type or location of the sink. For the medical USN simulation showed the Hurst parameter dependence on the interval between packet sending. For the 20 bytes packets and the intervals between packets less than 5 seconds increase of the Hurst parameter were shown. For the intervals grater then 5 seconds the Hurst parameter doesn't change. The R/S method estimation of the Hurst parameter is 0.66 in stationary state; the Higuchi method estimation is 0.76.

The Hurst parameter for tracking USN showed a weak dependence on the source packet rate and the OFF interval length (see Fig. 6). The Higuchi method showed an increase of the Hurst parameter for the OFF interval less then 50 seconds and the R/S method developed a decrease of the Hurst parameter for the OFF interval less then 50 seconds on the low rates (about 1 packet per second).

In the real systems packet rate is closer to 10 packets per second, therefore the stationary estimation of the Hurst parameter value for tracking USN may be taken equal to 0.78 according to both the R/S and the Higuchi methods.

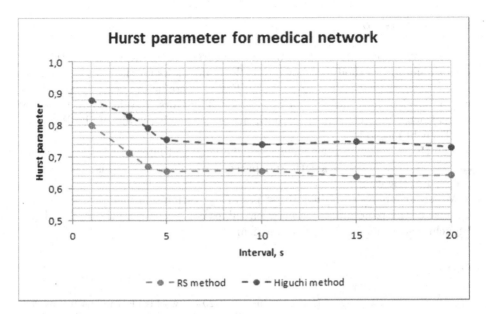

Fig. 5. Medical USN. Hurst parameter dependence on the length of interval between packets.

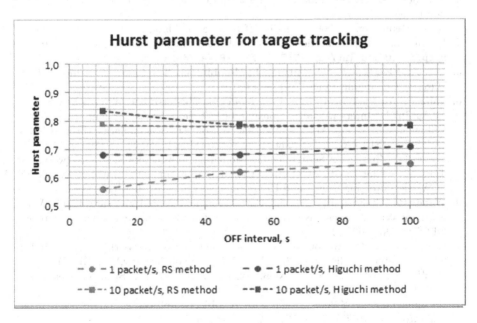

Fig. 6. Tracking USN. Hurst parameter dependence on the packet rate and OFF period length.

5   Conclusions

The USN traffic models for medical and target tracking applications investigation with simulation method has shown that the traffic flows for the fixed and hybrid sensor networks with CBR and ON/OFF source traffic are self-similar with a middle level of self-similarity in both cases. The Hurst parameter mean value estimations were determined for both models through the aid of the R/S and the Higuchi methods.

The Hurst parameter dependence on the length of interval between packets, as well as OFF interval length and packet rate observed in this study needs deeper investigation.

Moreover, analysis of other Hurst parameter estimation methods is required because of the large difference between values computed with the R/S and the Higuchi algorithms in the case of medical USN.

References

1. Sorensen, L., Skouby, K.E.: Use scenarios 2020 - a worldwide wireless future. Visions and research directions for the Wireless World. Outlook. Wireless World Research Forum 4 (2009)
2. Waldner, J.-B.: Nanocomputers and Swarm Intelligence. ISTE. Wiley & Sons, London (2008)
3. Recommendation Y.2060. Overview of Internet of Things. ITU-T, Geneva (2012)
4. Internet of Things Definition. European Union. ITU-T, IoT-GSI, Geneva (2012), http://www.itu.int/en/ITU-T/gsi/iot/Pages/default.aspx
5. Iera, A., Floerkemeier, C., Mitsugi, J., Morabito, G.: The Internet of Things. IEEE Wireless Communications 17(6), 8–9 (2010)
6. Tang, S.: An Analytical Traffic Flow Model for Cluster-Based Wireless Sensor Networks. In: 1st International Symposium on Wireless Pervasive Computing (2006)
7. Willinger, W., Taqqu, M., Sherman, R., Wilson, D.: Self-similarity through High-variability. IEEE/ACM Transaction on Networking 15(1), 71–86 (1997)
8. Wang, Q., Zhang, T.: Source Traffic Modelling in Wireless Sensor Networks for Target Tracking. In: 5th ACM International Simposium on Performance Evaluation of Wireless Ad Hoc Sensor and Ubiquitous Networks, pp. 96–100 (2008)
9. Wang, P., Akyildiz, I.F.: Spatial Correlation and Mobility Aware Traffic Modelling for Wireless Sensor Networks. In: IEEE Global Communications Conference, GLOBECOM 2009 (2009)
10. Koucheryavy, A., Prokopiev, A.: Ubiquitous Sensor Networks Traffic Models for Telemetry Applications. In: Balandin, S., Koucheryavy, Y., Hu, H. (eds.) NEW2AN/ruSMART 2011. LNCS, vol. 6869, pp. 287–294. Springer, Heidelberg (2011)
11. Recommendation Y.2062. Framework of object-to-object communication using ubiquitous networking in NGN. ITU-T, Geneva (2012)
12. Hurst, H.E.: Long-term storage of reservoirs: an experimental study. Transactions of the American Society of Civil Engineers 116, 770–799 (1951)
13. Higuchi, T.: Approach to an irregular time series on the basis of the fractal theory. Physica D 31, 277–283 (1988)

An Adaptive Codec Switching Scheme for SIP-Based VoIP

Ismet Aktas, Florian Schmidt, Elias Weingärtner,
Cai-Julian Schnelke, and Klaus Wehrle

Chair of Communication and Distributed Systems,
RWTH Aachen University
Ahornstr. 55, 52074 Aachen, Germany
`lastname@comsys.rwth-aachen.de`

Abstract. Contemporary Voice-Over-IP (VoIP) systems typically negotiate only one codec for the entire VoIP session life time. However, as different codecs perform differently well under certain network conditions like delay, jitter or packet loss, this can lead to a reduction of quality if those conditions change during the call. This paper makes two core contributions: First, we compare the speech quality of a set of standard VoIP codecs given different network conditions. Second, we propose an adaptive end-to-end based codec switching scheme that fully conforms to the SIP standard. Our evaluation with a real-world prototype based on Linphone shows that our codec switching scheme adapts well to changing network conditions, improving overall speech quality.

Keywords: VoIP communication, Codec Switching, SIP.

1 Introduction

The constantly changing dynamics of wireless and mobile environments are a great challenge for Voice-over-IP (VoIP) communications. Current VoIP software has only limited capabilities to deal with these dynamics. They typically support a number of codecs that differ in the optimal speech quality and required network parameters (jitter, packet loss rate, bandwidth). However, those VoIP clients typically negotiate one single codec for the entire duration of the call. While these codecs might be able to adapt themselves to a limited degree to changing network conditions such as available bandwidth, network delay or packet loss rate change in the meantime, the VoIP clients abide with their initial codec choice. Hence, they often apply a codec that is not well suited for the present network situation although better codec choices would be available.

To our knowledge, none of these VoIP clients implement an adaptive strategy to switch the session's speech codec upon changing network conditions.

In this paper we investigate how we can improve the speech quality of SIP-based VoIP calls using codec switching when the network conditions change. Specifically, this paper makes the following contributions:

S. Andreev et al. (Eds.): NEW2AN/ruSMART 2012, LNCS 7469, pp. 347–358, 2012.

- We first compare the speech quality of four standard speech codecs (GSM, PCMU, PCMA, Speex) given different network conditions using the MOS-LQO [7,2] metric (Section 2).
- We propose an adaptive codec switching scheme that is fully compliant with the Session Initiation Protocol (SIP). It allows specialized SIP VoIP clients to dynamically adjust their codec choice to the current network conditions (Section 3).
- We evaluate our codec switching scheme using a real-world prototype build around the Linphone [3] client software. Our evaluation shows that our scheme is well able to deliver an increased speech quality if the network conditions change during a VoIP session (Section 4).

We discuss important related work in Section 5. Section 6 concludes the paper with final remarks.

2   Sensitivity of Speech Codecs to Network Conditions

In this section we first discuss which networking conditions influence the speech quality of VoIP sessions in general. After a brief description of our evaluation environment, we analyze the influence of these different network conditions on the speech quality of four different free codecs: Speex (in its $8kHz$ version), GSM-FR and G.711, also known as Pulse Code Modulation (PCM) with its two variants PCMU and PCMA. In order to objectively evaluate perceived speech quality, we rely on the PESQ tool [7,2] that rates perceived quality with a MOS-LQO (mean opinion score – listening quality objective) score ranging from 1 (bad quality) to 5 (excellent quality).

2.1   Influencing Network Conditions

We now discuss the interplay of speech quality and network conditions. In this paper we regard (1) packet loss, (2) jitter, and (3) available bandwidth as the factors that define the current network condition. We opted for these parameters as we believe that they have the strongest effect on VoIP communications.

Packet loss strongly influences VoIP service quality because of the real-time streaming nature of VoIP traffic; lost VoIP packets are typically not retransmitted. Hence, knowing about how each codec can cope with packet loss is crucial for picking an adequate codec under unstable network conditions.

Similarly, the prevalence of **jitter**, that is, the variation of the end-to-end delay from packet to packet, also affects VoIP communication. VoIP's real-time properties require packets to arrive steadily. If the delay variation is too high, packets arrive too late to be of any use, and are discarded. Typically, streaming implementations include a jitter buffer that helps reducing the impact of this effect by buffering the packets for a short while before playing them back, effectively trading a reduction in packet loss for a higher delay. However, since delay has to be kept low for VoIP, this technique is limited, and if jitter increases beyond the bounds of the buffer, it negatively influences speech quality. We regard

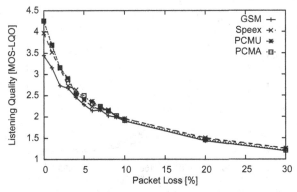

Fig. 1. Influence of packet loss on perceived speech quality for several voice codecs

jitter as an important network condition factor, as the abilities of the considered codecs to handle packet delay variations differ.

Finally, **bandwidth** is important as different codecs have different bandwidth requirements due to the fact that they employ different amounts of data compression. High bit rate codecs have larger bandwidth requirements, but usually provide higher quality.

2.2 Testing Environment

In order to analyze the influencing parameters we conducted several experiments. To assess speech quality, we used the PESQ test standardized by the ITU-T in the full-reference version. As reference file, we used the ITU-T test file u_afls03.wav (female voice speaking two sentences with a short pause between them) [10]. For VoIP communication, we employed the Linphone client as it is open source and already offers a fair set of commonly used speech codecs.

Our test setup consisted of two notebooks. Both notebooks were running Ubuntu 10.04 (LTS) and were connected to a router via $100\,MBit/s$ Ethernet. We chose a wired connection for these tests as we wanted to have full control over channel effects, and avoid additional uncontrollable environmental effects that would be witnessed in a WLAN connection.

All tests followed the same order. One notebook, the sender, initiated the call and transmitted the ITU-T test file via Linphone. The other notebook, the receiver, answered the call and recorded the audio output. We used netem [5] to insert jitter or packet loss into the connection in a controlled fashion, and employed traffic shaping via the token bucket filter [9] to reduce the available bandwidth. For each combination of codec and a certain packet loss/jitter/bandwidth, we repeated the experiment 100 times.

2.3 Codec Performance under Different Network Conditions

Figures 1, 2, and 3 show our experimentation results for the codecs GSM, Speex, PCMU (PCM with μ-law encoding), and PCMA (PCM with A-law encoding) under different network conditions.

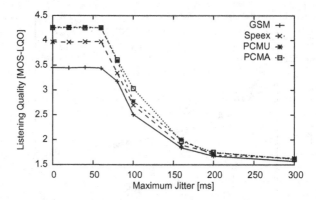

Fig. 2. Influence of delay jitter on perceived speech quality for several voice codecs

In Figure 1 the experimentation results for varying packet loss rates are shown that are manipulated with the help of netem [5]. The results for the different codecs are very close to each other with no codec having a real advantage over another. As the values above 10% packet loss are already too bad for a real communication, codec switching does not make any sense.

Jitter was also manipulated with netem. The performance was tested with a packet delay following a normal distribution and a maximum variation from $\pm 20\,ms$ to $\pm 100\,ms$ in $20\,ms$ steps and additionally for $\pm 160\,ms$, $\pm 200\,ms$ and $\pm 300\,ms$. The chosen delay variations also ultimately led to packet reordering. Figure 2 shows that the limit of the jitter buffer of all codecs is reached roughly between $60\,ms$ and $80\,ms$. Beyond this point, the speech quality degrades sharply for all codecs all codecs, and their MOS-LQO ratings converge.

In order to limit the bandwidth, we used the token bucket filter [9] traffic shaper.

Each audio codec was tested with upstream bandwidths from $10\,kbit/s$ up to $100\,kbit/s$ in $10\,kbit/s$ steps. The results can be seen in Figure 3. The divergence

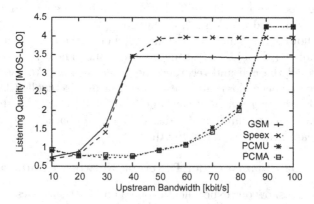

Fig. 3. Influence of available bandwidth on perceived speech quality for several voice codecs

between the value reached in 10 $kbit/s$ and the 20 $kbit/s$ test cases of the PCMU and the PCMA codecs may be due to the extremely low quality, at which output is so garbled due to data loss that quality assessment fails to properly evaluate the speech. From 20 $kbit/s$ to 30 $kbit/s$ the GSM codec performs slightly better than the Speex codec. Up from 40 $kbit/s$ until 90 $kbit/s$ the Speex codec outperforms all other investigated codecs. After 90 $kbit/s$ the PCMU and the PCMA codecs perform better than the other. Further tests are not necessary as neither the Speex codec nor the GSM codec are expected to outperform the PCMU or the PCMA codecs at higher bandwidths.

All in all the tests demonstrates that with the available set of codecs, bandwidth is the best characteristic to select a codec since it shows the most difference in quality. In all remaining cases the codecs behave similarly well, resulting in no need for a codec change. The experiments also show that the GSM codec can be dropped as a useful audio codec because it never outperforms the other codecs in any of the experimented network conditions.

Even though PCMA and PCMU always perform marginally better than Speex in the packet loss and jitter tests, we have to consider the low MOS-LQO value of the two PCM codecs at upstream bandwidths smaller than 90 kbit/s. Here the selection of Speex as a codec provides a better MOS-LQO value.

As a result, we draw the conclusion that especially switching between Speex and PCM[1] on available bandwidth is reasonable. In particular, we should use Speex for every bidirectional bandwidth between 0 kbit/s and 180 kbit/s while PCM above 180 kbit/s bandwidth. Note that, since we only investigated a limited set of codecs, this proposed switching decision is rather simple. The switching scheme proposed in the following will, however, also work with more sophisticated switching decisions and more codecs to switch between.

3 Adaptive Codec Switching

Based on the previous findings we now propose an adaptive codec switching scheme that dynamically switches the codecs of a SIP session in order to improve the speech quality. We first discuss the overall system design before we describe the switching scheme in further detail.

3.1 System Design

In todays VoIP clients typically the Session Initiation Protocol (SIP) is used to establish, modify and terminate sessions where two or several more participants are involved [8]. After a session is established, the Real-Time Transport Protocol is used to transport VoIP data to the participant. RTP is usually used in conjunction with its helper protocol Real-Time Control Protocol (RTCP) that provides periodic feedback about the the reception quality. A typical VoIP session creation and communication is depicted in Figure 4. The caller sends an

---

[1] PCMA or the PCMU codec curves are very similar in all cases, therefore from now on we use only the term PCM to indicate both

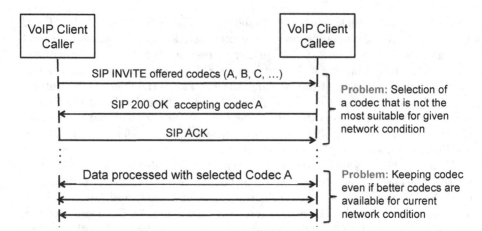

Fig. 4. A static codec selection scheme and its inherent problems

INVITE message with a set of usable codecs. If the callee agrees to the conditions, it sends the SIP 200 OK accepting code with the preferred codec in the message body. In a third step the caller sends out an ACK message to confirm the the start of call. After the call is established the whole conversation is encoded with the selected codec. There are two problems in a wireless and mobile environment where network conditions change rapidly and unpredictable. First, the selected codec at the beginning of the call may not be well suited to the network conditions at the beginning of the call. Second, network conditions may change, so there is also a need during a call to detect such changes.

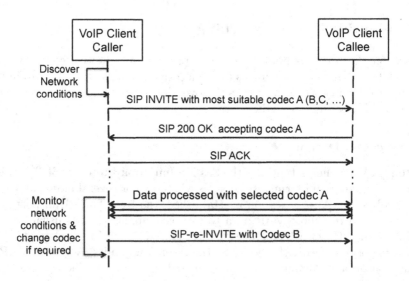

Fig. 5. Proposed codec selection scheme. Before the call is sent out, the available bandwidth is measured, and the optimal codec is chosen. During the call, network conditions are measured and codecs switched when beneficial.

In Figure 5 we present the design idea of our approach. Before a call, we measure the bandwidth and determine the adequate codec and send an INVITE message with the determined codec to the callee. The answer of the callee and the ACK of the caller remains same.

During a call we monitor the network conditions. If a high bandwidth consuming high quality codec is used, we check if the current packet loss given by RTCP reports raises above a certain threshold to decide whether we change to a low bandwidth consuming low quality codec. If we are currently using a low bandwidth consuming low quality codec, we measure the currently available bandwidth regularly. If it is above a certain threshold, we change to a high bandwidth consuming high quality codec. To coordinate the change with the callee, we used the re-INVITE message defined in the SIP standard. This message facilitates adding, removing, or modifying a session. Our codec change constitutes such a modification. The detailed description on how we selected these conditions for the available set of codecs that we have is explained in the following sections.

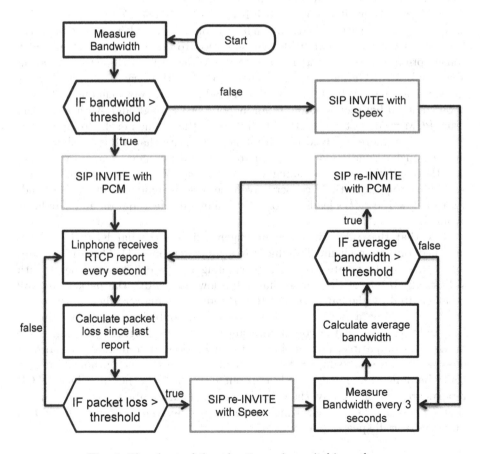

Fig. 6. Flowchart of the adaptive codec switching scheme

3.2   Switching Scheme

As discussed, bandwidth is the most relevant network condition for our set of available codecs. In Figure 6 we graphically show the decision graph for our codec selection scheme which runs fully automatic and without user interaction.

In a first step we have to ensure that we use the best performing codec for the current bandwidth when we start the call. Since Linphone does not offer a bandwidth measuring possibility, we used WBest [11]. Although this causes longer initiation time (around 1 second) for a call, we believe that such a short time is not annoying for a user if it is at the beginning of a call. If the available bandwidth exceeds our threshold ($180\,kbit/s$), we send out a SIP INVITE message with PCM as a codec. If it is smaller than our threshold, we offer only the low bandwidth consuming codec Speex.

Recognizing a decrease in the bandwidth is fairly easy. If the codec needs more bandwidth than we have available, this ultimately leads to packet loss. RTCP reports already provide information about packet loss. If the reported packet loss increases above our threshold of 10%, we switch from the high bandwidth consuming codec to the low bandwidth consuming.

An increase of available bandwidth is harder to discover. Zero percent packet loss does not necessarily mean that we have enough bandwidth to spare to switch from a codec with low bandwidth consumption to one with a high bandwidth consumption. So we once again have to use WBest to get the currently available bandwidth. We measure the available bandwidth every 3 seconds as every measurement with WBest creates an overhead to our network channel. This is a good trade-off between reacting to an increase fast enough and overhead. When using PCM we do not run WBest at all because there is no benefit to finding out whether even more bandwidth is available. To ensure that a bursty, short-term increase in bandwidth does not lead to a premature codec change, we calculate the bandwidth over a sliding window history of three measurements. We switch from Speex to PCM conservatively because switching to PCM at a bandwidth below $180\,kbit/s$ leads to considerable quality degradation and should be avoided.

Note that the codec switching intelligence lies in the hands of the caller's modified Linphone client. This has three main reasons. (1) There is no need for both sides to create overhead by running bandwidth measurements with WBest. (2) If one of the parties has a too low bandwidth the packet loss will be recognized by the caller via the RTCP reports no matter whether the caller or the callee has a bad connection. (3) A caller-only implementation means that only one side of the communication has to be changed, the changes are backwards-compatible, and deployment can be done incrementally because it does not require any support from the callee. However, one optimization that could be done at the callee's side is to increase the frequency at which RTCP reports are sent. That way, the caller is informed earlier about packet loss rates that suggest a codec switch, which increases responsiveness of our scheme and increases overall speech quality.

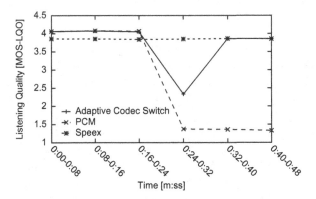

Fig. 7. MOS-LQO values for decreasing bandwidth. At 24s, bandwidth was reduced from 200 kbit/s to 65 kbit/s, and our scheme switched from PCM to Speex.

4   Evaluation

In our evaluation we investigate the speech quality achieved by our codec selection scheme. We compare our scheme with a static use of either Speex or PCM. To highlight the effects of the switch in either direction, we present two evaluation cases: a bandwidth increase and a bandwidth decrease situation.

In the decreasing case, the bandwidth is limited to $200\,kbit/s$ initially, and is further reduced to $65\,kbit/s$ after 24 seconds. Conversely, in the increasing case, the available bandwidth is increased from $65\,kbit/s$ to $200\,kbit/s$ 24 seconds into the experiment. We use the same ITU-T test file as in Section 2.3. We loop that test file six times to create a 48-second test file out of the 8-second sample.

Note that each case was tested 20 times with each approach (Speex, PCM and our codec switch scheme). The total MOS-LQO value for the whole test period of 48 seconds measured by the PESQ tool is not fair to compare, since the longer the test file is the lesser influence is of the switching gap from one codec to another. Therefore, we decided to divide each record to 8 second chunks and compare those chunks with our original 8 second test file. This has the further advantage that it also shows the perceived quality over the course of the experiment, instead of one aggregated result.

The results from the decreasing case are shown in Figure 7. As expected, the speech quality of the Speex codec stay constant throughout the test, because the bandwidth limitation to $65\,kbit/s$ is still above Speex's requirements. On the other hand, PCM shows a strong degradation of quality after the bandwidth reduction. The effect of our codec switching scheme can clearly be seen. Note also how it correctly chooses PCM as initial codec. The temporary degradation between 24 and 32 seconds can be attributed to two factors. (1) The bandwidth decrease is detected due to frame losses, which reduces the listening quality in the time before the switch takes place. (2) Linphone's current implementation reacts to a re-INVITE codec switch with a small playback gap of about $200\,ms$, which also decreases the perceived quality.

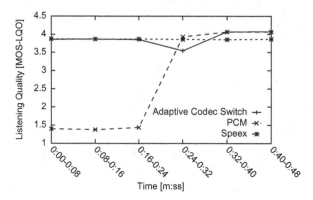

Fig. 8. MOS-LQO values for increasing bandwidth. At 24s, bandwidth was increased from 65 kbit/s to 200 kbit/s, and our scheme switched from Speex to PCM.

The results from the increasing case are shown in Figure 8. Again, Speex's speech quality stays constant. PCM benefits from the increased available bandwidth, which leads to a strong quality improvements after the 24-second mark. Our codec-switching scheme correctly decides on the better codec to use at any given point in time by choosing Speex as initial codec, and switching to PCM after the bandwidth change at 24 seconds. The slight degradation between 24 and 32 seconds can be attributed to Linphone's playback gap, which temporarily decreases perceived quality.

To conclude our evaluation, our test show that our codec switching scheme selects the specified codec properly at the beginning and during the communication. In addition, we can see how it improves listening quality compared to a static codec choice, except for the short time of the switch itself. Note that we do not expect to change codecs very frequently, so these evaluation results overemphasize the temporary quality loss during the switch; in a real setup with long conversations and only occasional codec switches, the overall quality improvement will strongly outweigh the short degradations.

5   Related Work

Related work on adaptive codec switching typically focuses on adaptation to degrading network conditions, and only discusses adaptation to improving conditions briefly or not at all. Furthermore, to the best of our knowledge, none of these approaches provides a solution for choosing the optimal codec at the beginning of the call.

In [1] the authors suggest to use packet delay as an indicator to select codecs. Their reasoning is to detect network congestion that way, and to preemptively switch codecs before prohibitively high packet loss occurs. However, their purely analytical approach does not focus on speech quality as a metric; therefore it is not clear whether such a switching approach actually would improve quality.

In [6] packet loss is used as an indicator to switch the codec in degrading as well as improving conditions. However, how and when to infer from low packet loss that additional bandwidth is available is not discussed. The adaptation to improving conditions is not evaluated either, so this question stays open. Similarly, the authors of [12] use packet loss as an indicator for both degrading and improving channel conditions. Furthermore, they propose a handover scheme between different types of network (e.g., WiFi and WiMAX) that also takes signal strength into account. Again, no evaluation for adaptation to improving network conditions is presented. Both [6] and [12] employ a SIP re-invite technique similar to ours.

The authors of [4] propose an adaptation that combines the goals of quality and security. They continuously monitor the MOS via a no-reference scoring algorithm and then decide on which codec to use, whether to introduce additional forward error correction, and how much security overhead they can introduce without compromising quality. However, the authors do not fully address the increasing bandwidth case. Moreover, their design has neither been implemented nor tested, so it is unclear how well their approach would work in reality.

6 Conclusion and Future Work

In this paper, we analyzed the speech quality of several standard VoIP codecs for different network conditions. The results of this analysis showed that bandwidth is a defining metric for quality.

Based on that, we designed an adaptive coded switching scheme that fully conforms to the SIP standard and integrated it into the Linphone [3] VoIP client. Depending on available bandwidth, our adaptive codec switching scheme performs three tasks: (1) choosing the currently best performing codec before the actual communication starts, (2) changing to a low bandwidth consuming low quality codec when the packet loss increases, and (3) changing to a high bandwidth consuming high quality codec when the bandwidth increases.

Our evaluation shows that our solution produces improvements in the perceived listening quality compared to a static codec choice at the beginning of the call.

In the future, we plan to investigate the integration of further open codecs such as G.726 or Internet Bit Rate Codec (iLBC). These codecs have low bandwidth consumption which makes them a good alternative to Speex. In addition, we aim at evaluating the effects of further network metrics such as packet or bit error rate on codec performance. More codec choices together with more metrics may lead to an extension of our switching scheme with more sophisticated strategies that further improve listening quality.

References

1. Casetti, C., De Martin, J.C., Meo, M.: Framework for the analysis of adaptive voice over ip. In: 2000 IEEE International Conference on Communications, pp. 821–826 (June 2000)

2. ITU-T P.862: Perceptual evaluation of speech quality (pesq): An objective method for end-to-end speech quality assessment of narrow-band telephone networks and speech codecs (2001)
3. Linphone, http://www.linphone.org
4. Mazurczyk, W., Kotulski, Z.: Adaptive voip with audio watermarking for improved call quality and security. Journal of Information Assurance and Security 2(3), 226–234 (2007)
5. netem, http://www.linuxfoundation.org/collaborate/workgroups/networking/netem
6. Ng, S.L., Hoh, S., Singh, D.: Effectiveness of adaptive codec switching voip application over heterogeneous networks. In: 2005 2nd International Conference on Mobile Technology, Applications and Systems, pages 7 (November 2005)
7. Pesq, http://www.pesq.org/
8. Rosenberg, J., Schulzrinne, H., Camarillo, G., Johnston, A., Peterson, J., Sparks, R., Handley, M., Schooler, E.: SIP: Session Initiation Protocol. RFC 3261 (Proposed Standard) (June 2002); Updated by RFCs 3265, 3853, 4320, 4916, 5393, 5621, 5626, 5630, 5922, 5954, 6026, 6141
9. tbf, http://linux.die.net/man/8/tc-tbf
10. Testfile, http://www.itu.int/rec/T-REC-P.862-200511-I!Amd2
11. Wbest: a bandwidth estimation tool for IEEE 802.11 wireless networks, http://web.cs.wpi.edu/~claypool/papers/wbest/
12. Yee, Y.C., Choong, K.N., Low, A.L.Y., Tan, S.W.: Sip-based proactive and adaptive mobility management framework for heterogeneous networks. J. Netw. Comput. Appl. 31, 771–792 (2008)

Stop the Flood –
Perimeter Security- and Overload-Pre-evaluation in Carrier Grade VoIP Infrastructures

Michael Hirschbichler, Joachim Fabini, Bernhard Seifert, and Christoph Egger

TU Wien, Institute of Telecommunications
Favoritenstr. 9-11/E389
A 1040 Wien, Austria
{michael.hirschbichler,joachim.fabini,bernhard.seifert,
christoph.egger}@tuwien.ac.at

Abstract. With the upcoming introduction of the Session Initiation Protocol to carrier grade telecommunication infrastructures, the threat of attacks is increasing massively. Multiple types of unsolicited communication, like high and low rate Denial-of-Service attacks as well as Spam over Internet Telephony driven by Botnets will be an upcoming risk for all telecommunication operators.

In this document, we introduce an enhanced Session Border Controller which is able to detect high-rate DoS attacks and which will mark all forwarded requests with a value indicating the "quality" of the request. This value, which we denote as "dropability", reflects the effort the system has already invested for this request. This dropability-value depends amongst other presented factors on the spam-probability and the economic- or QoS-effect of this request.

This introduced value supports overloaded core-components to decide with minimum processing effort, which requests to drop first and which requests have severe effects on the customers perception or the economic income of the carrier.

Keywords: SIP, DoS, Spam, Carrier Grade Networks: Overload Control.

1 Introduction

Standardization bodies are working on solutions both against Spam over Internet Telephony (SPIT) and for implementing overload control communication mechanisms. In complex carrier grade environments with multiple hops and components, there is a strong requirement to keep overload at the perimeter in order to keep the core clear of unnecessary load.

In this document, we merge the ideas of IETF standardization activities and extend these by developing an extended perimeter security component named "Session Border Controller - Advanced" (SBC-A). This multistage-component

S. Andreev et al. (Eds.): NEW2AN/ruSMART 2012, LNCS 7469, pp. 359–370, 2012.

prevents from flooding denial-of-service (DoS) attacks on one hand and, on the other hand, it marks all requests for later overload-control dropping algorithms. By focusing all qualification mechanisms on the entry-point of the core infrastructure, this component unloads the core and supports prioritization of important requests during high-load periods.

In the first part of this document, we introduce the SBC-A and a new Session Initiation Protocol (SIP)-header X-dropability which is added by the SBC-A to communicate the "quality" of a request to the protected core. For example, INVITE-requests of authenticated and trusted customers have a high quality (and a low X-dropability-value), probable SPIT-requests of unknown callers have a low quality. In overload state, the affected components are able to drop low-quality requests earlier than high-quality requests by considering the X-dropability-header.

In the second section, we show, how this X-dropability value can be integrated in current overload-control communication standards and standard drafts.

The last section summarizes the concept and presents further steps verifying our approach.

With the presented technique, the processing load on carrier grade core components can be reduced to the benefit of handling more goodput.

2   Related Work

SIP, as underlying protocol of the presented overload control technique is specified by Schulzrinne et. al. in RFC 3261[12]. The Session Border Controller and its provided functionality is discussed in RFC 5853[3].

In RFC 5390[10] J. Rosenberg defined different causes of overload and compared their impact on SIP-operated infrastructure.

For solving the overload threat, ETSI/TISPAN has introduced the highly complex transprotocol overload control protocol *GOCAP* in standard TS 283. 039.2-4 [1]. A more generic approach is the informational IETF standard "Design Considerations for Session Initiation Protocol (SIP) Overload Control" RFC 6357[4]. Based on this standard and improved by this paper, the IETF sip-overload working group develops a hop-to-hop overload control protocol for SIP components[2]. This protocol communicates overload states to neighbor upstream hosts. The used dropping and rate-limiting algorithms are proposed in the early-state draft [7].

Noel and Johnson compare in [6] multiple overload control algorithms with the finding to throttle the overload as close to the source as possible. In the here discussed carrier grade infrastructure, this location is the core perimeter where we will place the SBC-A.

Preventing (low-rate) DoS-attacks on SIP application layer and the need of SIP- and RTP-aware *perimeter security* is presented by Ormazabal et. al. in [8]. Using the Hellinger distance for detecting stealth flooding is discussed in [15]. Methods for detecting unsolicited communications are defined in RFC 5039[11] and the marking of SIP-messages with spam-score headers are proposed in the (outdated) IETF-draft of Wing et. al. in [16].

3   Session Border Controller - Advanced

The term SBC is unspecific and not standardized but meanwhile used by many VoIP providers as a (often) proprietary border security function. At this edge node, the network policies are usually enforced using the SBC. The RFC 5853[3] summarizes these common functionalities and highlights three SBC-specific areas: " a) perimeter defense(...) b) functionality not available in the endpoints (...) and c) traffic management (...)". We extend both area "a)" and "c" by adding overload control-preevaluation and -marking and name this extended SBC *Session Border Controller - Advanced (SBC-A)*.

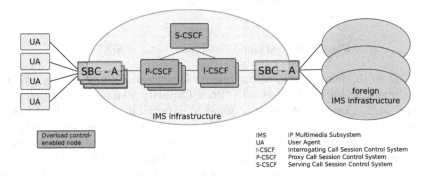

Fig. 1. Multiple SBC-A' protecting an oc enabled IMS infrastructure both from accessnet as also from peering partners

The SBC-A consists of multiple nested stages (see figure 2) with an increasing grade of packet inspection and an increasing load per request (*lpr*). We first define two categories, *Unknown Relationship (UR)* and *Known Relationship (KR)*, whereas the latter is split into *Trusted Relationship* and *Untrusted Relationship*. An incoming request passes the two categories (*UR* and *KR*) and the three nested analysis stages:

1. Denial of Service (DoS)-detection and -protection
2. Unsolicited Communication (UC)-detection
3. Overload Specific Request Qualifying (OSRQ)

In the next subsections, we describe these three stages and show the results in overload-control marking.

3.1   Denial-of-Service–Detection and –Protection

To keep the overall load in the next steps low, we block incoming high-rate DoS-attacks first, using techniques proposed and discussed in [8] (figure 2 [1]).

Due to the high complexity of SIP-operated infrastructure, successful blocking of high-rate attacks does not stop the problem of overload in general.

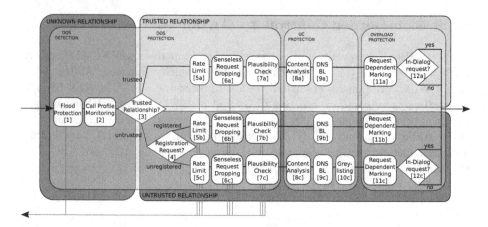

Fig. 2. Detailed step-by-step processing of SBC-A

RFC 5390[10] defines six different reasons for overload, which can be grouped into a) operator initiated overload ("Poor Capacity Planning", "Dependency Failures" and "Component Failures") and b) external initiated overload ("Avalanche Restart of clients", "Flash Crowd of multiple users simultaneously creating a call" and "DoS"[10]). From this list, only DoS from one single source can be blocked by simple high-rate attack preventing mechanisms. All other reasons for overload stem from multiple sources, which create in sum the overload. With the proposed algorithm, the system is capable to mark suspicious and/or unimportant requests for earlier dropping than other requests.

For this reason, the next step compares the current incoming traffic profile with already observed and recorded daily call-profiles (see figure 3 and figure 2 [2])). If there is a significant difference between the current call amount and this profile, all requests passing this SBC-A are marked with an increased X-dropability-header and a factor, by which the current profile exceeds the expected profile. Requests, which arrived during a timespan with common load will also be marked with a X-dropability-header and a value of "0".

Using the Hellinger distance as proposed in [14] and [15] will additionally assist to mark stealthy flooding requests.

The presence of a X-dropability-header within the signaling flow notifies the other SIP components about an SBC-A within the signaling path.

In listing 1.1, we present the RFC 3261 INVITE-request extended by the dropability factor 2

```
INVITE sip:bob@biloxi.com SIP/2.0
Via: SIP/2.0/UDP pc3.atlanta.com;branch=z9hG4bKnashds8
X-dropability: 2
To: Bob <sip:bob@biloxi.com>
From: Alice <sip:alice@atlanta.com>;tag=1928301774
Call-ID: a84b4c76e66710
```

Listing 1.1. INVITE request (w/o SDP) with added X-dropability-header

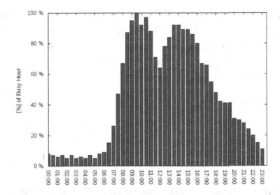

Fig. 3. Example request profile (INVITE requests) in a carrier grade VoIP-infrastructure. The diagram represents the call-setup distribution over a whole working day. Equivalent profiles are used for all other relevant SIP requests to detect abnormal request load.

In opposite to the high-rate flood protection, the requests are not dropped (or canceled).

As the irregular increased load can not clearly be allocated to evil attacks, the requests are instead marked and forwarded to the next processing stage. With this approach, we prevent false-positives and the protected core can block suspicious requests only in case of overload.

In the next step of DoS-protection, the SBC-A decides, whether the incoming request is part of a *trusted* or an *untrusted relationship*. Here, an incoming request is assumed as *trusted*, when the sending socket of the request is already actively registered (fig. 2 [3]).

By differentiating between trusted and untrusted relationships, the system can handle new requests distinctly and offer them various priorities. In our architecture, we propose to limit the allowed amount of system load for processing untrusted requests in high-load-situations to a fixed, configurable value. All requests exceeding this limitation shall be terminated with a 5xx-final response (fig. 2 [5a-c]).

As a result, the requests from trusted relationships (which are more valuable to the operator) have guaranteed processing resources even in high-load-situations.

Most of the "untrusted" requests are initial registration requests of well-known customers. Hence we define the next differentiation that depends on the type of request. Requests that are received from an untrusted socket and that are REGISTER-requests must be handled with higher priority than other requests (fig. 2, [4]).

After this point, we propose that the operator includes simple request-dependent filter rules: For instance, SUBSCRIBE-, NOTIFY- and OPTIONS-requests from unregistered sockets can simply be canceled with a "403 Forbidden" right there, as they will never result in a positive response from the core network (fig. 2, [6a-c]).

The next step in DoS-protection is the *plausibility check*. Following predefined regular expression-patterns, we decide whether requests are compliant to RFC 3261 and operator-specific security policies (fig. 2, [7a-c]). Depending on the security or local regulatory settings, we either drop suspicious requests immediately or increase the dropability parameter (see listing 1.2).

```
INVITE sip:bob@biloxi.com SIP/2.0
..
X-dropability: 3
...
```

Listing 1.2. increased X-dropability-header

This plausibility step is located subsequent to the trusted-relationship delimitation, as we want to assure for performance reasons, that operations with a high *lpr* must be executed after operations with low *lpr*.

After the DoS-detection and protection stage, the system has divided all requests into a preferred trusted and a non-preferred untrusted relationship and marked all "suspicious" requests with a X-dropability-value larger than 0.

3.2   Unsolicited Communication Detection-Marking

The next stage in the SBC-A pipeline is the *Unsolicited Communication* (or *SPIT*)-analysis. For highest transparency, we focus mainly on non-intrusive SPIT analysis where the procedures differ between trusted relationship, untrusted relationship–REGISTER-request and untrusted relationship–other-requests.

From this stage on, we do not drop suspicious requests at the SBC-A (due to the risk of getting false positive results) but we increase the X-dropability-header-value by a value according to the evaluated SPIT-grade.

In the outdated IETF-draft[16], Wing et. al. propose an additional SIP-header for marking SPIT-suspicious messages. We decide to merge this header with our X-dropability-header as both are aiming at the same goal. Here – as a side-effect to overload-control – the X-dropability-header could also be used by the callee to decide whether he wants to accept a specific request or not. In the following we detail on three variants of UC detection:

- **UC marking for requests from trusted relationships**
  In this paper, we propose to use two explicit and one implicit UC-detecting techniques: First, we analyze the content and the payloads of all requests for terms like "Rolex", "Viagra" or "Cyalis", equivalent to mail-spam-filtering. If such terms are found, the X-dropability-header-value is increased(fig. 2, [8a,c]).

  As second approach, we adopt the DNS blacklist technique presented in [5] and compare all IP-addresses in the requests with these lists. Although the system knows that the socket is trustworthy regarding authentication and authorization in SIP universe, it cannot guarantee, that the customer's infrastructure is not misused as SPIT-proxy (fig. 2, [9a-c]).

The third UC-detecting technique is the whitelist-test proposed in RFC 5539. This test is implicitly solved by using the trusted socket. All requests arriving from a known socket have been tested against the home subscriber server before and are implicitly more trustworthy than other requests.

- **UC marking for REGISTER-requests from untrusted relationships**
  The REGISTER-request is the most intrusive request when creating low-rate DoS attacks. Inherently, a REGISTER request creates the highest load in an infrastructure as it traverses a high number of components to the location database (e.g., in IMS the Home Subscriber Server (HSS)).

  A REGISTER request is not able to transfer SPIT messages at all, but we can use SPIT-qualification methods to verify the plausibility of an incoming REGISTER-request. When using the DNS-whitelist and -blacklist method, we can check if a REGISTER request arrives from a socket which has sent an successful initial REGISTER request before. Additionally, we can check, if a REGISTER request arrives from a socket, which is on the black-list for sending unsolicited messages before.

  According to the result of these DNS-lookups, the `X-dropability`-header-value is increased.

- **UC marking for all other requests from untrusted relationships**
  These requests, arriving from an unregistered host are the most untrustworthy requests.

  During our real-live measurements on the A1overIP-core[1], we noticed that only about 1% of all valid call-setup requests are from a foreign domain directed to the local domain.

  This small number of requests must be inspected carefully, as this is the main entrance for possible SPIT and (next to REGISTER flooding) the main threat for low-rate attacks to the core infrastructure.

  Here, the proposed SBC-A uses content analyzing techniques (in addition to DNS-whitelist and -blacklist analyses and regarding the request type). Additionally, we propose to move away from non-intrusive tests to the intrusive greylisting-test[9] for INVITE-requests (fig. 2, [10c]). Using greylisting, a call-setup attempt is canceled by the SBC-A with a "486 Busy Here"-response, pretending that the callee is currently in another call. The caller ID is stored temporarily and if the caller retries, the call is forwarded to the called party. This technique based on Turing tests is used regularly in Email-spam protection infrastructures.

Splitting the request analysis into three categories supports the system to handle requests of differing quality distinctly and helps – even in case of local SBC-A system-overload – dropping the untrustworthy and invaluable requests first.

3.3   Overload Specific Request Qualifying (OSRQ)

The IETF draft "Session Initiation Protocol (SIP) Overload Control"[2] proposes an overload control mechanism, where the upstream neighbor reduces the load

---

[1] A product of A1 Telekom Austria, the largest austrian mobile and fixed net operator, http://www.a1.net

Table 1. The `X-dropability`-header is increased by the value of the column "Rank" to enable the upstream neighbor of an overloaded component to drop "invaluable" requests first

Request from *Trusted* Socket		Request from *Untrusted* Socket	
Request	**Rank**	**Request**	**Rank**
ACK	1	REGISTER	9
BYE	2	ACK	10
INVITE	3	BYE	11
REGISTER	4	INVITE	12
MESSAGE	5	MESSAGE	13
UPDATE	6	UPDATE	14
NOTIFY	7	NOTIFY	15
SUBSCRIBE	8	SUBSCRIBE	16

sent to the overloaded downstream component. The drop-rate or alternatively the rate limitation is defined by the overloaded component and communicated to the upstream neighbor.

By default, there is no standardized differentiation between the different request-types. The fact that for example, a dropped ACK-request as part of an INVITE dialog may produce more load through retransmission is only considered informational. Instead, the upstream host is free to decide which requests should be dropped and which should be forwarded.

Our proposal is to support this upstream host deciding, which requests are "cheap" to drop and which requests are "valuable".

Hereby we propose – after "DoS detection and protection" and "UC detecting" – to qualify the type of request as additional input value for our `X-dropability`-header. For this reason, we suggest in table 1 a ranking of requests (as defined in RFC 3261 ff., fig. 2, [10a-c]) and observe, if they are part of an already established dialog (fig. 2, [12a,c]).

The sorting of requests in table 1 considers the request-priority. The most important requests are the in-dialog-requests "ACK" and "BYE". "ACK" concludes a nearly completed call-setup, whereas the BYE-request finishes calls and reduces the workload on an infrastructure. The INVITE-request is expected by the customer to be handled in soft-realtime for QoS-reasons and must also be handled with high priority.

The REGISTER-requests keep an existing registration alive. If they are dropped, the customer premises equipment (CPE) is unreachable for incoming requests, which decreases customer satisfaction.

The remaining requests are text-based requests and can be ranked equally: they do not have to be handled in realtime and delays of multiple seconds are acceptable. Here, we propose to prefer the MESSAGE-request because of its revenue generation when the message is sent over the providers SMS gateway to the CS environment.

For requests arriving from an untrusted socket, the REGISTER request must be handled prioritized: new REGISTER-requests arrive from potential customers and must be ranked higher than other messages arriving from unregistered sockets.

On all requests, which will be forwarded to the core network, we propose to add the rank-value from table 1 to the X-dropability-header. Considering the INVITE-request from listing 1.1, we increase the X-dropability-header by 3 if the request arrives from a registered socket, respectively by 12 if the request arrives from an unregistered socket.

3.4   Forwarding to the Next Hop

After the three stages of analyzing and marking in SBC-A, the initial INVITE-request of listing 1.1 appears now like in listing 1.3:

```
INVITE sip:bob@biloxi.com SIP/2.0
Via: SIP/2.0/UDP sbc-a.biloxi.com;branch=z9hG4bKd84ks2
Via: SIP/2.0/UDP pc3.atlanta.com;branch=z9hG4bKnashds8
X-dropability: 14
To: Bob <sip:bob@biloxi.com>
From: Alice <sip:alice@atlanta.com>;tag=1928301774
Call-ID: a84b4c76e66710
```

Listing 1.3. INVITE-request after SBC-A analysis

This request is forwarded to the next SIP-downstream hop of the carrier grade infrastructure for further SIP processing.

4   Integrating SBC-A Marked Requests into Existing Overload Control Infrastructure

The standard RFC 3261 defines only minimum overload- and congestion-control. Overloaded components should cancel incoming requests by responding with "503 Service Unavailable"-responses and optional retry timers. The informational RFC 6357 proposes to avoid local overload control and to locate the actuator responsible for blocking the overload situation at least one hop upstream to the overloaded system.

For satisfying this need, the current IETF draft "Session Initiation Protocol (SIP) Overload Control"[2] introduces the "overload control" parameter. This "oc"-parameter, added to the Via-header of a SIP message, communicates the desired traffic-reduction to the neighbor upstream node (see listing 1.4). The current draft does not define the supported algorithms in detail.

```
SIP/2.0 180 Ringing
Via: SIP/2.0/TLS p1.example.net;branch=z9hG4bK2d4790.1;
     oc=20;oc-algo="loss";oc-validity=1000;oc-seq=1282321615.782
     ...
```

Listing 1.4. a oc-enabled 180 Ringing-response defining a loss-rate of 20% over a timespan of 1000ms

Instead, this draft proposes in 5.10.1 that the upstream neighbor should prefer requests, which are for example, in-dialogue-requests or marked with *Resource-Priority*-flags as defined in RFC 4412[13]. This evaluation and decision-finding on each upstream host is producing additional load in a per-se overloaded infrastructure. We propose to use the new introduced X-dropability-header instead to keep this decision outside of the core infrastructure at the perimeter-security SBC-A.

The new introduced draft "Session Initiation Protocol (SIP) Rate Control"[7] extends [2] by the request-limitation algorithms, where we propose to include the X-dropability-header in one of three ways:

1. on loss-based algorithms: we propose to drop requests with higher X-dropability-values first until the wanted loss-rate is reached
2. on rate-based algorithms: like in loss-based algorithms, the requests with high X-dropability-values shall be dropped until the expected rate is reached
3. X-dropability-value–compliant oc-communication: as loss- and rate-based algorithms are extensive to be calculated (especially when dropping requests non-uniformly) we propose to add an algorithm to [7], where the overloaded host explicitly communicates a maximum allowed X-dropability-value. All requests with X-dropability-values greater than the maximum allowed value shall be dropped at the upstream neighbor. For archiving this task, we further propose to define a X-dropability-compliant oc-Via-parameter as in listing 1.5.

```
SIP/2.0 180 Ringing
Via: SIP/2.0/TLS p1.example.net;branch=z9hG4bK2d4790.1;
   oc=12;oc-algo="x-dropability";oc-validity=1000;
   oc-seq=1282321615.782
   . . .
```

Listing 1.5. signaling a X-dropability compliant overload-control algorithm to the next upstream node. In this example the overloaded hosts signals to drop all requests with a X-dropability-value of larger or equal "12" over a timespan of 1000ms

5   Advantages of Pre-evaluation and Summary

In case of overload in a core element, there is no way around blocking and dropping of requests to reduce the load to an acceptable amount. The preevaluation supports the decision process, which requests are better to be dropped, and which requests are needed to keep existing calls or registrations alive, resp. which are needed to fulfill the customers need of high QoS in the best possible way.

The presented approach is not completely aware of false-positive decisions, but compared to uniform dropping, the preevaluation keeps the number of lost important requests lower as requests marked with a high X-dropability-value are dropped earlier. High value requests (with a low dropability-value) are instead "protected" from being dropped at an early stage high-load-situation.

As [2] already presents basic ideas of request dependent request dropping, this draft does not bring aspects such as unsolicited communications or suspicious low-rate DoS-attacks into account.

Our approach stops flooding DoS-attacks and marks requests for possible UC and low-rate DoS as well as it qualifies SIP requests by their (economic- and QoS-) quality. It takes the *oc*-analyzing- and -deciding–load away from the (high-loaded) core components to a single, dedicated perimeter-security component, the SBC-A.

6   Outlook and Future Work

This proposed technique is the first conceptual step. In the next steps, we will use our testlab and simulator to compare uniform- and request dependent-dropping with our multistage approach. Here, we expect interesting chain-reactions and side-effects. For these simulations, we will use carrier grade user profiles provided by our project partners.

Our solution currently focuses primarily on request-qualification, but in further work, we will consider also response-flooding attacks. Based on Kamailio[2], the rate-limit-, a DNS-blacklist- and SpitAssassin[3]-module, we will implement the SBC-A as a reference implementation.

References

1. ETSI. NGN Congestion and Overload Control; Part 4: Overload and Congestion Control for H.248 MG/MGC. ES 283 039-4, Telecommunications and Internet converged Services and Protocols for Advanced Networking (TISPAN) (April 2007)
2. Gurbani, V., Hilt, V., Schulzrinne, H.: Session Initiation Protocol (SIP) Overload Control. Internet-Draft draft-ietf-soc-overload-control-07, Internet Engineering Task Force (January 2012) (work in progress)
3. Hautakorpi, J., Camarillo, G., Penfield, R., Hawrylyshen, A., Bhatia, M.: Requirements from Session Initiation Protocol (SIP) Session Border Control (SBC) Deployments. RFC 5853, Internet Engineering Task Force (April 2010)
4. Hilt, V., Noel, E., Shen, C., Abdelal, A.: Design Considerations for Session Initiation Protocol (SIP) Overload Control. RFC 6357, Internet Engineering Task Force (August 2011)
5. Hirschbichler, M., Egger, C., Pasteka, O., Berger, A.: Using E-Mail SPAM DNS Blacklists for Qualifying the SPAM-over-Internet-Telephony Probability of a SIP Call. In: Third International Conference on Digital Society, ICDS 2009, pp. 254–259 (February 2009)
6. Noel, E., Johnson, C.R.: Novel overload controls for SIP networks. In: 21st International Teletraffic Congress, ITC 21 2009, pp. 1–8 (September 2009)
7. Noel, E., PhilipWilliams, P.: Session Initiation Protocol (SIP) Rate Control. Internet-Draft draft-noel-soc-overload-rate-control-02, Internet Engineering Task Force (December 2011) (work in progress)

---

[2] http://www.kamailio.org

[3] http://www.spitassassin.org

8. Ormazabal, G., Nagpal, S., Yardeni, E., Schulzrinne, H.: Secure SIP: A Scalable Prevention Mechanism for DoS Attacks on SIP Based VoIP Systems. In: Schulzrinne, H., State, R., Niccolini, S. (eds.) IPTComm 2008. LNCS, vol. 5310, pp. 107–132. Springer, Heidelberg (2008)
9. Quinten, V.M., van de Meent, R., Pras, A.: Analysis of Techniques for Protection Against Spam over Internet Telephony. In: Pras, A., van Sinderen, M. (eds.) EUNICE 2007. LNCS, vol. 4606, pp. 70–77. Springer, Heidelberg (2007)
10. Rosenberg, J.: Requirements for Management of Overload in the Session Initiation Protocol. RFC 5390, Internet Engineering Task Force (December 2008)
11. Rosenberg, J., Jennings, C.: The Session Initiation Protocol (SIP) and Spam. RFC 5039, Internet Engineering Task Force (January 2008)
12. Rosenberg, J., Schulzrinne, H., Camarillo, G., Johnston, A., Peterson, J., Sparks, R., Handley, M., Schooler, E.: SIP: Session Initiation Protocol. RFC 3261, Internet Engineering Task Force (June 2002)
13. Schulzrinne, H., Polk, J.: Communications Resource Priority for the Session Initiation Protocol (SIP). RFC 4412, Internet Engineering Task Force (February 2006)
14. Sengar, H., Wang, H., Wijesekera, D., Jajodia, S.: Detecting VOIP floods using the hellinger distance. IEEE Transactions on Parallel and Distributed Systems 19(6), 794–805 (2008)
15. Tang, J., Cheng, Y.: Quick detection of stealthy SIP flooding attacks in VOIP networks. In: 2011 IEEE International Conference on Communications (ICC), pp. 1–5 (June 2011)
16. Wing, D., Niccolini, S., Stiemerling, M., Tschofenig, H.: Spam Score for SIP. Internet-Draft draft-wing-sipping-spam-score-02, Internet Engineering Task Force (February 2008) (work in progress)

Queuing Model for Loss-Based Overload Control in a SIP Server Using a Hysteretic Technique*

Pavel Abaev, Yuliya Gaidamaka, and Konstantin E. Samouylov

Telecommunication Systems Department,
Peoples' Friendship University of Russia,
Ordzhonikidze str. 3, 115419 Moscow, Russia
{pabaev,ygaidamaka,ksam}@sci.pfu.edu.ru
http://www.telesys.pfu.edu.ru

Abstract. In this paper, we develop a mathematical model of a load control mechanism for SIP server signaling networks based on a hysteretic technique developed for SS7 from ITU-T Recommendation Q.704. We investigate loss-based overload control, as proposed in recent IETF documents. The queuing model takes into account three types of system state – normal load, overload, and discard. The hysteretic control is made possible by introducing two thresholds in the buffer of total size B – the overload onset threshold H and the overload abatement threshold L. We denote the mathematical model using the modified Kendall notation as an $M|G|1|\langle L, H \rangle|B$ queue with hysteretic load control. We also develop an analytical model for the case of an $M|M|1$ queue and a simulation model for an $M|D|1$ queue. We investigate the return time from an overload state as the target performance measure of overload control in a SIP server, and provide numerical examples in order to examine the difference between the $M|M|1$ and $M|D|1$ systems.

Keywords: SIP server, hop-by-hop overload control, loss-based overload control, hysteretic control, return time, queuing model.

1   Introduction

In this paper, we investigate the problem of overloading in a SIP server signaling network [1,2]. Our purpose is to build and analyze queuing models to estimate the parameters of overload control mechanisms and methods [3]. We also apply the hysteretic load control technique to the case of loss-based overload control, which relates to an explicit overload control mechanism [4].

First, we describe hysteretic load control using a model consisting of four components: Monitor, Actuator, Control Function, and the System under control (SIP Processor). Second, after some simplifying assumptions, we build and analyze a two-flow queuing model with hysteretic control. For an $M|M|1$ system,

---

* This work was supported in part by the Russian Foundation for Basic Research (grants 10-07-00487-a and 12-07-00108), and by Rosobrazovanie (project no. 020619-1-174).

S. Andreev et al. (Eds.): NEW2AN/ruSMART 2012, LNCS 7469, pp. 371–378, 2012.

we obtain the distribution function for one of the critical parameters, i.e., the mean return time. Finally, we present a numerical example for the mean return time of $M|M|1$ and $M|D|1$ queues, using a simulation in the latter case. The goal of our case study is to illustrate the control mechanism that minimizes the mean return time.

2   System Model for Loss-Based Overload Control

Two overload control mechanisms are distinguished in [3]: explicit and implicit overload control. The latter is used in the absence of a response from a downstream SIP server, and uses packet loss as an indication of overload. In an explicit overload control mechanism, a downstream server uses an explicit overload signal to indicate that it is reaching its capacity limit. Upstream neighbors receiving this signal can adjust their transmission rate accordingly to a level that is acceptable to the downstream server.

In this paper, we build a queuing model for an explicit overload control mechanism, namely loss-based overload control [4]. Note that another mechanism under intensive study in IETF is rate-based overload control [5]. We investigate one of four generic SIP server configurations [3], the so-called "multiple sources" configuration shown in Fig. 1, in which the downstream server receives traffic from K upstream servers. Each of these servers can contribute a different amount of traffic, which can vary over time.

Fig. 2 shows the system model of loss-based overload control implemented between each "upstream – downstream" server pair.

The model identifies components of loss-based overload control that we propose to implement using a hysteretic technique. We develop the model in accordance with the RFC 6357 recommendations on design considerations for explicit SIP overload control, and include the following components: SIP Processor, Monitor, Control Function, and Actuator.

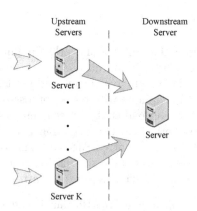

Fig. 1. Model of multiple sources (cooperation between servers)

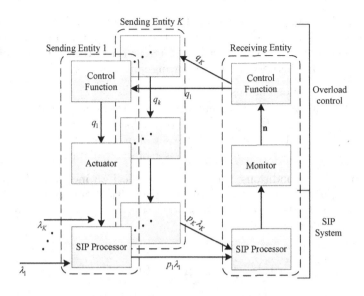

Fig. 2. System model for a loss-based overload control schema

The Processor is protected by an overload control mechanism and includes a buffer where SIP messages are queued according to the FIFO order.

The Monitor measures the Processor load in the Receiving Entity (RE) and reports the buffer occupancy $\mathbf{n} = (n_1, \ldots, n_K)$ to the Control Function, where n_k is the number of messages received from the k-th Sending Entity (SE).

The Control Function uses the buffer occupancy to determine whether an overload is likely to occur according to the hysteretic load control algorithm, and identifies the limitation required to adjust the load sent to the Processor on the RE. Thus, the Control Function in the RE sends the dropping probability to the SE.

Note that the Control Function in the RE realizes two algorithms. The first determines whether an overload will occur, and the second calculates the dropping probability $0 < q_k \leq 1$ for the throttling of traffic at the SE. According to RFC 6357, the Control Function at the SE is empty and simply passes q_k along as feedback to the Actuator. The Actuator implements the throttling algorithm for traffic forwarded to the RE.

In the next section, under the assumption of a stationary Poisson input, we construct a model of a loss-based overload control algorithm in a SIP server using a hysteretic technique. Without loss of generality of the model, we assume that $K = 2$.

3   Queuing Model of Loss-Based Overload Control

We now build and analyze a mathematical model of the hysteretic overload control system described in the previous section. The model is represented by a single-server queuing system with a finite buffer of capacity B, as shown in Fig. 3. Note that the case of $B = \infty$ was studied in [7]. Two Poisson customer flows reach the system with an intensity of $\lambda_i\,(s, n)$, $i = 1, 2$, the graph of which is shown in Fig. 4, where we use the following notation:

$s \in \{0, 1, 2\}$ – overload status;
n – buffer occupancy, i.e., total number of customers in the buffer;
$\mathcal{X}_0 = \{s = 0,\, 0 \le n \le H - 1\}$ – set of normal load states;
$\mathcal{X}_1 = \{(s, n) :\ s = 1,\, L \le n \le B - 1\}$ – set of overload states;
$\mathcal{X}_2 = \{(s, n) :\ s = 2,\, H + 1 \le n \le B\}$ – set of discard states.

Under normal load conditions, the intensity of the i–th incoming flow is equal to $\lambda_i > 0$ and, if the system is in an overload condition, the flow rate decreases such that $\lambda_i' = p_i \lambda_i$, $0 < p_i < 1$. If the system is in a discard state, the intensity of the incoming flow is equal to zero.

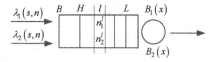

Fig. 3. Queuing model of $M|G|1|\,\langle L, H \rangle\,|B$ with hysteretic load control

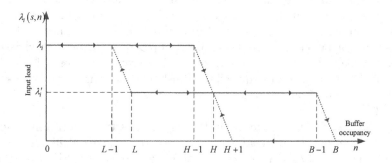

Fig. 4. Function of the intensity $\lambda_i\,(s, n)$ of the i–th flow

Both types of customer are served by a single server on a FCFS basis, and let $B_i\,(t)$ be the generic service-time distribution function of the i–th flow. Let $\left(n_1^l, n_2^l\right) \in \{(0, 0), (0, 1), (1, 0)\}$ indicate the state of the l-th position of the buffer, i.e., $n_i^l = 1$ if the l-th position is occupied by the i-th customer, and $n_i^l = 0$ otherwise. The total number of customers of both types can then be calculated by the formula:

$$n = n_1 + n_2 = \sum_{l=1}^{B} n_1^l + \sum_{l=1}^{B} n_2^l, \qquad (1)$$

where n_i is the number of the i-th customer in the buffer.

We assume that the customer being served retains a place in the queue, which is released when the service is terminated. We implement hysteretic overload control of the input flows through two thresholds: an abatement threshold L, and an onset threshold H [6].

In the simplest case of $B_1(t) = B_2(t) = 1 - e^{-\mu t}, t \geq 0$, the queuing system can be considered as a Markov process $X(t)$ over the state space:

$$\mathcal{X} = \mathcal{X}_0 \bigcup \mathcal{X}_1 \bigcup \mathcal{X}_2. \tag{2}$$

Thus, the transition diagram of the system can be pictured as in Fig. 5, where $\lambda = \lambda_1 + \lambda_2$ and $\lambda' = \lambda'_1 + \lambda'_2$.

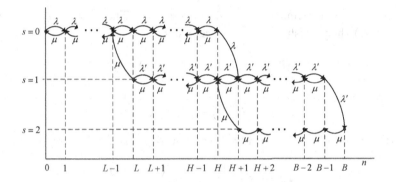

Fig. 5. Transition diagram of Markov process $X(t)$

Let random variable τ measure the time from when the process $X(t)$ reached the overload set $\mathcal{X}_{12} = \mathcal{X}_1 \bigcup \mathcal{X}_2$, i.e., the state $(1, H+1)$, for the first time until it reaches the normal load set $\mathcal{X}_0$, namely the state $(0, L-1)$. τ is called the return time to the normal load set. Below, we propose a method for calculating the probability distribution function of the random variable τ for an $M|M|1| \langle L, H \rangle |B$ system.

Let $\widehat{X}(t)$ be the truncation of process $X(t)$ to the set $\widehat{\mathcal{X}} = \mathcal{X}_1 \bigcup \mathcal{X}_2 \bigcup \{(0, L-1)\}$, and let $\widehat{p}_{sn}(t) = P\left\{ \widehat{X}(t) = (s, n) \right\}$, $(s, n) \in \widehat{\mathcal{X}}$, $\widehat{\mathbf{p}}(t) = (\widehat{p}_{sn}(t))_{(s,n) \in \widehat{\mathcal{X}}}$, and $\widehat{\mathbf{P}}(t) = (\widehat{p}_{(s,n),(r,m)}(t))_{(s,n),(r,m) \in \widehat{\mathcal{X}}}$, where $\widehat{\mathbf{p}}(t)$ is the row vector of state probabilities of Markov process $\widehat{X}(t)$ at the moment $t \geq 0$ and $\widehat{\mathbf{P}}(t)$ is the transition probability matrix in the interval $[0, t)$. The state transition diagram for the process $\widehat{X}(t)$ is shown in Fig. 6.

It is known that matrix $\widehat{\mathbf{P}}(t)$ can be written in the following form:

$$\widehat{\mathbf{P}}(t) = e^{\widehat{\mathbf{\Lambda}}t} = \sum_{n \geq 0} \frac{\widehat{\mathbf{\Lambda}}^n t^n}{n!}, \tag{3}$$

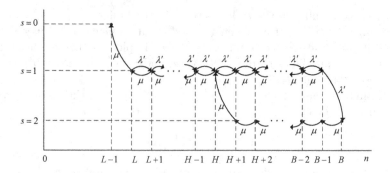

Fig. 6. State transition diagram for $\widehat{X}(t)$

where $\widehat{\Lambda}$ is the infinitesimal operator of process $\widehat{X}(t)$. The vector $\widehat{\mathbf{p}}(t)$ for Markov process $\widehat{X}(t)$ then satisfies the following equations:

$$\widehat{\mathbf{p}}^T(t) = \widehat{\mathbf{p}}(0)\,\widehat{\mathbf{P}}(t) = \widehat{\mathbf{p}}(0)\,e^{\widehat{\Lambda}t}, \tag{4}$$

$$\frac{d}{dt}\widehat{\mathbf{p}}^T(t) = \widehat{\mathbf{p}}^T(t)\,\widehat{\Lambda}. \tag{5}$$

As part of the problem being solved, the initial probability vector $\widehat{\mathbf{p}}(0)$ can be written as:

$$\widehat{p}_{(s,n)}(0) = \begin{cases} 1, & (s,n) = (1, H+1), \\ 0, & \text{otherwise}, \end{cases} \tag{6}$$

and then, the distribution function $F_\tau(t)$ of the random variable τ is given by:

$$F_\tau(t) = \widehat{p}_{(0,L-1)}(t). \tag{7}$$

Thus, the problem of finding the characteristics of the random variable τ (return time to the set of normal load states) is reduced to the calculation of the distribution function $F_\tau(t)$ by equations (4), (5), and (7), while the initial probability vector $\widehat{\mathbf{p}}(0)$ is given by (6). The mean return time $\overline{\tau}$ is then given by:

$$\overline{\tau} = \int_0^\infty t\frac{d}{dt}\widehat{p}_{(0,L-1)}(t)\,dt.$$

4   Case Study

To estimate the mean service rate μ of SIP messages, we have taken into account that each session involves the exchange of seven SIP messages. We assume that the processing time of an INVITE is 10 ms and the processing time of a non-INVITE message is 5 ms. Taking into account that a basic session is composed of one INVITE and six non-INVITE messages, the average processing time of a SIP message is about 5 ms, hence $\mu = 200\,s^{-1}$.

The problem is stated as follows: minimize the mean return time with respect to the choice of the two thresholds, L and H, such that requirements $R1 - R3$ are satisfied [8]. Formally:

$$\bar{\tau}(L, H) \to \min;$$
$$R1: \ B(\mathcal{X}_1) \leq \gamma_1;$$
$$R2: B(\mathcal{X}_2) \leq \gamma_2;$$
$$R3: \tau \geq \gamma_3.$$

For a given dropping probability $q \in \{0.3, 0.4, 0.5, 0.6\}$, we now seek to solve the problem of choosing the two threshold values. Let us also consider minimizing the mean return time $\bar{\tau}$ such that the offered load $\rho = \frac{\lambda}{\mu} = 1.2$, buffer capacity $B = 100$, $\gamma_1 = 0.2$, $\gamma_2 = 10^{-4}$, and $\gamma_3 = 450$ ms. Using the above formulae, we developed an algorithm for solving the optimization problem. Note that for the optimum solution obtained by this algorithm, requirements $R1$ and $R2$ are always binding, making the mean control cycle time as high as possible. The results of calculations with the above-defined input data for exponential (M) and deterministic (D) service times are presented in Fig. 7.

The graph shows that the mean return time for the $M|M|1$ system is about twice that for the $M|D|1$ system. Taking into account that six out of seven SIP messages have almost the same length, we recommend the use of the $M|D|1$ system for analysis, i.e., choose the thresholds according to the dashed curve.

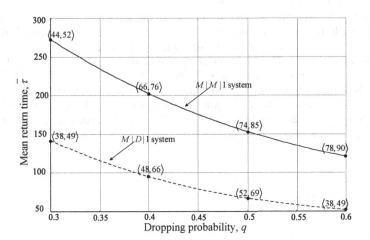

Fig. 7. Mean return time for the $M|M|1$ and $M|D|1$ queues. Values of L and H are also shown.

5   Conclusion

In this paper, we developed a hysteretic control mechanism to solve the problem of loss-based overload control in a SIP signaling network. The mechanism was modeled as a $M|G|1| \langle L, H \rangle |B$ queue, with two input flows being throttled depending on the hysteretic control thresholds. The performance measures of the control system were determined as a function of input load, dropping probability, abatement, and onset thresholds. In the case of an $M|M|1$ queue, we obtained an analytical form for the distribution function of the return time to a normal state. In the case of an $M|D|1$ queue, we analyzed performance measures using a simulation model. A case study showed that the $M|D|1$ queue is preferable to the $M|M|1$ queue for the performance analysis of hysteretic control systems. Future study will be devoted to the development of models for rate- and window-based overload control methods.

References

1. Rosenberg, J., Schulzrinne, H., Camarillo, G., et al.: SIP: Session Initiation Protocol. RFC3261 (2002)
2. Rosenberg, J.: Requirements for Management of Overload in the Session Initiation Protocol. RFC5390 (2008)
3. Hilt, V., Noel, E., Shen, C., Abdelal, A.: Design Considerations for Session Initiation Protocol (SIP) Overload Control. RFC6357 (2011)
4. Gurbani, V., Hilt, V., Schulzrinne, H.: Session Initiation Protocol (SIP) Overload Control. draft-ietf-soc-overload-control-08 (2012)
5. Noel, E., Williams, P.M.: Session Initiation Protocol (SIP) Rate Control. draft-ietf-soc-overload-rate-control-02 (2012)
6. Abaev, P.O.: Algorithm for Computing Steady-State Probabilities of the Queuing System with Hysteretic Congestion Control and Working Vacations. Bulletin of Peoples' Friendship University of Russia (3), 58–62 (2011)
7. Roughan, M., Pearce, C.E.M.: A martingale analysis of hysteretic overload control. Advances in Performance Analysis: A Journal of Teletraffic Theory and Performance Analysis of Communication Systems and Networks (3), 1–30 (2000)
8. Abaev, P., Gaidamaka , Y., Pechinkin, A., Razumchik, R., Shorgin, S.: Simulation of overload control in SIP server networks. In: Proc. of the 26th European Conference on Modelling and Simulation, ECMS 2012, pp. 533–539 (2012)

Applying MIMO Techniques to Minimize Energy Consumption for Long Distances Communications in Wireless Sensor Networks

Edison Pignaton de Freitas[1,2,4], João Paulo C. Lustosa da Costa[1],
André Lima F. de Almeida[3], and Marco Marinho[1]

[1] Department of Electrical Engineering, University of Brasília, Brasília, Brazil
[2] Institute of Informatics, Federal University of Rio Grande do Sul, Brazil
[3] Teleinformatics Engineering Department, Federal University of Ceará, Brazil
[4] Brazilian Army Center of Technology, Brazilian Army, Brazil
epfreitas@inf.ufrgs.br, jpdacosta@unb.br, andre@gtel.ufc.br,
marco@aluno.unb.br

Abstract. This paper explores the usage of cooperative multiple input multiple output (MIMO) technique to minimize energy consumption used to establish communications among distant nodes in a wireless sensor network (WSN). As energy depletion is an outstanding problem in WSN research field, a number of techniques aim to preserve such resource, especially by means of savings during communication among sensor nodes. One such wide used technique is multi-hop communication to diminish the energy required by a single node to transmit a given message, providing a homogeneous consumption of the energy resources among the nodes in the network. However, it is not the case that multi-hop is always more efficient than single-hop, even that it may represent a great depletion of a single node's energy. In this paper a cooperative MIMO transmission technique for WSN is presented, which is compared to single-hop and multi-hop transmission ones, highlighting its advantages in relation to both. Simulation results support the statement about the utility in applying the proposed technique for energy saving purposes.

Keywords: Wireless Sensor Networks, Cooperative Multiple Input Multiple Output, Energy Efficiency.

1   Introduction

Wireless sensor networks (WSN) are been used in a number of emerging applications representing an important technology for the future [1]. However, a paramount concern in relation to the usage of WSN is the energy consumption. Wireless sensor nodes are usually resource constrained platforms, driven by batteries, which limits their energy budget. Additionally, these sensor nodes are usually deployed in areas that are difficult to be accessed, thus making impracticable the replacement of such energy resources. As a result, to overcome such problem, a smart energy resource management is a must.

S. Andreev et al. (Eds.): NEW2AN/ruSMART 2012, LNCS 7469, pp. 379–390, 2012.
© Springer-Verlag Berlin Heidelberg 2012

Considering that the most energy consuming task in the sensor nodes is communication [2], efficient communication mechanisms are highly desirable to reduce the energy depletion in wireless sensor networks. A number of proposals address this problem, such as alternative routing protocols [3], energy aware broadcast [4], among others. A common aspect of these approaches is the exploration of multi-hop communication to spread the energy depletion among the nodes in the network, so that no single node suffers a great decrease in its energy budget due to expensive long distance single-hop transmissions. However, multi-hop does not represent a "silver bullet" to solve the problem, as there are cases in which even a single-hop transmission can perform better than a multi-hop one.

To address the problem of long distance communications in WSN, this paper presents a cooperative multiple input multiple output (MIMO) strategy, in which a number of nodes cooperate to send/receive data aiming at an efficient usage of their energy resources. Theoretical and simulation comparisons are performed providing evidences of the value in applying the proposed techniques to address the energy consumption problem for long distance communications in WSN.

The remaining text is structured as follows: Section 2 presents background context and motivations. In Section 3 a theoretical analysis of energy consumption for single-ho and multi-hop transmissions is presented. Section 4 describes the proposed cooperative MIMO strategy is described, while Section 5 is dedicated to the presentation of simulation results and comparisons with single-hop and multi-hop alternatives. Related works are discussed in Section 6, and concluding, Section 7 summarizes the paper and provides directions for future works.

2   Motivation and Background

2.1   WSN Communications

Energy consumption is the paramount problem that still hinders a larger usage of WSN nowadays [2]. Due to the constrained energy budget that the sensor nodes count with, a careful usage of this resource in each individual node has to be taken into account in order to enlarge the lifespan of the entire network. As all distributed systems, WSNs have their basic functionalities highly dependent on the communication among their nodes. However, as wireless communications are very costly in terms of energy consumption, it leads to an impasse about the usage of the communications. The solution of this impasse has to consider an efficient usage of the communication in order to minimize the waste of energy.

Wireless sensor networks usually present a planar or a hierarchical architecture [5]. In the first one, sink nodes disseminate information in the network, which are transmitted from node to node according to the type of the information being disseminated, such as queries to specific locations for instance, and receive the replies by similar multi-hop communications from the nodes that provide the required information. Hierarchical-based WSNs restrict the more expensive communications to special nodes that exchange messages among them and are responsible for a number of other nodes, as representatives. Examples of such WSNs are clustered-based WSNs, in which those special sensor nodes are called cluster-heads, and they can be more powerful nodes that are in charge for long range communications with other cluster-heads and sink nodes, representing a group of sensor nodes under their responsibility (cluster members). Other configurations of clustered WSNs are possible, in which the cluster-heads and the cluster members are equally powerful, but differences such as remaining available resources, geographical positioning among other criteria can be used to select a given sensor node as the cluster-head [6]. Figure 1 presents an example of these two network architectures.

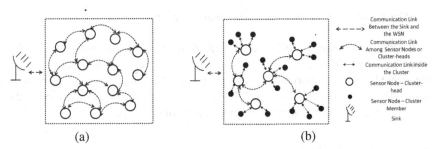

Fig. 1. WSN architectures: (a) Planar; (b) Hierarchical

Regardless of the WSN architecture, the sensor nodes need to communicate with each other, and their corresponding nodes can be close or far from them. In case of short range communication with close neighbour nodes, the problem related to energy consumption is not so significant, but the impact of long range communication for the energy resources consumption have to be considered. Usually, in order to avoid the energy depletion of a single node in an expensive single-hop transmission to another node far from the sender node, multi-hop communication is used, so that the sensor nodes do only perform short range transmissions. This is the case, for example, when a sensor node, or a group of them represented by a cluster-head, has to send replies from a query to the sink, as presented in Figure 2.

Although multi-hoping is considered an efficient communication solution for WSN, there are cases in which single hop represents a better alternative in terms of energy consumption [7]. Such cases are shown in Section 3.1. On the other hand, in several scenarios, the usage of single hop for long distance communication may compromise the entire network lifetime. This is the case, for instance, when the single hop communication results in an unpaired depletion of the energy resources of individual nodes, which is highly undesired.

Despite the specific problem of the energy consumption itself, the above described single hop and multi-hop schemes with fixed data rate may incur in several errors during communication, which require retransmissions, then increasing even more the energy cost associated to communication.

Observing the basic characteristics of a WSN, in which several nodes in a neighbourhood provide similar data, alternative solutions can be created. One of them explores the concept of hierarchical WSN presented above, in which different types of sensor aggregation or sensor fusion techniques are used [8]. In spite of the benefits of such techniques, they may require several communications among the cluster members, depending on the agreement protocol that is used, thus increasing the energy consumption due to communication.

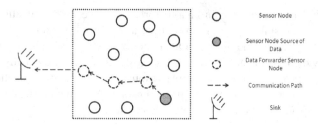

Fig. 2. Multi-hop data communication from a data source sensor node towards the sink

Besides single and multi-hop communication in WSN, cooperative multiple input multiple output (MIMO) schemes are also important alternative solutions that can be considered. As neighbouring sensors need to send the same or even not the same, but data at the same time, cooperative MIMO can be used, as briefly described in the next subsection, and further explored in next sections of this paper.

2.2   Cooperative MIMO

The cooperative MIMO communication considered in this work is based on two steps. The first step estimates the cooperative MIMO channel by using pilot signals. Once the channel is estimated, the new information can be transmitted. In Section 4 this cooperative MIMO approach for communication in WSN is detailed, while the results obtained with this alternative solution are compared with those obtained with single and multi-hop in Section 5.

3   Single-Hop and Multi-hop Transmissions in WSN

From energy efficiency point of view, multi-hop are preferable in relation to single-hop transmissions in WSN. This is due to the minimization of the energy consumed by a single node and due to the more evenly depletion of the energy resources among the sensor nodes along the communication path. However, besides the aspect related to the uniform energy consumption among the sensor nodes, if only the total amount of energy spent in a given communication is considered, the advantage in using multi-hop instead of single-hop is not true for all cases, and depends on the distances between the source and destination nodes, and among the nodes between themselves, as presented and discussed in [7].

According to [7], the energy spent on a communication between two nodes, say i and j, can be divided in two terms, one for the transmission and another for the reception, respectively, according to (1) and (2):

$$E_t(i,j) = \alpha \cdot f_{i,j}, \tag{1}$$

$$E_r(j,i) = \beta \cdot f_{i,j}, \tag{2}$$

where $f_{i,j}$ stands for the bit rate, β is a constant and α is a parameter that depends on the distance between the nodes as follows:

$$\alpha = \begin{cases} a + b \cdot d_{i,j}^{\gamma}, & if \ d_{min} \leq d_{i,j} \leq d_{max} \\ a + b, & if \ 0 \leq d_{i,j} \leq d_{min} \end{cases}, \tag{3}$$

where a and b are constants corresponding to the energy consumption per transmitted bit, $d_{i,j}$ is the distance between nodes i and j, γ is a decay factor according to the propagation model, d_{min} and d_{max} are respectively the minimum and the maximum distances for the communication range.

To study the difference between the single and multi-hop transmission, assume an integer k so that $d_{max} = k \cdot d_{min}$, and consider two nodes, i and j, in a WSN that are d_{max} apart from each other. Node i can transmit to node j either via a single hop transmission, as its range reaches the destination node, or via multi-hop, using the sensor nodes between them, which by their turn are separated from each other by a d_{min} distance. Figure 3 illustrates this scenario.

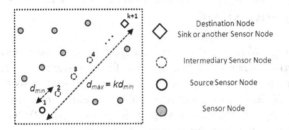

Fig. 3. Distances between communicating nodes and intermediary nodes

The cost in terms of the energy consumed by the single hop transmission of one bit can be expressed by the sum of the energy consumed in the transmission and the reception, and by using (1) – (3), it possible to come to the following expression:

$$E_t(1, k+1) + E_t(k+1, 1) = 2a + b \cdot d_{max}^2 = 2a + b(k \cdot d_{min})^2. \tag{4}$$

On the other hand, for the multi-hop alternative, the total energy consumed for the same one bit communication is:

$$\sum_{i=1}^{k} E_t(i, i+1) + \sum_{j=2}^{k+1} E_r(j, j-1) = k(a + b \cdot d_{min}^2) + k \cdot a = 2ka + bkd_{min}^2. \tag{5}$$

From [7], the condition that makes the single-hop alternative outperforms the multi-hop one is given by:

$$k \leq \frac{2a}{b \cdot d_{min}^2}. \tag{6}$$

Assuming a pair of nodes apart from each other by a distance D, from [8] it is known that a condition to achieve the minimum energy consumption in a multi-hop communication between these two nodes is that the distances between the intermediary nodes be identical (d). Assuming n between them, $d = \frac{D}{n}$. Moreover, the optimum number of hops between a sender and receiver node is given by:

$$n_{opt} = \sqrt{\frac{b}{2a}} \cdot D. \tag{7}$$

The term $\sqrt{\frac{b}{2a}}$ in (7) derives $\sqrt{\frac{2a}{b}}$, which is called characteristic distance (d_{char}), and condition for the optimum number of hop n_{opt}, results in minimal consumed energy is $d = d_{char}$.

From the analysis of these results, the study in [7] concludes that the single-hop communication is more efficient that the multi-hop one when the distance between the communicating nodes is less than d_{char}. Observing the situation illustrated in Figure 3, this means that if $d_{max} < d_{char}$ the single-hop communication is more energy efficient than the multi-hop one.

4   Energy Efficient Transmissions in WSN Based on Cooperative MIMO

Taking advantage of the cooperative nature of the sensor networks operation, cooperative MIMO can be introduced in these systems to provide a reliable communication, by diminishing the bit error rate (BER), possibly requiring less energy resources. Instead of a conventional arrangement of multiple antennas in a single sensor as in the traditional MIMO, in cooperative MIMO system, multiple sensors cooperate to transmit and receive data. Figure 4 presents an example in which two clusters of sensors establish a communication as a MIMO system. In the figure it is also possible to observe another possible situation in which a cluster of sensors establish communication with a single sensor similar to a single input multiple output/ multiple input single output (SIMO/MISO) system. Notice that the term "cluster" used to describe the situation in Figure 4 has not the same meaning that it has in the part of the paper that is used to explain the hierarchical WSN above (Section 2). In Figure 4 there is no difference among the sensor nodes, and the term "cluster" is used to identify a group of cooperating sensor nodes.

Using this approach, if two sensors near to each other cooperate to transmit information, and two sensors on a far cluster cooperate to receive data, the efficiency is effectively doubled, as two symbols can be transmitted over the same time slot.

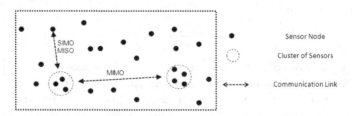

Fig. 4. Cooperative MIMO communication between clusters of sensors and SIMO/MISO communication between a cluster an individual sensor

Figure 5a presents an example of a MIMO systems composed of Q transmitter sensors and P receiver sensors. The communication channel between the i-th transmitter sensor to the j-th receiver sensor is given by $h_{i,j}$.

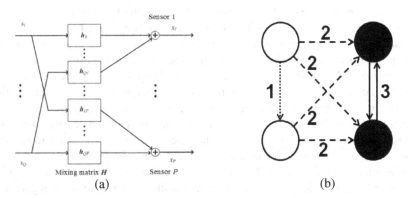

(a)   (b)

Fig. 5. Examples of: (a) cooperative Q by P MIMO system; (b) 2 by 2 MIMO system

Considering the MIMO system in Figure 5, the output of the j-th sensor $x_j(n)$ is given by:

$$x_j(n) = \sum_{i=1}^{Q} h_{i,j} s_i(n),\tag{10}$$

where $s_i(n)$ is the transmitted symbol at the n-th time instant by the i-th source transmits.

Considering the Q input signals, the P output signals, and the mixing or channel matrix H, we can rewrite (10) in the matrix form as follows:

$$X = HS.\tag{11}$$

As mentioned in Subsection 2.1, the Bit Error Rate is going to be computed in two steps.

In the first step given X and S, we can solve (11) and find $\hat{H}$, which is an estimate of H. The symbol matrix S contains the pilot signals. For low Signal-to-Noise Ratio (SNR), the estimation of H can be severely degraded.

After the channel estimation step, the Q transmitter sensors send unknown symbols, which are estimated by using $\hat{H}$ and (11). Then, from this step, $\hat{S}$ is obtained, which are used to compute the Bit Error Rate (BER).

Figure 5b presents an example of 2 by 2 MIMO system in which the numbers represent the steps that are described as follows: (1) The transmitting sensors exchange the information that needs to be transmitted; (2) Both sensors transmit different symbols at the same time slot; (3) The receiving sensors exchange the received information so that the original symbol sequence can be obtained.

The next section presents the performance evaluation of the aforementioned cooperative MIMO scheme in a WSN in comparison to single-hop and multi-hop.

5   Experimental Results

5.1   Simulation Setup

The cooperative MIMO channels are generated by using the IlmProp [13], which is a flexible geometry-based multi-user MIMO channel modeling tool. The evaluation of the efficiency of the cooperative MIMO communications for the WSN is done by measuring the Bit Error Rate (BER). Simulations are made to determine the BER for the cooperative MIMO, the single-hop and the multi-hop techniques. A random pair of sensors that are about 100 m apart from each other is selected presenting the possibility for multi-hop communication via other nodes that are 30 m apart from each other, i.e. 30 m distance for each hop. The single-hop approach involves only the sensor that has the data that needs to be transmitted and the sensor that will receive this data. Both sensors are 100 m away as mentioned above, so communication requires a very high power at the output of the antenna, since the free space attenuation is very high. 30 pilot symbols are transmitted to estimate channel gain and the rotation it causes on the symbol constellation. After 1000 data symbols are transmitted, the BER is estimated comparing the transmitted symbols with the received ones.

For the multi-hop approach the process is very similar, but the distance the data needs to travel is split in three, with sensors cooperating along the way to transmit the data over smaller distances, thus requiring less power. 30 pilot symbols are transmitted between each pair of sensors so the channel gain can be estimated, after that 1000 data symbols are transmitted similarly as in the single hop approach. It is worth to highlight that for a sensor start transmitting the data to the next sensor it first needs to receive the entire data string and decode it, which may imply in a significant delay. The BER is estimated comparing the transmitted symbols with the symbols received at the last sensor, i.e. last hop.

Direct communication between the two nodes is used, then a scenario with 3 hops is simulated, and then the MIMO scenario using a 2 × 2 configuration. The sensors are distributed according a random pattern following Poisson distribution in two dimensions, for all simulation runs. Following this distribution, 50 sensors are displaced in an area of 400 × 400 m^2. Figure 6a presents the simulated scenario without multipath components and also exemplifies each solution. Note that the selected sensors are varying during the simulations.

The first simulation considers only the line of sight (LOS); the second simulation considers 10 obstacles that provide 10 different multi-path components, in which each obstacle represents a reflection of the signal that presents different amplitude and phase from the LOS signal, resulting in heavy interference. The simulation conditions are the same for all experiments, the sensors' position are preserved, only the multipath components changed from one simulation to the other. For the cooperative MIMO case, the symbols to be transmitted are first transmitted to the cooperating sensors of the transmit cluster. Perfect synchronization among sensors is assumed.

The channel coefficients are estimated by means of a set of 30 pilot symbols, followed by the transmission of 1000 data symbols. The carrier frequency is 2.4GHz and flat fading over the transmission bandwidth is assumed. The simulation ranges from -10 to 10 dB SNR and 1000 independent Monte Carlo runs are assumed for each SNR. The results for the energy cost are based on the characteristics of the Berkeley Mica2 Mote [9].

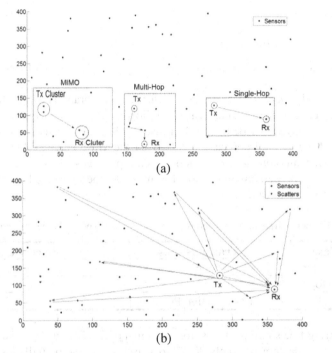

Fig. 6. Simulated scenario: a) LOS only exemplifying the Cooperative MIMO, Multi-hop and Single-hop solution; b) LOS and NLOS with 10 multipath exemplifying only the Single-hop solution

5.2 Results and Discussion

Figure 7 presents the average BER as function of the SNR. Direct single-hop communication requires a high output power at the transmitter for the signal to be properly decoded. Multi hop communication requires much less transmitting power across the transmitters, but grows increasingly costly as the number of hops increases, due to the fact that receiving can cost up to twice as much power as transmitting a low power signal. In the MIMO case, the same signal can be transmitted by two or more sensors, and decoded by one or more sensors, signals can be transmitted at a lower power, thus requiring less energy. This further improves energy efficiency over the network when long distance transmissions need to be made and direct transmission is not possible. This impossibility can be either due to a high number of multi path components or due to the fact that there are no intermediary nodes available to relay the packages across the network.

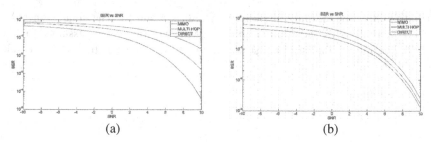

Fig. 7. BER vs. SNR results: a) LOS only; b) LOS and NLOS with 10 multipaths

Comparing the energy spent by the 3 different configurations (single-hop, multi-ho and MIMO) to achieve the same BER, the following results are acquired. Considering the LOS case only, for a BER of 10^{-6} a SNR of 12.5 dB is necessary for the direct case, 10 dB for the multi-hop case and 7 dB for the MIMO case. Table 1 presents the amount of transmissions and receptions of data packages of the same size that are needed to perform the communication 100 m across the network. Figure 8a presents the relative energy consumption.

Table 1. Number of transmissions and receptions for each type of the performed simulations

Case	Transmissions	Receptions
MIMO (2 × 2)	6	6
Multi-Hop	3	3
Single-Hop	1 (Very High Power)	1

Considering free space path loss, for the signal of 2.4 GHz used in the simulations, the results provide approximately 80 dB loss over 100m and 69.6 dB loss over 30 m, a loss approximately 11 times higher for a distance only 3.3 times longer.

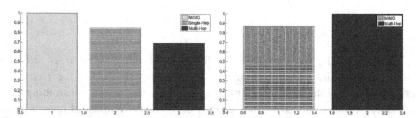

Fig. 8. Relative Energy Consumption: a) First comparison among MIMO, single-hop and multi-hop; b) Second comparison between MIMO and Multi-hop

At first impression (Figure 8a) the 2 × 2 MIMO case seems to have the highest cost, but if two sensors in the transmit cluster and two at the receive cluster are allowed to cooperate, the proposed cooperative MIMO approach presented in Section 4.2 can be used. Now, with two symbols being transmitted over the same time slot, the energy cost per symbol is decreased by 50%. This configuration can achieve ranges much longer than the direct communication (single-hop), with a lower BER. As the distance increases, the amount of multi-hops necessary to get the data package to its destiny becomes increasingly costly, and if the data that needs to be transmitted is sensitive to delay, it might not be possible to use to multi-hop approach. Comparing the costs for the multi-hop and MIMO alternative to get a message across a distance in which 5 multi-hops are needed, the results show that MIMO is more energy efficient than multi-hop, as can be observed in Figure 8b.

In this case the energy efficiency of the MIMO case surpasses the efficiency of the multi-hop approach, which becomes increasingly costly especially due to the number of receptions that need to be performed. Receptions can cost twice as much energy in transmissions of a low power signal than in transmissions of high power.

Transmitting data over the direct approach is the best option only if the distance is smaller than d_{char} (see Section 3), or the multi-hop approach is not an option and the number of multi-paths of the signal is minimum.

The multi-hop approach is the best option if the data is not sensitive to delay and if the number of multi-hops necessary does not make the transmission cost bigger than the MIMO approach. It is also the case to use it if the MIMO approach is not available due to the absence of other sensors in the neighborhood to cooperate with in the reception.

A SIMO approach, a special case of MIMO (i.e. 1×2), is an option in case the multi-hop approach is not available, the number of multi-paths of the signal is too much for the single-hop and there is no sensor available to cooperate at the reception.

6   Related Work

The use of MIMO presents enhancement in the energy efficiency of WSN [10][11]. However, MIMO is also used for spatial diversity and multiplex gain. In [12] techniques for energy efficient communication aiming at to diminish the total energy transmission and the energy consumed in the processing performed by the circuit for MIMO and SISO systems were proposed. In [10] the increase in the overhead involved in the training of the MIMO systems is studied. In [11] the efficiency in the cooperative transmission of space-time block codes, STBC, is analysed. Besides the innovative usage of cooperative MIMO proposed in our paper, we presented a comparative study considering single and multi-hop alternatives, which was not performed so far to the best of authors' knowledge.

7   Conclusions and Future Work

This paper presents a study comparing single-hop, multi-hop and cooperative MIMO-based transmissions over WSNs. Simulation results were presented, discussed and compared to theoretical ones. As observed by the analysis of the obtained results, the cooperative MIMO approach becomes the best option as the number of multi-hops increase too much, (using the MICA2 power model assumed on this paper, when 4 hops are necessary, the MIMO approach costs 8.7% more power, for number of hops higher than that, the cooperative MIMO approach starts to be more power efficient than the multi-hop approach) or if the data is sensitive to delay. Our results corroborate that cooperative MIMO is a powerful option for data transmission over long distances across the WSNs.

Directions of future work are the investigation of possible enhancements in the proposed cooperative MIMO technique, in particular by considering sensor nodes' mobility, which was not addressed in the present paper.

Acknowledgements. The authors thank the Research Financing Agency of Federal District in Brazil (FAPDF) for the provided support to develop this research.

References

1. Business Week: 21 ideas for the 21st century, August 30, pp. 78–167 (1999)
2. Mini, R.A.F., Loureiro, A.A.F.: Energy in Wireless Sensor Networks. In: Garbinato, B., Miranda, H., Rodrigues, L. (eds.) Middleware for Network Eccentric and Mobile Applications, pp. 3–24. Springer (2009)
3. Goyal, D., Tripathy, M.R.: Routing Protocols in Wireless Sensor Networks: A Survey. In: Second International Conference on Advanced Computing & Communication Technologies, pp. 474–480 (2012)
4. Durresi, A., Paruchuri, V., Barolli, L., Raj, J.: QoS-energy aware broadcast for sensor networks. In: Proceedings of 8th ISPAN, p. 6 (2005)
5. Akyildiz, I.F., Su, W., Sankarasubramaniam, Y., Cayirci, E.: A survey on sensor networks. IEEE Communications Magazine 40(8), 102–114 (2002)
6. Boyinbode, O., Le, H., Mbogho, A., Takizawa, M., Poliah, R.: A Survey on Clustering Algorithms for Wireless Sensor Networks. In: 13th International Conference on Network-Based Information Systems, pp. 358–364 (2010)
7. Chen, C., Ma, J., Yu, K.: Designing Energy-Efficient Wireless Sensor Networks with Mobile Sinks. In: Sensys 2006. ACM (2006)
8. Nakamura, E.F., Loureiro, A.A.F., Frery, A.C.: Information fusion for wireless sensor networks: Methods, models, and classifications. ACM Computing Surveys (CSUR) 39(3), Article 9, 1–55 (2007)
9. Hill, J., Culler, D.: A wireless embedded sensor architecture for system-level optimization, 12 p. Technical Report, UC Berkeley (2002)
10. Jayaweera, S.K.: Energy analysis of MIMO techniques in wireless sensor networks. In: 38th Annual Conf. on Information Science and Systems, Princeton, NJ, USA (2004)
11. Li, X., Chen, M., Liu, W.: Application of STBC-encoded cooperative transmissions in wireless sensor networks. IEEE Sig. Proc. Letters 12(2), 134–137 (2005)
12. Cui, S., Goldsmith, A.J., Bahai, A.: Energy-efficiency of MIMO and cooperative MIMO techniques in sensor networks. IEEE Journal on Select. Areas Communications 22(6), 1089–1098 (2004)
13. Del Galdo, G., Haardt, M., Schneider, C.: Geometry-based channel modeling in MIMO scenarios in comparison with channel sounder measurements. Advances in Radio Science - Kleinheubacher Berichte 2, 117–126 (2004)

Namimote: A Low-Cost Sensor Node for Wireless Sensor Networks

Ivan Müller[1], Edison Pignaton de Freitas[2,3,4], Altamiro Amadeu Susin[1],
and Carlos Eduardo Pereira[1,2]

[1] Electrical Engineering Department, Federal University of Rio Grande do Sul, Brazil
[2] Institute of Informatics, Federal University of Rio Grande do Sul, Brazil
[3] Electrical Engineering Department, University of Brasília, Brasília, Brazil
[4] Brazilian Army Technological Centre, Rio de Janeiro, Brazil
{ivan.muller,altamiro.susin}@ufrgs.br,
epfreitas@inf.ufrgs.br, cpereira@ece.ufrgs.br

Abstract. Namimote is a wireless sensor node developed in the context of the Namitec project that aims to provide a low cost and multi-purpose sensor node platform for wireless sensor networks. This paper describes the Namimote architecture, highlighting its features and main characteristics. A case study is also presented, in which the performance of the sensor node is evaluated.

Keywords: Wireless sensor networks, Wireless sensor node, LR-PAN transceiver.

1   Introduction

Wireless sensor networks (WSN) are being used in a number of emerging applications representing an important technology for the future [1]. Different parameters determine the efficiency of a WSN to perform their sensing tasks, specially the communication protocols used by the sensor nodes to exchange data and the sensor node architecture, which needs to be designed to address the specific requirements of WSN operation, such as robustness and energy consumption awareness. A paramount concern in relation to the usage of WSN is related to energy consumption. Wireless sensor nodes are usually resource constrained platforms, driven by batteries, which limits their energy budget. Additionally, these sensor nodes are usually deployed in areas that are difficult to be accessed, thus making virtually impracticable the replacement of such energy resources. As a result, to overcome such problem, a smart energy resource management is a must.

Considering that the most energy hungry task in the sensor nodes is communication [2], efficient communication mechanisms are highly desirable to reduce the energy depletion in wireless sensor networks. A number of proposals try to address this problem in different abstraction levels, such as alternative routing protocols [3][4] in the network level, and energy efficient architectures in the node level [5], among others.

S. Andreev et al. (Eds.): NEW2AN/ruSMART 2012, LNCS 7469, pp. 391–400, 2012.

Usually, WSN are deployed using a great number of nodes, which demands the development of low cost sensor node platforms that address the specific WSN issues, such as those mentioned above. Low cost sensor node platforms allow the deployment of large scale WSN, which is the existing demand nowadays. To get an idea of the wide coverage of WSN subject, the following applications can be mentioned:

i) In-vehicles networks: the cabling used in vehicles results in an increased final cost, and cause a large increase in weight. This weight leads to extra fuel consumption and reduced charged capacity. Added to this, the cost of copper used in cable manufacturing has been increasing due to of global reserve reduction;

ii) Home automation: Easy installation of sensor nodes and actuators aiming at the automation of residential and commercial buildings is important because of the cost savings afforded by these networks. Without the need to use ducts to accommodate communication cables, the WSN can be installed on demand;

iii) Industry: The cost of cabling is a big incentive for many plant managers to consider the use of wireless control and monitoring systems, since they can promote reductions of 20% to 80% of the cost of installation. Moreover, the network can be installed on demand, according to the growth or modification of the plant;

iv) Assistive technology: the increasing development of home automation as a facilitator in the performance of domestic activities provides convenience, comfort and safety for users. These benefits can be extended to people with some kind of special need, increasing their autonomy and quality of life. The integration of WSN in this type of automated environment aims to provide flexibility in installing these systems;

v) Disaster management support: in the event of an earthquake, WSN nodes can be deployed around the site, launched from an airplane, for example. Acoustic sensors can be employed in order to capture voices. The ad-hoc network formed by the sensor nodes can be used when local telecommunications infrastructure are damaged;

vi) Environmental monitoring: The quality of air, water, and noise levels can be continuously checked in large cities by means of WSN. The detection of the presence of pollutants such as hydrocarbons in water is another important application. Habits of animals particularly sensitive to climate change can be monitored in order to rapidly detect changes to reverse the impacts;

vii) Precision agriculture: WSN are employed in agriculture to maximize productivity, to quantify the losses and control activities in the field. The applications vary from soil moisture monitoring for irrigation purposes to animals tracking and tracing in the field.

Other applications that can be cited are lighting control, traffic control, border monitoring, fire detection, landslides and monitoring of structures such as buildings and bridges.

In a near future, WSN will be ubiquitous, when the sensor nodes will be tiny and cheap enough for employment on a global scale.

Because of such importance, comprehensive study and development of WSN have been considered a strategic issue for many countries. For this reason, many universities and companies around the world are developing and / or marketing their motes. Intel, Berkeley, Oracle, Yale and Digi are some examples. The development of a sensor node for applications in research on WSN is therefore a strategic matter. Technology dependence in this area induces unnecessary expenses and lack of knowledge. The usage of ready to use modules, non-free software and other constrains leads to technology dependence. In order to conduct different WSN researches such as energy consumption optimization, packet routing strategies, techniques for data dissemination and intelligent sensing, it is critical to have total control of the employed technology. In this paper, a WSN node called Namimote is presented. It is a complete device for WSN, including on board sensors, MCU, battery management, mass storage and radio transceiver in a small 35 by 55 mm board. With this device, it is possible to develop several different researches because the entire platform is accessible. The development of hardware, software and applications are presented and discussed in this paper.

The remaining text is structured as follows: Section 2 presents related work in the area, bringing information about some wireless sensor nodes available nowadays. In Section 3 the Namimote is presented, describing its characteristics and usability. Section 4 describes a case study in which the Namimote was successfully used and evaluated, while Section 5 summarizes the paper and provides directions for future works.

2   Related Works

Several companies and universities are developing their own sensor nodes for the above mentioned matters. In [5] a customizable wireless sensor node called FemtoNode is presented. It contains a customizable ASIC and a wireless communication interface, which are configured according to application requirements. The node uses the FemtoJava processor [6], a stack-based microcontroller that natively executes Java byte-codes. It implements an execution engine for Java in hardware, through a stack machine that is compatible with the Java Virtual Machine specification. It includes a wireless transceiver form Texas Instruments CC2420, which utilizes the IEEE802.15.4 standard communication protocol targeted to wireless sensor network applications with a low data rate. A module adapter described in VHDL, implements the interface with the wireless transceiver. A great advantage offered by the FemtoNode is the possibility to customize the sensor node to a specific WSN application, resulting in a very efficient platform targeting that application. However, this same feature can be seen as a drawback, considering that it makes the sensor node not general purpose, which may hinder its usage to other applications.

SunSPOT (Sun Small Programmable Object Technology) [7] is an embedded platform developed by Sun Microsystems Laboratories to program Java applications for WSN. It is a small programmable wireless device that is able to communicate and interface with other system parts on a network level using the IEEE 802.15.4 MAC and PHY layers over a Wireless Personal Area Network (WPAN). The SunSPOT is based on a 32-bit 180 MHz ARM920T core that includes 4 MBytes of flash memory and 512 KBytes of RAM memory. SunSPOTs are easy to program, as it uses Java as programming language, but it has the drawbacks of having very short communication range, and it is relatively expensive, which prohibits its massive usage in large scale deployments.

IMote 2 (Intel Mote 2) [8] is a sensor node platform designed for highly demanding WSN applications. It is equipped with a PXA271 XScale low power processor and a wireless MMX coprocessor. It has 256kB SRAM, 32 MB flash and 32 SDRAM memory. A 802.15.4 radio transceiver from ChipCom (CC2420) provides its wireless communication support with a nominal range of about 30 meters, which can be extended with an external antenna. The great plus of this sensor node is the processing power and the ability to accelerate multimedia operations, by using its MMX coprocessor. This feature makes this node well suited to multimedia applications in WSN. A drawback is the short range distances, besides the high cost and power.

The Berkeley Mica Motes (Mica 2 and Mica 2 dot) [9] are wireless sensor nodes created at University of California Berkeley and further developed and commercialized by the Crossbow company. They are equipped with an Atmel Atmega 128L processor, having 4K RAM and 128K Flash memory capacity. A ChipCom CC1000 provides the wireless link for short communication range (around 15 to 20 meters). The Mica motes provides a convenient platform for experimental purposes for an affordable cost and the flexibility to add/remove components by adding boards on top the main board that accommodate the basic components. Again, its main drawback is the short range.

There are a number of sensor node platforms available on the market, developed by both universities and companies with different designs, targeting different WSN applications. A very complete survey of these platforms can be found in [10]. Namimote differs from those presented here in several aspects. The most relevant is the wide range of usage, based on its intrinsic characteristics as well as the modularity. For instance, the sensor node can act as a base station to connect several nodes or it can acquire data and store locally, in the case of a communication failure. Some of these features can be found in certain degree in some of the sensor nodes mentioned above. However, for the best of our knowledge, Namimote is more complete and presents lower final cost when compared with commercial versions.

3   Namimote Development

The Namimote is the developed sensor node for the Namitec project, part of a major investment from Brazilian government for nano and micro technologies [11]. It is the effort of the WSN group of the project, to develop a sensor node based on off-the-shelf components, to be the future basic design for an entire WSN SoC development. The Namimote block diagram can be seen in Figure 1. The central control of the node is the microcontroller unit (MCU) which has an embedded IEEE 802.15.4 transceiver. The system is supplied by a 3.7 Volts Lithium battery whose charge is obtained by a standard USB port. The charge is monitored by the MCU and a dedicated integrated circuit. There are three onboard sensors: luminosity, temperature and three-axes accelerometer. A general purpose input/output (GPIO) with digital and analog ports is provided for any type of expansion. The RF port of the sensor node feed a power amplifier (PA) to provide long range links. The features of the sensor node are explained in details in the next subsections.

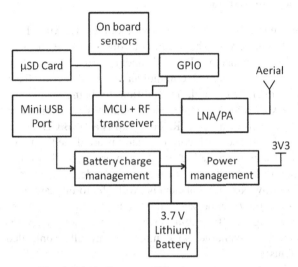

Fig. 1. Block diagram of Namimote sensor node

3.1   System-on-a-Chip

The employed microcontroller unit is part of a system-on-a-chip that has a 2.4 GHz IEEE 802.15.4 transceiver included. The main reasons for its choice are the 32-bit ARM 7 core, the integrated transceiver, several peripherals and hardware acceleration modules for MAC and security. The device can be programmed through a standard JTAG port but can also accept the firmware by the serial port or by the RF transceiver. There are 128 kB of flash memory for the firmware but it cannot run from there. The firmware is mirrored in the 96 kB RAM that must be shared between data and code. The remaining Flash area can be used for non-volatile data. Also, it has 80 kB of ROM, whose code is the IEEE 802.15.4 MAC library and other utilities.

3.2   Onboard Memory

The Namimote has two storage units to retain acquired data locally. They can be used independently in the case it has no RF connection or in other situations, according to the application. The AT25DF321 32 Mbits serial flash memory is a fast storage unit that presents low power consumption. The other memory unit is a retractable micro SD card. The advantage of this type of memory is the ability to retain large data in contrast with the serial flash. On the other hand, the consumption of a micro SD card is much higher and the information is not transferred as quickly as in the serial flash.

Still, it is a good choice to collect large amount of data during a long time, in a typical data logger application. It is also possible to store data in MCU's non-volatile (NVRAM). This is done usually for configuration data, which is retrieved during the system start-up.

3.3   Communication System

The MC13224 SoC has several communication modules: USART, I^{2}C, SPI, as well as the IEEE 802.15.4 transceiver. The Namimote does not use the I^{2}C port, but it is available in GPIO for expansions. The SPI module is used by the SoC to access the NVRAM and is used for internal communication with serial flash memory, the accelerometer, the micro SD card, in addition to being available on the GPIO expansion buses. Two USART modules are available. One is employed for the USB communication, by means of a FT232R. This chip is a USB CDC compliant and it is powered by the USB port itself. RF communications are done through a 2.4 GHz ISM band front-end connected to the MCU. It allows more than 1 km links, depending on environmental conditions, positioning of the sensor nodes, output power and antenna type. Also, a LNA (Low Noise Amplifier) is available, to increase even more the link range. The LNA sensitivity can be selected (0 or 10 dB) and the PA presents 20 dB of gain, typically providing 22 dBm output power. As more sensitivity and output power are used, less battery is available, so it must be carefully controlled by means of firmware adjustments.

3.4   Power Management

The Namimote is powered by a 3.7 V Lithium-ion battery. The use of this type of battery is justified by the increased energy density in relation to other types. However, Li-ion batteries are very sensitive to high temperatures and do not tolerate deep discharges. High recharge currents may cause overheating and eventually lead to explosions. Therefore it is necessary to monitor and control the recharge as well as the discharge processes. Being supplied by the USB port (5 Volts), an integrated circuit charges the battery monitoring and controlling its voltage and current. If the battery is deeply discharged, the controller enables a soft-start appropriate for the situation. The charging process is autonomous, i.e. do not dependent on the firmware. On the other hand, the discharge process must be controlled by software, by monitoring the battery voltage. The voltage is frequently monitored and the system enters in a power down mode when it reaches a 3 Volts minimum.

To ensure a constant supply voltage to all components, Namimote uses a high efficiency buck-boost voltage regulator. It is a DC-DC converter that provides an average voltage of 3.3 V for the circuit. The step-up / step-down converter switches an internal transistor at 1 MHz frequency. The input voltage can vary from 2.5 to 5.5 V which permits the use of different primary power sources from energy harvesting applications, for instance. The power consumption is a trade off, because it is dependent on radio cycling and output power. At 22 dBm, an instantaneous current of 140 mA appears at the supply circuit, which drains the battery fast. So, it is imperative to manage radio cycling and power to obtain high autonomy.

3.5   Onboard Sensors

The onboard sensors measure temperature, light, and acceleration. The temperature sensor has also one internal 2.56 V precision reference voltage for the MCU 12-bit analog-to-digital converter. The temperature sensitivity is 10mV / °C and it can operate in the range of -40 °C to +125 °C (although the remaining circuit cannot). At first, the sensor does not require calibration, especially in applications where high precision is not important. The manufacturer reports an error of less than 1.2 °C at 25 °C. The IC typically consumes 150 µA and can be disabled by the MCU when it is idle, then going down to 1 µA.

The employed light sensor encapsulates a photodiode with a current amplifier. The output voltage is proportional to the incident light and can vary from 1 to 1000 lux approximately. The maximum efficiency is around 600 nm light wave length. The sensor settling time is approximately 100 ms. During this period, the signal is unstable and any acquired information should be discarded.

The accelerometer is a capacitive triaxial MEMs device that can measure tilt or vibrations in the range of ± 2, 4, 8 and 16 g. It has an internal AD converter and thus provides data output in digital form. The resolution is fixed to 10 bits always keeping 4 mg/LSB. The chip has also an internal 32 level FIFO memory that can be used for temporary data storage in order to reduce the MCU operations. The communication with the MCU is through the SPI bus. The sensing measurement bandwidth is 1600 Hz and higher clock rate 5MHz. The current consumption is typically 145µA when operating at high data rates and can be reduced to 1µA when in standby.

3.6   PCB Layout Considerations

The 2.4 GHz ISM band of IEEE 802.15.4 demands careful circuit and PCB design. At this range, even a short length PCB track could have enough inductance to lead the circuit to instability [12, 13]. Mismatched impedances produce standing waves, and consequently reduced circuit performance. Output impedance is function of width, thickness and length of PCB tracks as well as the dielectric constant of the material. RF PCB tracks must be as short as possible and placed at component side in order to minimize vias whose inductance is relevant in 2.4 GHz circuits. Copper pours are placed everywhere to fulfill large spaces but they are avoided in small ones because they can act as RF radiation/absorption pads. Passive components are selected taking into account their resonance frequency because after the resonance point, capacitors acts like inductors and vice-versa. The 4-layer Namimote PCB is designed taking into account the RF signals at the front end.

3.7   Final Prototype

Prototypes are produced to validate the design. Several tests are performed and reveal the feasibility of the sensor node. Tests included not only functionality but power consumption, noise produced by the energy management circuits, spectral occupancy, bit error rate and achieved range. These parameters are considered the most important, that validate the hardware design. The other ones are details to be normally corrected in the final production version. Figure 2 shows one of the prototypes. It is possible to see all the main components, the GPIO for piggybacking boards and the UMCC connector for the antenna. Other versions included a PIFA antenna that can be detached from the board and replaced by an external one, as needed.

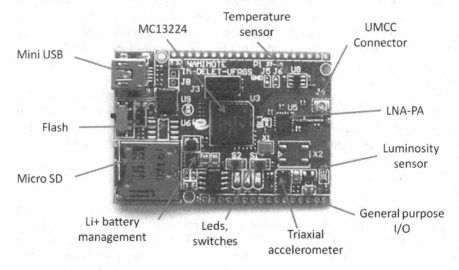

Fig. 2. Prototype of Namimote, with its main components highlighted

4   Case Study

To evaluate the sensor node, a demonstration firmware and PC application software are developed. The firmware provides communication between a master radio, connected to a PC and five slave radios, in a simple star topology. The collected data is sent to the PC through USB connection from the master radio. The simple MAC from Freescale is used, to perform the evaluation of the sensor node. Other stacks are tested, including plain IEEE 802.15.4, Zigbee [14] and even WirelessHART [15,16], an industrial grade protocol. Although not tested, other solutions are available to be used within the Namimote's MCU, such as IP-centric Contiki [17][18].

To evaluate the wireless sensor performance, a C++ software is developed to run on a PC, together with the embedded software. The software shows the up and down link between two peer devices by means of link quality indicators along the time. Figure 3 shows the graphical interface, developed for five slave radios. The software communicates with the master radio through an emulated serial port and plots the collected data of the link between two peer devices as well as the sensors information. The provided data are: the sensor node's temperature, luminosity, three-axes resultant acceleration, link quality indication and battery voltage. All data are plotted in time scaled graphs that can be stored and retrieved. Adjustable parameters include RF power level and channel, passed by the RF link to the slave radios. Other parameters are also available, such as scan speed and lost packets counter.

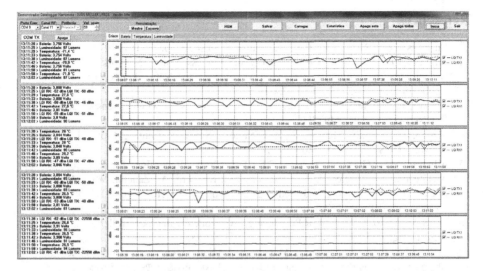

Fig. 3. GUI demo software. On left part, five text memo objects where collected data are showed. On right part, five graphs, one for each slave node. These may contain information about the link quality, battery voltage, temperature and luminosity, according to the selected page in the top of the GUI.

The obtained data are presented in four different graphs and saved on the hard disk in XML format after two hours of collection. It takes about 35 MBytes of memory on hard disk for 7 days of data collection, depending on the scan rate. The user interaction is all done through the graphical interface without the need for adjustments or setup. With this software and firmware demonstrators, it is possible to evaluate the prototypes, their functionality and behavior.

Although a full application can be developed based on this software and firmware, much more can be done with Zigbee or Contiki. However, for evaluation purposes, the developed test systems are sufficient to verify the feasibility of the design and even to develop simple applications.

5   Conclusion and Future Work

This paper presented the Namimote, a multi-purpose low cost sensor node for WSN. The proposal of this platform is a hardware that provides access to all its components, allowing the nodes adaptation and customization according to the user needs. The sensor node is depicted in the text through the components description and design details. Several tests were done and revealed the feasibility of the proposal. Future work include the use of Namimote in different WSN applications such as air monitoring, integration with mobile sensor nodes, animal tracking and monitoring, and a complete WirelessHART battery operated field device.

References

1. Business Week: 21 ideas for the 21st century, August 30, pp. 78–167 (1999)
2. Mini, R.A.F., Loureiro, A.A.F.: Energy in Wireless Sensor Networks. In: Garbinato, B., Miranda, H., Rodrigues, L. (eds.) Middleware for Network Eccentric and Mobile Applications, pp. 3–24. Springer (2009)
3. Akkaya, K., Younis, M.: A survey on routing protocols for wireless sensor networks. Ad Hoc Networks 3, 325–349 (2003)
4. Magoo, C., Shefali, Sharma, S.: Comparative Study of Routing Protocols in Sensor Network. IJCST 3(1), 678–680 (2012)
5. Allgayer, R.S., Götz, M., Pereira, C.E.: FemtoNode: Reconfigurable and Customizable Architecture for Wireless Sensor Networks. In: Rettberg, A., Zanella, M.C., Amann, M., Keckeisen, M., Rammig, F.J. (eds.) IESS 2009. IFIP AICT, vol. 310, pp. 302–309. Springer, Heidelberg (2009)
6. Ito, S.A., Carro, L., Jacobi, R.P.: Making Java work for microcontroller applications. In: IEEE Design and Test of Computers, pp. 100–110 (2001)
7. SUN. Sun SPOT Programmer's Manual. Sun Microsystems Laboratories/Oracle (2010), `http://sunspotworld.com/docs/Yellow/SunSPOT-Programmers-Manual.pdf` (accessed in: June 2011)
8. INTEL. Intel Mote 2 Engineering Platform Data Sheet. Intel Coorporation (2006), `http://ubi.cs.washington.edu/files/imote2/docs/imote2-ds-rev2.0.pdf` (accessed in: December 2011)
9. Hill, J., Culler, D.: A wireless embedded sensor architecture for system-level optimization, 12 p. Technical Report, UC Berkeley (2002)
10. Wireless Sensor Node Hardware: A Review. In: Proceedings of the 2008 IEEE Sensors Conference, pp. 621–624 (2008)
11. INCT Namitec, `http://namitec.cti.gov.br`
12. Gupta, K.C.: Microstrip Lines and Slotlines. Artech House Publishers (1996) ISBN 0-89006-766-X
13. Bowick, C.: RF Circuit Design. Newnes Publishing (1982) ISBN 0-7506-9946-9
14. ZigBee Alliance, `http://www.zigbee.org`
15. Song, J., et al.: WirelessHART: Applying Wireless Technology in Real-Time Industrial Process Control. In: Real-Time and Embedded Technology and Applications Symposium, pp. 377–386 (2008)
16. Zhu, X., Han, S., Mok, A., Chen, D., Nixon, M.: Hardware Challanges and Their Resolution in Advancing WirelessHART. In: 9th IEEE International Conference on Industrial Informatics, INDIN (July 2011)
17. Contiki OS, `http://www.contiki-os.org`
18. Using the Freescale MC1322x Series ARM7 Processor with integrated 802.15.4, `http://mc1322x.devl.org/`

Fast Restoration of Connectivity
for Wireless Sensor Networks

Nourhene Maalel[1,2], Mounir Kellil[1], Pierre Roux[1], and Abdelmadjid Bouabdallah[2]

[1] Communicating Systems Laboratory
DRT/LIST/DIASI/LSC
Gif sur Yvette, France
name.surname@cea.fr
[2] Heudiasyc Laboratory,
UMR CNRS 6599, UTC
Compiegne, France
bouabdal@hds.utc.fr

Abstract. Node failures represent a fundamental problem in wireless sensor networks. Such failures may result in partitioned networks and lose of sensed information. A network recovery approach is thus necessary in order to ensure continuous network operations. In this paper, we propose CoMN2 a scalable and distributed approach for network recovery from node failures in wireless sensor networks. CoMN2 relies on a new concept called network mapping which consists in partitioning the network into several regions of increasing criticality. The criticality is set according to the energy, the traffic distribution and the deployment of nodes. Using this network mapping, our solution CoMN2 ensures the continuous network activity by efficiently swapping nodes from low critical area to highly critical area when required. Simulation results prove the effectiveness of our approach and show that the obtained improvement in terms of lifetime is in the order of 40%.

Keywords: WSN, fault tolerance, sensor failure.

1 Introduction

Because of the unprecedented prospects they offer and the idea of eliminating human intervention, wireless sensor networks (WSNs) have perceived a tremendous attention from the research community in recent years [1]. Upon their deployment, WSN form a network composed of a set of sensor nodes and one or several sink nodes. Sensors role is to make measurement then to relay it as routers from sensor to sensor toward a sink node. We do also consider the presence of specific entity called mobile node which has the ability of processing data and making the appropriate decision. WSNs encounter many challenges [1] such as energy efficiency and limited battery. Our contribution focuses on the issue of sustainable connectivity.

Uneven energy distribution and consumption or accidental failures due to harsh environments may cause sensors to die arbitrarily [3]. These damaged nodes generate constrained region which provoke the network partitioning and thus hamper the

S. Andreev et al. (Eds.): NEW2AN/ruSMART 2012, LNCS 7469, pp. 401–412, 2012.
© Springer-Verlag Berlin Heidelberg 2012

network activity. On the other hand, the data traffic is typically concentrated on the sensor nodes surrounding the sink [4-5]. Consequently, those bottleneck nodes around the sink exhaust their batteries faster than other nodes which lead to the isolation of the sink from the rest of the network. Therefore, the network may get partitioned into disjoint areas (referred to as holes) due to node failure or energy depletion. If the network gets useless when the first holes occur, the remaining resources are wasted. Consequently, maintaining network connectivity is a crucial concern to ensure application operation.

In this paper, we exploit the non-uniform consumption, traffic distribution and deployment of wireless sensors to elaborate an efficient dynamic recovery approach. We present CoMN2 a protocol restoring WSNs Connectivity using a Mobile Node in a Mapped Network. CoMN2 introduces a new concept called *Network Mapping* which consists in partitioning the network in several regions of increasing criticality. For instance a dense area with low sensing activity will have lower criticality than nodes in a sparse area highly solicited.

Through this network mapping, our solution CoMN2 ensures the continuous network activity by swapping nodes from low critical area to high critical ones when required. The swapping is performed using a mobile node which is aware of the sensor status. Basically, the algorithm is triggered depending on the *Zone* of the failed node in our mapped network. This solution minimizes the energy consumption by inhibiting the recovery process as long as the damaged node is not vital for the network. Above and beyond that, CoMN2 proceeds to a periodic network mapping update in response to unpredictable network dynamics.

CoMN2 design fulfills the following points that we consider as requirements for our context:

- In view of the autonomous and unsupervised characteristics of WSN, the network recovery from node failure should be performed in a distributed manner.
- The process should be rapid and lightweight to preserve the WSN responsiveness to detected events.
- The overheads should be minimized in order to consider the scarce resources of WSNs.

The remainder of this paper is organized as follows: the next section gives an overview of the problem statement. The details of the algorithms and analysis are given section 3. Section 4 presents the previous approaches then section 5 provides the simulation results and finally section 6 concludes this paper.

2   Problem Statement

Maintaining inter-node connectivity is a crucial concern in WSNs. Many applications like disaster management require efficient and highly reliable nodes collaboration to efficiently assess damage and identify safe escape paths which emphasize the importance of this concern. Basically, the failure of a sensor in critical region may create 'holes' (Figure 1) and induces a partitioned network, so a reduction of the network operation efficiency. A particular case of this point is the HotSpot problem: Sensor nodes around the sink are extremely solicited to forward the sensed data (from the entire network nodes) toward the sink. In spite of the large deployment of sensors to tolerate possible node failures, we have to face the problem of isolation of the sink node caused by the depletion of the energy of sensor nodes surrounding it. This may result in the failure of the whole network whereas the remaining nodes are perfectly functional.

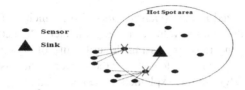

Fig. 1. Node isolation in the hotspot area

On the other hand, sensor nodes activity may be non-uniform in an hybrid network. For instance, regular traffic is expected for monitoring a given parameter whereas some nodes may be further involved in relaying data toward the sink. This non-uniform activity distribution depletes the battery of some sensors faster than the rest of the network.

Nevertheless, some areas are covered by excessive sensors which create redundant sensing region because closest nodes may have similar data [3]. Table 1 presents the relationship between the number of neighbors and the percentage of redundancy. As shown above, with 11 neighbors, the probability of complete redundancy is almost 92.28%, and the percentage of redundancy area exceeds 99%.

Table 1. Number of neighbors Vs Percentage of redundancy extracted from [3]

Number of neighbors	Probability of complete redundancy	Percentage of the redundant area
5	31.22-37.01%	91.62%
7	64.29-65.21%	96.89%
9	82.97-83.09%	98.85%
11	≈ 92.28%	99.57%

Thus, the loss of a sensor in a redundant area does not impact the rest of the network because its neighbors are capable to achieve exactly the same role.

Keeping in mind all these aspects, our problem statement can be formulated as follows: given a set of nodes,

- How can we recover from node failure?
- When should we replace the damaged node and which nodes are vital for the functioning of the network?
- How can we handle unpredictable dynamics of the networks?

To deal with the first issue, CoMN2 proposes to recover from node failure by swapping nodes using a mobile node. This solution raises another issue which is the selection of the node to move. This problem is resolved thanks to a new concept proposed in this paper which is the network *Mapping*.

Upon the detection of a damaged node, CoMN2 considers four characteristics:

- The battery level
- The number of neighbors
- The location of the node (Hotspot area or not)
- The activity of the node

These issues are summarized in the *ZoneAffectationTable* which will be presented in the next section. Depending on the value of these parameters, the network is mapped in three zones of increasing rank expressing the criticality of sensors of each zone. This mapping is periodically updated to consider the network dynamism.

The detailed CoMN2 algorithm is explained in the next section.

3   CoMN2 Operation

3.1   Overview of the Mechanism

In the following scenario, we are considering a hybrid WSN with static sensors, along with a sink and a mobile node. The mobile node is resource-rich node equipped with high processing capabilities and longer battery life. We assume that the sensors are randomly scattered over the network and that every node is aware of its location and its remaining energy.

The communication between sensors and the mobile node is performed thanks to the sink. All the control messages as initially sent to the sink which transfers it to the mobile node. We assume that the mobile node notifies the sink of its new position each time it moves.

As mentioned above, the crash of some nodes may stir up the division of the network into disjoint segments and leads to the formation of holes. To avoid this, our solution uses a mobile node carrying out the network restoration by switching a failed node with another from a redundant area (less critical). Furthermore, our solution handles simultaneous sparse node failure by classifying nodes according to their importance through an innovative process called network Mapping (partitioning). This method consists in organizing the network into three zones of increasing rank depending on the criticality of nodes belonging to each area. The detailed mechanism will be presented in the following part.

3.2   Network Model

Our idea of mapping (Figure 2) the network emanates from the observation that sensors have no uniform activity in the network as explained in section II. The key plan is to hierarchically divide the sensor field into zones of increasing importance according to the node's activity.

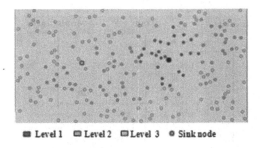

■ Level 1   ▣ Level 2   ▣ Level 3   ● Sink node

Fig. 2. Network Mapping

In fact, the more the node is solicited, the faster its energy depletes and consequently it will die first. On the other side, some high density zones are monitored by excessive sensors what means that the failure of a node of that zone does not impact its functioning as stressed in section 3. That's why we decide to give priority to connectivity in specific zones instead of favoring connectivity of the entire network.

We divide the network into three zones of growing rank. Our basic assumption is that the nodes around the sink are the most actives [4-5] and that they belong to the first zone. Since we are considering a heterogeneous network with nodes assigned to diverse applications, some nodes may be very solicited for sensing or reporting operations. In view of their importance, the sensors of these specific zones belong to the first zone too. We define the third zone as the less critical one containing the set of nodes with more than 10 neighbors (section 3). The remaining nodes constitute the second zone. Finally we obtain a network's cartography based on how critical is each zone. This network cartography will be dynamic: it will be updated contingent on the activity of nodes.

3.3   Protocol Operation

Initial Affectation of Nodes through Zones: The first affectation of nodes through zones exploits their position and their neighborhood repartition. The mobile node will maintain a table referred to thereafter as *ZoneAffectationTable* containing five parameters {*Id, Pos*, NN, RZ, *Zone*} where:

- *Id* is a unique identifier of each node.
- *Pos* designate the local position of the node.
- *NN* is the number of neighbors.
- *RZ* is a binary parameter designating risky zones: 1 for risky one/ 0 for no risky. Initially, RZ is set to 0 for all nodes. It will be set to 1 during the update process when the   mobile node will notice the high activity of some nodes.
- *Zone* indicates the zone of affection. It can be 1, 2 or 3 with 1 designating the most critical zone.

To establish *ZoneAffectationTable,* each node should send *NN* and *Pos* to the sink which will relay it to the mobile node (Figure 3). The whole process is kept local by requiring every node to maintain a list of only its close neighbors: we restrain our list to the 1-hop and 2-hop ones only.

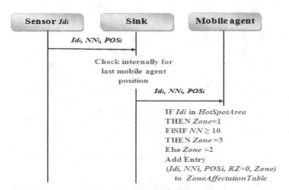

Fig. 3. Initialization phase

After receiving this information from nodes, the mobile node would proceed to the zone affectation depending on the number of neighbors around each node:

$$Zone = \begin{cases} 3 \text{ IF } NN \geq 10 \\ \\ 2 \text{ Otherwise} \end{cases} \qquad (1)$$

Nodes around the sink constitute the special case of the *HotSpotArea*. They are automatically affected to zone 1.

Update of the Network Mapping: CoMN2 pursues a dynamic approach by updating the mapping of the network throughout the exchange of specific messages between nodes. This reorganization involves the refresh of the *ZoneAffectationTable* (Figure 4) and more precisely of *NN* and *RZ*.

NN is kept up to date by sending periodic *HEARTBEAT* messages between neighboring nodes. The absence of *HEARTBEAT* response from a neighboring node F would mean that F failed then a message is automatically sent to the sink which will notify the mobile node. Upon receiving this notification, the mobile node modifies the value of the corresponding *NN* in the *ZoneAffectationTable*.

Furthermore, we define *Tobs* as the time that the sink waits before sending a periodic activity report to the mobile node for analysis. This report aims to determine the most solicited nodes by quantifying the number of operation (including sensing and routing) per node: each sensor is equipped with a counter which is incremented after every action like relaying or sensing data. The result of this counter is periodically sent to the sink. Keeping in mind that this report must characterize the network and is used to identify the most active zones, special attention is paid to well choose *Tobs* which should be long enough. Once the account is relayed by the sink, the mobile node calculates the average number of operations per node (the mean value) and sets RZ to 1 for the most solicited nodes: with a number of operations greater than the mean value. Thanks to this parameter, a node initially belonging to zone 3 can be classified as zone1 if it's highly active. Additionally, as soon as a node's energy drops below a critical level, it sends a *DISTRESSMESSAGE* containing its *ID* to the mobile node.

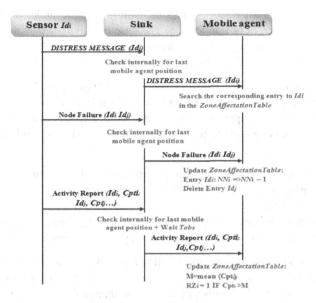

Fig. 4. Update phase

Initiation of the CoMN2 Recovery Process: To expound how CoMN2 Algorithm performs, let's consider these two cases with a first classical one of a single alert message and a second case showing how our proposed algorithm handles simultaneous node failure.

Before giving rise to the recovery process upon detecting a failure, the mobile node waits during a period of *Thold* to check if there is another alert.

• Single Alert

For the case of single alert, the mobile node should decide whether to initiate the recovery or not. This choice is based on the zone to which the node belongs. Actually, the network mapping aims to avoid restoring connectivity at the cost of additional energy consumption when we can do without this node: The protocol tussles to minimize the movement of the mobile node in order to save its energy. In fact, despite its high battery compared to the sensors, the energy of the mobile node is not unlimited. Execution of restoration is triggered only if the node belongs to zone 1 or 2 which are the most important regions. If the dying node is from zone 3, we suppose that it has a sufficient number of close redundant neighbors to replace it. The mobile node checks at first *ZoneAffectationTable* to determine the *Zone* of the node then decide to *rescue* it or not.

• Simultaneous Alert

Let's consider the case of 2 alerts received during *Thold* from node A in zone 1 and node B in zone 3. The mobile node proceeds to the comparison of the zone of each node using *ZoneAffectationTable* then decides to rescue the node with the most important zone (which is node A in our example). If all the alerts arise from zone 3, no recovery process is triggered because the failure of these nodes does not impact the efficiency of the network application. The case of simultaneous alerts with the same importance is handled by a FIFO mechanism (First In First Out). The mobile node continues its current task then deals with the second affected node.

Recovery Execution: The recovery solution is based on the swap of nodes and the replacement of the broken ones with the help of the mobile node. For the beginning of the network recovery approach, the mobile node exploits the *Network Mapping* to find the most suitable node to move. The choice of the best candidate is based on two criteria:

- A minimum change in the network topology. This condition can be achieved by choosing a node from a redundant area (zone3) which will not involve the reorganization of the network.

- The nearest one given the position of the failed node and the actual position of the mobile node in order to minimize the overhead of the recovery process.

The mobile node searches the closer node of zone 3 in the *ZoneAffectationTable* given the position of the failed node and his current position. This way, the inter-node connectivity is re-established without involving a change of the network topology. Since the spare node reaches the failed node location, it becomes responsible for carrying out its tasks including both routing and the application level. The choice of the optimal path to reach the failed node is out of the scope of this paper.

4   Related Works

How to deal with node failure is a challenging concern that has been widely studied in the past years. For instance, [12] proposed a power extension algorithm consisting on the increase of the transmission power to extend the communication range so that a node can reach further nodes when its next hop fails. This algorithm handles the connectivity aspect but does not provide a solution to ensure the sensing function of the failed node. Another proposed approach in [13] uses the neighbor information to construct a new path and take the routing role of the failed nodes. This work assumes that sufficient number of neighbors surround the failed node. Yet, this is not always true because of the random deployment of sensors in some scenarios such as hostile environments and hazardous zones.

Many researches investigate the use of mobile nodes [6-8] [14-15] to restore connectivity in mobile networks. In [7], the author suggests to use a mobile base station in order to load balance the charge around the sink by changing the nodes located close to it. Unlike our proposed approach, this algorithm restrains its efficiency on nodes surrounding the sink and involves an extra message overhead to update the location of the mobile base station. Another solution referred to as Coverage Conscious Connectivity Restoration algorithm [15] (C3R) relocates one or multiple neighbors of the failed node to recover from the damage. Each neighbor temporarily moves to substitute the failed node what leads to intermittent connectivity. The assumptions in the above work are strict because there is no guarantee to have multiple neighbors available.

Controlled mobility is a framework operating in context aware mobile devices with intelligent ability permitting to determine their future location on the basis of some conditions like battery level or data gathered. Authors in [14] exploit controlled mobility to propose an algorithm for sensor nodes relocation. The algorithm assumes WSNs with cluster topology and allows relocation of sensor nodes between clusters based on predefined utility function. Despite its interest, this approach arises clustering limitations such cluster head deployment and maintaining coverage in clusters.

5   Experimental Evaluation

This section describes the simulation experiments to assess the effectiveness of the proposed approach. We compared CoMN2 to a Mobile Sink solution and to a basic topology without restoration.

5.1   Baseline Approach

We use two metrics to evaluate the performance of CoMN2: the network lifetime and the failure rate.

-The network lifetime is a generic term depending on the considered scenario. In our case, we define it as the time for the first node to become unable to reach the sink (no route available because of the failed nodes). Our goal is to increase the network lifetime.

-The failure rate indicates the rate of failed nodes. It serves as a measure of the fault tolerance of the network: The higher the failure rate is at the end of the network life, the more fault-tolerant the network is.

A java simulator has been developed to evaluate CoMN2. The simulation experiments involve randomly generated WSN topologies with varying number of nodes from 100 to 800 in a network field with dimensions of 10000mx20000m. The location of the sink is selected randomly. For each topology set-up (network size, protocol) simulation is run 10 times and the average performance is reported. During each run, packets are generated and sent successively. The sending sensors are randomly selected and the path to the sink is created using least-cost routing in terms of number of hops. We assume the energy distribution is uniform in the network: all nodes have initially the same battery level. Nodes' battery-level decrease at each operation until attaining a critical threshold equivalent to 20% of the initial battery level. Once the threshold reached, CoMN2 is activated if the node belongs to zone 1 or 2 and the node is considered as failed.

As explained in section 4, many researches propose solutions with Mobile Sink. Our simulation uses a configuration where the sink moves periodically in the network to load balance the traffic. It chooses a location where nodes in the neighborhood (nodes around the new position of the sink) have a high level of battery available.

5.2   Performance Comparison

We conduct different experiment by varying the network density. To test the scalability of our approach, we increased progressively the number of nodes for a fixed network field dimension and transmission range. The results are depicted in Figure 5. As can be seen from the figure, our algorithm significantly increases the network lifetime compared to the basic configuration (without restoration) and to the mobile sink for all network sizes. This rise is more significant for dense network which is attributed to the fact that the number of the back-up nodes used to replace the failed nodes in CoMN2 is higher in a dense network.

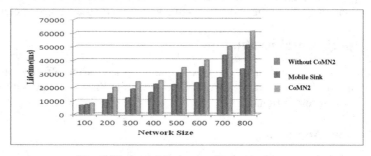

Fig. 5. Network Lifetime vs. Network size

Table 2 summarizes the increase of the network lifetime for the different configurations. We observe an average increase of more than 11% compared to the Mobile Sink and 40% compared to the network without CoMN2 which assess the effectiveness of our approach.

Table 2. Percentage of the network lifetime increase with CoMN2

Network Size	% No CoMN2	%Mobile Sink
100	17,41	12,19
200	46,81	23,20
300	50,04	22,15
400	35,43	11,19
500	37,10	11,54
600	41,79	12,23
700	46,15	12,83
800	45,46	16,93

Moreover, we tested the performance of CoMN2 in term of fault tolerance. Curves in Figure 6 shows that the fault tolerance is considerably raised thanks to CoMN2: the network tolerates a higher percentage of failed nodes (~35% vs ~10% without CoMN2).

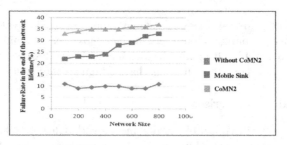

Fig. 6. Fault tolerance vs. Network Size

This increase is due to the location of the failed nodes which differs as shown in Figure 7. We notice that unlike CoMN2, failed nodes are scattered around the sink for the first figure which shortens the network lifetime and provokes the hotspot problem.

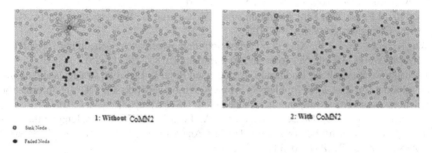

1: Without CoMN2   2: With CoMN2

○ Sink Node

● Failed Node

Fig. 7. Failed node location for the same failure rate

In fact, the curves of failure rate vs. time for a network with and without CoMN2 follow the same slope (Figure 8) in the beginning of the simulation. Nevertheless, CoMN2 maintains nodes around the sink alive by swapping them with nodes far from it which increase the lifetime of the network. However, the hotspot area around the sink remains vulnerable to node failure (highly solicited) which increase its failure rate hence the higher slope of CoMN2 from t=500.

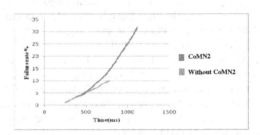

Fig. 8. Failure rate vs. Time

6   Conclusion and Future Works

Network mapping based on node resources represents an attractive paradigm in the design of wireless sensor network. In this paper, we presented a distributed and dynamic recovery protocol CoMN2 which handles node failure in WSNs. As it is unknown whether such crash causes a network partitioning or not, we provide a technique based on network mapping deciding if a node should be replaced or not. Simulations show that CoMN2 achieve more than 40% improvement of the network lifetime.

For the future work, we plan to reduce the network recovery delay of our approach by investigating the optimal routing path to follow to join the sink.

References

1. Akyildiz, I.F., Kasimoglu, I.H.: Wireless Sensor and actor networks: Research Challenges. Elsevier Ad hoc Network Journal 2, 351–367 (2004)
2. Akyildiz, I.F., Su, W., Sankarasubramaniam, Y., Cayirci, E.: Wireless sensor networks: a survey. Computer Networks 38 (2002)
3. Wu, K., Gao, Y., Li, F., Xiao, Y.: Lightweight deploymentaware scheduling for wireless sensor networks. Mobile Networks and Applications 10, 837–852 (2005)
4. Jaichandran, R., Anthony Irudhayara, A., Emerson Raja, J.: Effective strategies and optimal solutions for Hot Spot Problem in wireless sensor networks (WSN). In: Information Sciences Signal Processing and their Applications (ISSPA), pp. 389–392 (2010)
5. Wang, J., de Dieu, I.J., De Leon Diego Jose, A., Lee, S., Lee, Y.-K.: Prolonging the Lifetime of Wireless Sensor Networks via Hotspot Analysis. In: Applications and the Internet (SAINT) 10th IEEE/IPSJ, pp. 383–386 (2010)
6. Vasanthi, V., Romen Kumar, Ajith Singh, Hemalatha, M.: A detailed study of mobility model in wireless sensor networks. Journal of Theoretical and Applied Information Technology 33(1) (2011)
7. Luo, J., Hubaux, J.P.: Joint mobility and routing for lifetime elongation in wireless sensor networks. In: INFOCOM, vol. 3, pp. 1735–1746 (2005)
8. Basagni, S., Carosi, A.: Controlled sink mobility for prolonging wireless sensor networks lifetime. Journal of Wireless Networks 14, 831–858 (2008)
9. Vemulapalli, S., Akkaya, K.: Mobility-based Self Route Recovery from Multiple Node Failures in Mobile Sensor Networks. In: WLN, Denver (2010)
10. Efrat, A., Har-Peled, S., Mitchell, J.S.B.: Approximation algorithms for two optimal location problems in sensor networks. In: Broadnets, Boston, MA (October 2005)
11. Wang, Y.-C., Hu, C.-C.: Efficient Placement and Dispatch of Sensors. IEEE Trans. on Mobile Computing, TMC-0278-1006.R1 (2007)
12. Liang, C., Huang, X., Deng, J.: A fault tolerant and energy efficient routing protocol for urban sensor networks. In: InfoScale 2007: Proceedings of the 2nd International Conference on Scalable Information Systems, pp. 1–8. ICST, Brussels (2007)
13. An, D., Cam, H.: Route recovery with one-hop broadcast to bypass compromised nodes in wireless sensor networks. In: Wireless Communications and Networking Conference, WCNC, Kowloon, pp. 2495–2500 (March 2007)
14. Denkovski, D., Mateska, A., Gavrilovska, L.: Extension of the WSN Lifetime through Controlled Mobility. In: The Seventh International Conference on Wireless On-demand Network Systems and Services IEEE/IFIP WONS, Kranjska Gora (2010)
15. Tamboli, N., Younis, M.: Coverage-Aware Connectivity Restoration in Mobile Sensor Networks. In: ICC on Communications, Dressen (2009)

FDAP: Fast Data Aggregation Protocol in Wireless Sensor Networks

Sahar Boulkaboul[1,2], Djamel Djenouri[1], and Nadjib Badache[1]

[1] CERIST Research Center, Algiers, Algeria
[2] Mira Univeristy, Bejaia, Algeria
{bsahar,ddjenouri,badache}@mail.cerist.dz

Abstract. This paper focuses on data aggregation latency in wireless sensors networks. A distributed algorithm to generate a collision-free schedule for data aggregation in wireless sensor networks is proposed. The proposed algorithm is based on maximal independent sets. It modifies DAS scheme and proposes criteria for node selection amongst available competitors. The selection objective function captures the node degree (number of neighbors) and the level (number of hops) contrary to DAS that simply uses node ID. The proposed solution augments parallel transmissions, which reduces the latency. The time latency of the aggregation schedule generated by the proposed algorithm is also minimized. The latency upper-bound of the schedule is $17R+6\Delta+8$ time-slots, where R is the network radius and Δ is the maximum node degree. This clearly outperforms the state-of-the-art distributed data aggregation algorithms, whose latency upper-bound is not less than $48R+6\Delta+16$ time-slots. The proposed protocol is analyzed through a comparative simulation study, where the obtained results confirm the improvement over the existing solutions.

1 Introduction

Data aggregation has been proposed as an energy/bandwidth efficient paradigm for routing in sensor networks. The key idea behind this concept is to combine data from different source nodes, at the aim of eliminating existing redundancies. A router combines several packets from neighboring nodes into a single packet before relaying towards the sink. This is different from traditional packet-based routing, and it clearly minimizes the number of transmissions and reduces power consumption. The major shortcomings of data aggregation is the increased end-to-end latency due to aggregator timeout on each hop which is the time an aggregator waits for its children to submit their data before making the aggregation and relaying the final packet. This problem is tackled in this paper and a new solution is proposed. It consists in a distributed approach that takes into account the network topology and the number of neighboring nodes (node degree) during the construction of the aggregation tree. The approach also increases parallel transmissions during data aggregation, while avoiding collisions. The proposed solution is compared with a state-of-the art protocol, namely DAS [1], which

S. Andreev et al. (Eds.): NEW2AN/ruSMART 2012, LNCS 7469, pp. 413–423, 2012.

we implemented using the Python environment. The remainder of this paper is organized as follows, In Section 2 the related work is outlined. Section 3 provides problem formulation and the notations. The proposed distributed scheduling algorithm is introduced in Section 4, and it is analyzed in Section 5. Section 6 presents the simulation results, and finally Section 7 draws the conclusion and the perspectives.

2   Related Work

Several in-network aggregation protocols considering the end-to-end delay (latency) reduction have been proposed. In cascading timeout [2] , nodes located in the same level of a data aggregation tree have the same aggregation time, regardless of the number of child nodes. Therefore, the nodes with more children will loose some of the childrens data; and the nodes in the same level will send the data at almost the same time, which leads to traffic congestion. ATC [3] proposes an adaptive timing control that distributes the aggregation time. That is, the nodes with more children are allocated a longer waiting time, for the purpose of maximizing the level of data aggregation. However, the timeout problem persists at the aggregation nodes. In ZFDAT [4], the timeout is calculated for each sub-tree in the cluster, which is based on the packet transmission delay and the cascading delay.Some other algorithms have considered jointly the transmission time and the energy consumption. In [5], a generalized and combined SPT-MST algorithm was proposed, which constructs DAC (Data Aggregation enhanced Convergescast) tree for all values of the data growth factor (α).This parameter gives an indication on the rate at which the data grows as it moves upstream on the tree. The data latency was reduced by re-structuring the energy-efficient tree using β-constraint, which puts a soft limit on the maximum number of children a node can have in a DAC tree. In [6], Galluccio et al. provide a closed form expression for the end-to-end delay and the network lifetime distributions in a wireless sensor network that applies data aggregation. Authors of [3],[4]and [2] mention the collision problem, but leave it to the MAC layer. Solving this problem at the MAC layer incurs a large amount of energy consumption and time latency during aggregation. Some new solutions consider collision during aggregation. A typical data aggregation algorithm consists in two phases; tree construction and aggregation scheduling. The most relevant work on aggregation scheduling are [[7],[8],[9],[10] and [1]].Chen et al [7] proved that the minimum data aggregation time (MDAT) problem is NP-hard. They designed an algorithm named SDA (Shortest Data Aggregation) based on the shortest Path Tree. They proved a latency bound of $(\Delta - 1)$R, where Δ is the maximum degree of the network graph. Qan [9] propose a time-efficient data aggregation algorithm that aggregates delay-constrained data within a given time deadline in clustered wireless sensor networks. Another aggregation scheduling algorithm was proposed by Huang et al [8], which has a latency bound of $23R + \Delta - 18$, where R is the network radius and Δ is the maximum degree. Unfortunately, the generated schedules are not collision-free in many cases. Xu et al. [10] presented a collision

free data aggregation scheduling with a latency bound of $16R + \Delta - 14$, where
R is the network radius. All the algorithms mentioned above are centralized.
Centralized algorithms are generally unpractical,not scalable, and even infeasi-
ble for some applications. Yu et al. [1] propose a distributed algorithm named
DAS. It has a latency bound of $48R+6\Delta+16$, where R is the network radius and
Δ is the maximum degree. It adopts the concept of Connected Dominating Sets
(CDS) [11] to construct the aggregation trees. DAS also proposes an adaptive
strategy for updating the schedule when nodes break down, or new nodes join
the network. However, it uses nodes' ID to select a node among its competitors.
This criterion does not help to reduce latency.New criteria are proposed in this
paper to increase parallel transmissions and reduce the latency.

3   Notation and Problem Definition

A Wireless sensor network with sink node s can be represented as a unit disk
graph $G =(V,E)$, where V denotes all the sensor nodes in the network. An
edge $(u,v) \in E$ indicates that u lies in v's transmission area. Each node can
send/receive data to/ from all directions.The transmission range of any sensor
node is a unit disk (circular region with a constant radius) centered at the sensor.
A data aggregation schedule is a sequence of sender sets $\{S1, S2, ...\}$ satisfying
the following conditions:

1) $S_i \cap S_j = \emptyset, \forall i \neq j$;
2) $\cup_{i=1}^{l} S_i = V - \{s\}$;
3) Data are aggregated from S_k to $V \setminus \cup_{i=1}^{k} S_i$ at time slot k, for all $k = \{1..l\}$
All the data are aggregated to the sink s in l time slot. The number l is called
the data aggregation latency. The distributed aggregation scheduling problem is
to find a schedule $S1, S2,..., Sl$ in a distributed way, such that l is minimized.
The following notations are used in the rest of the paper.

$Level_i$: Number of hops to the sink.
$Degree_i$: Number of i's neighbors.
ID_i: The unique identifier of node i.
$CH(u)$: u's children set.
Given a couple of 3-tuple variables $(Level_i, Degree_i, ID_i)$ and
$(Level_j, Degree_j, ID_j)$,the weight function $W(Level, Degree, ID)$ satisfies
$W(Level_i, Degree_i, ID_i) > W(Level_j, Degree_j, ID_j)$ if any of the following conditions is fulfilled:

1) $Level_i > Level_j$.
2) $Level_i = Level_j$ and $Degree_i < Degree_j$.
3) $Level_i = Level_j$ and $Degree_i = Degree_j$ and $ID_i > ID_j$.
The DAS (*Distributed Aggregation Scheduling*) problem is defined as follows.
Given a graph $G = (V, E)$ and the sink node $s \in V$, find a data aggregation
schedule with minimum latency. In the next section, a distributed algorithm
with latency, $17R+ 6\Delta+8$, is proposed, where R is the network radius, and Δ is
the maximum degree.

4   Ditributed Data Aggregation Scheduling

The proposed distributed algorithm, we call FDAP (Fast Data Aggregation Protocol), improves the DAS protocol by modifying the construction of the tree of aggregation, and adding other factors for priority ranking of nodes among its competitors in the scheduling aggregation phase. The proposed approach allows FDAP to reduce the end-to-end aggregation delay, by increasing the parallel transmissions while avoiding collisions. The protocol involves several steps that are described in the following.

4.1   Distributed Aggregation Tree Construction

Topology Center: In this phase, a tree is constructed by running a distributed algorithm. For a given graph $G = (V, E)$ and a sink node $s \in V$, the node $v_c \in V$ that is located at the center of the network represented by of G is selected as the root of the tree; it is called the network center in G, and it is defined by [10], $v_c = argmin_v\{max_u d_G(u, v)\}$, where $d_G(u,v)$ denotes the distance between u and v in G, as the minimum number of hops from u to v. The network center node is supposed to be known and static, similarly to the previous works such as [10]. After the network center v_c gathers the aggregated data from all the nodes, it forwards the aggregation to the sink node using ordinarily point-to-point routing.

Node Rank in CDS: To construct an aggregation tree, a connected dominating set (CDS) is used, which serves as the virtual backbone of the sensor network the distributed algorithm by Wan et.al. in [11] is used to generate the CDS, with a modified ranking priority for node coloring. the new priority considers the number of neighboring nodes, which accelerates data aggregation. This algorithm runs in two phases, they build, respectively, an MIS (*Maximal Independent Set*), and a dominated tree. During the first phase, the root node v_c is chosen to build a spanning tree. It is assumed that each node knows its level and neighbors. Initially, each node that has the smallest rank among its neighbors is colored black. It broadcasts a message showing its status, When a node receives the message for the first time, it will be colored gray. Then, the latter broadcasts its status.If a node receives the messages from all its neighbors with lower ranks than its rank, then it will be marked black. As soon as a leaf node is marked, it transmits a MIS-COMPLETE message to its parent node. Each internal node receiving the MIS-COMPLETE message passes it to its ascendancy, the process is repeated until the root reached. The second phase is to build a dominant tree, say T, where all nodes in this tree form a CDS Initially, the dominant tree is empty. The root is the first node that joins the tree. When a node joins T, it sends an invitation message to all blacks nodes with two hops, (that are not in T) to join the tree. The invitation message is relayed by the gray nodes. every black node joins T upon receiving the invitation message for the first time, and the gray node relays the message. The process repeats until all nodes join the tree. Consequently, the CDS will contain all nodes that belong to the tree. A black node will be a dominator node, and a grey node (*connector node*) will connect two black nodes.

Grey Nodes Reduction: In this phase. The number of connectors that a dominator will use to connect to all dominators within 2-hops is bounded. Based on lemmas proved in [12], and [10], we try to remove the redundant connectors to ensure that each dominator uses at most 12 connectors to connect itself to all dominators in lower level located within 2-hops. The idea of Algorithm1 is as follows: For each grey node, find all of its black neighbors with higher level; and assure that removing a connector will not disconnect any of the 2-hop dominators from a dominator.

Algorithm 1. Reducing the number of connectors
Input: $G=(V,E)$; Data aggregation tree T;
Output: Reduced data aggregation tree T1.
1: **For** each grey node u
2: Select $NLS(u)$(u's black Neighbors set,dominator node with Superior Level)
3: Send message to $NLS(u)$one-hop neighbors with higher levels
4: **For** each node v upon receiving message from u
5: **If** $\mid NLS(u) \mid \geq \mid NLS(p_T(v)) \mid$ **then**
6: Remove edge $(v, p_T(v))$ from E_T;
7: Add edge (v, u) into E_T;
8: $CH(u) \longleftarrow CH(u) - v$
9: **If** $CH(u) = \emptyset$**then**u color itself white(leaf node)

4.2 Distributed Aggregation Scheduling

A schedule of a node u in a sensor network is defined as the time slot allocated to u to send its data. The SCHDL algorithm -used by DAS- determines the schedules for all the nodes in a distributed way. The competitor set, say CS,is defined by,$CS(u) = N(p(u)) \cup (\cup_{v \in N(u) \setminus Ch(u)} Ch(v)) \setminus \{u, p(u)\}$, where $p(u)$, $Ch(u)$, and $N(u)$ are v's parent in T1, u's children set in T1, and u's one-hop neighbor set except $p(u)$,respectively.Each node u maintains the following information:

1) $CS(u)$: u's competitor set in T1.
2) $NCH(u)$:The number of u's children that have not been scheduled, which is initialized to the number of u's children.
3) $ts(u)$:the time slot assigned to u, which is initialized to 1.
4) $RCS(u) = \{v/v \in CS(u)$ and $NCH(v) = 0\}$. This is u's ready competitor set, initialized to null set. Each node, $v \in RCS(u)$, is ready to make its schedule.
5) $SCH(u)$: u's schedule state, initialized to false. If $SCH =$ true, it denotes that u has already been assigned a timeslot $ts(u)$.
6) RY,NRY: ready(respectively not ready)node state for schedule execution.
7) Weight Function: decides the scheduling priority.

The pseudocode of RSCHDL(Reduce scheduling) is given in Algorithm 2.

An example is illustrated by Figure1. It shows the latency when the level is considered for choosing a node among its competitors to make a schedule.

Algorithm 2. RSCHDL

Input: A network G, and the reduced data aggregation tree T1 ;
Output: $ts(u)$ for every node u.
1:**While** (not SCH) **do**
2: **For** each leaf node u
3: u sends a $RY(u)$ containing u's ID to all the nodes in $CS(u)$ and receives RY or NRY from $CS(u)$.
4: **If** node u receives RY or NRY messages from all the nodes in $CS(u)$**then**
a: **If** $RCS(u) \neq \emptyset$ **then** **If**$(u.level, u.degree, u.ID) > (v.level, v.degree, v.ID)$ For each $v \in RCS(u)$ //***checks if u's priority is the highest in $RCS(u)$ **then** u sends COMPLETE$(u, ts(u))$ to all the nodes in $CS(u) \cup \{p(u)\}$ and sets SCH to true.
b: **If** $RCS(u) = \emptyset, u$ sends COMPLETE$(u, ts(u))$to all the nodes in $CS(u)$ and sets SCH to true.
5: **For** each node u, upon receiving RY from v
6: **If** $v \in CS(u)$ **then** u adds v to its $RCS(u)$
7: **If** $NCH(u) \neq 0$
8: u sends NRY to v
9: **For** each node u, upon receiving NRY from v
10: **If** $v \in CS(u)$ **then**
11: u records NRY reception from v
12: **For** each node u, upon receiving COMPLETE$(v, ts(v))$from v
13: u deletes v from $RCS(u)$;
14: **If** $v \in CH(u)$, **then** $NCH(u) \longleftarrow NCH(u) - 1$;
15: $ts(u) \longleftarrow \max\{ts(u), ts(v) + 1\}$ //***$ts(u) > ts(v)$.
16: **If**$NCH(v_c) = 0$ **then** the topology center transmits aggregated data using the shortest path from v_c to s.

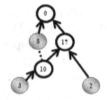

ID	(Level, ID)
S1= {8,2}	S1= {3,2}
S2= {3}	S2= {8,10}
S3= {10}	S3= {17}
S4= {17}	

Fig. 1. Latency with level

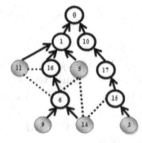

(Level ,ID)	(Level,Degree,ID)
S1= {14}	S1= {3,9}
S2= {3,9}	S2= {11,14,15}
S3= {5,6,15}	S3= {5,6,17}
S4= {16,17}	S4= {10,16}
S5= {10,11}	S5= {1}
S6= {1}	

Fig. 2. Latency with level and degree

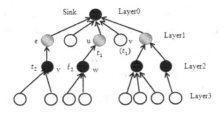

Fig. 3. Analysis on an aggregation tree

The final data aggregation schedule is given as follows: Four time slots are needed when using ID-only in DAS. In the proposed approach, only three time slots are needed. This can be explained by the fact that data aggregation is performed from the deepest layer until layer 0 (the sink). The leaf node with superior level is selected to accelerate data aggregation. Figure 2 shows the impact of adding a degree in the proposed solution. Only 5 time-slots care needed when adding the level (ID,Level). Since a node with a lower degree has less competitors, parallel transmissions are augmented.

5 Performance Analysis

The worst case performance of FDAP with respect to time latency (upper bound), say T, is analyzed. To calculate T, the latest schedule for each layer should first be calculated. Note t_i the latest schedule in layer i as shown in figure3. Two cases can be distinguished,

1. The Latest Scheduled Node Is Gray. Let w denote the u's child that has the schedule (t_2'),and y is the node that has the latest schedule t_2, such as $t_2' \leq t_2$. Each gray node in layer 1 receives data from all its children in interval $t_2 + 1$. Two situations may be distinguished during aggregation, i) when all the white nodes have sent their data, and ii) some white nodes have not sent their data to the sink at the beginning of the time slot $t_2 + 1$. In the first situation, the gray nodes compete only with gray nodes. Each gray node must have at least one black child, given that the tree construction phase chooses gray nodes to interconnect black nodes. According to the lemmas proved in [[10],[12]], if each black node two-hop neighbor includes at most 12 neighbors, then $t_1 \leq t_2 + 12$. In the second situation, the latest schedule of all the white nodes in layer 1, say t_{w1}, must be less than the largest size, Z, of the white nodes competitor sets (We refer to the lemmas in [[12],[10] and [[1], pp6]). After time slot t_{w1}, only layer 1 gray nodes' data remain unsubmitted, thus less than 12 time slots are needed. Therefore, the total latency T, can be bounded by: $T \leq Z + 12$.

2. The Latest Scheduled Node Is White. Similarly to the second situation in case one, the total latency is less than Z. In summary, the following Inequality yields.

$$t_1 \leq \begin{cases} 12 + t_2, all \ submissions \ end \ at \ t_2 + 1; \\ 12 + Z, \quad otherwise. \end{cases} \tag{1}$$

Now we calculate t_2 wich is the latest schedule of black node in layer 2. In the unit disk graph (UDG) model, a gray node has at most 5 black neighbors [8].

$$t_2 \leq \begin{cases} 5 + t_3, \text{ all submissions end at } t_3 + 1; \\ 5 + Z, \quad \text{ otherwise.} \end{cases} \quad (2)$$

Generally speaking, the formula for computing t_{2k} and t_{2k-1} is given by,

$$t_{2k-1} \leq \begin{cases} 11 + t_{2k}, \text{ all submissions end at } t_2 + 1; \\ 11 + Z, \quad \text{ otherwise.} \end{cases} \quad (3)$$

$$t_{2k} \leq \begin{cases} 5 + t_{2k+1}, \text{ all submissions end at } t_2 + 1; \\ 5 + Z, \quad \text{ otherwise.} \end{cases} \quad (4)$$

The difference between layer 1 and the other layers is due to the consideration of the number of gray nodes that are competing with the gray node u. 11 nodes are accounted for layer 1 instead of 12. We get,

$T = t_1 \leq 12 + t_2$

$\leq 12 + 5 + t_3 \leq ...$

$\leq \underbrace{(12 + 5) + (11 + 5) + ... + (11 + 5)}_{k-1} + t_{2k-1}$

$\leq \underbrace{(12 + 5) + (11 + 5) + ... + (11 + 5)}_{k-1} + 11 + t_{2k}$

$\leq \underbrace{17 + 16 + ... + 16}_{(m-1)/2} + t_m (m \text{ is odd}) \; t_m \leq Z$

or $\leq \underbrace{17 + 16 + ... + 16}_{m/2-1} + 11 + t_m \; (m \text{ is even})$

$T \leq 8m + Z - 4$, where m ($m \leq 2R$) denotes the the deepest layer number in the aggregation tree and Z ($Z \leq 6\Delta + 12$) the largest size of node's competitor set [1]. T is a latency upper-bound for the aggregated data to reach v_0. The added parameter R is the upper-bound of the time required for v_0 to send the data via the shortest path tree towards the sink. We obtain $T \leq 17R + 6\Delta + 8$.

6   Simulation Results

We implemented DAS and RDAL using PYTHON [[13],[14]] for the simulation comparison. In the simulation scenarios, a number of sensor nodes has been randomly deployed in a two-dimensional square region where all nodes have the same transmission range. The sink's position was also random. The following results (Figure 4, and Figure 5) are presented with a confidence interval of 95%. FDAP and DAS have been tested and compared in two different cases. For the first case, the network topology has been randomly generated with different number of nodes, while keeping the network density (Δ) unchanged (set to 10). The delay performance (in terms of number of time slots) of the two methods-FDAP and DAS- is illustrated in Figure 4. The average aggregation latency is

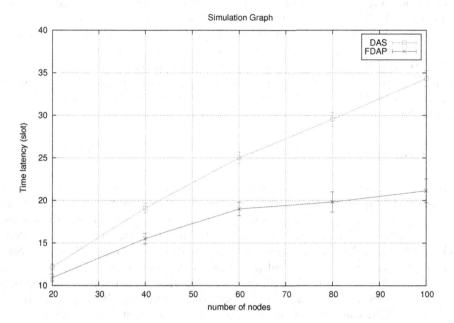

Fig. 4. Latency with different number of nodes

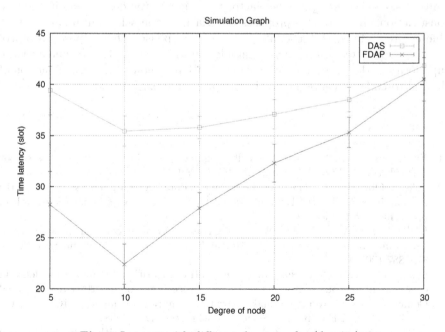

Fig. 5. Latency with different degree nodes (density)

proportional to N (number of nodes). The pattern of those curves matches with our theoretically estimated latency bound. The bigger the number of nodes, the better the improvement of FDAP over DAS. For the second case, the number of nodes was to 100, to compares the number of slots needed to aggregate data when the degree varies. Figure 5 confirms large difference between the two protocols, notably for low and average density. Simulation results demonstrate lower latency for FDAP over DAS, which is due to assigning parallel transmissions according to both the level and degree order in.

7   Conclusion

In this paper, latency of data aggregation in wireless sensor networks has been considered. Amongst the solutions proposed in the literature for minimizing data aggregation latency, the protocol DAS has been chosen for its advantages. It is a distributed collision-free solution, which takes into account node addition/deletion. DAS uses the connected dominating set concept (CDS) to construct the aggregation tree, and proposes a collision-free scheduling algorithm by introducing the concept of competitor sets. This helps reducing aggregation latency. However, it uses the node ID as a criterion for selecting a node amongst its competitor, which is not based on any optimization strategy. The new protocol, FDAP, takes advantage of DAS, and it reduces the latency bound to 17R $+6\Delta+8$, vs. 48R$+6\Delta+16$ for DAS. FDAP increases the numbers of parallel transmissions by adding another factor to designate the dominators in the tree construction phase. It also proposes criteria for node selection amongst available competitors in the scheduling aggregation phase. The proposed protocol has been compared with DAS by simulation. The results clearly demonstrate superiority of the proposed protocol over DAS and show significant reduction of the end-to-end aggregation delay in various scenarios.

References

1. Yu, B., Li, J., Li, Y.: Distributed data aggregation scheduling in wireless sensor networks. In: INFOCOM, pp. 2159–2167 (2009)
2. Solis, I., Obraczka, K.: The impact of timing in data aggregation for sensor networks communications. In: IEEE International Conference on Communications, vol. 6, pp. 3640–3645 (June 2004)
3. Li, H., Yu, H., Yang, B., Liu, A.: Timing control for delay-constrained data aggregation in wireless sensor networks: Research articles. Int. J. Commun. Syst. 20(7), 875–887 (2007)
4. Quan, S.G., Kim, Y.Y.: Fast data aggregation algorithm for minimum delay in clustered ubiquitous sensor networks. In: Proceedings of the 2008 International Conference on Convergence and Hybrid Information Technology, ICHIT 2008, pp. 327–333. IEEE Computer Society, Washington, DC (2008)
5. Upadhyayula, S., Gupta, S.K.S.: Spanning tree based algorithms for low latency and energy efficient data aggregation enhanced convergecast (dac) in wireless sensor networks. Ad Hoc Netw. 5(5), 626–648 (2007)

6. Galluccio, L., Palazzo, S.: End-to-end delay and network lifetime analysis in a wireless sensor network performing data aggregation. In: Proceedings of the 28th IEEE Conference on Global Telecommunications, GLOBECOM 2009, pp. 146–151. IEEE Press, Piscataway (2009)
7. Chen, X., Hu, X., Zhu, J.: Minimum Data Aggregation Time Problem in Wireless Sensor Networks. In: Jia, X., Wu, J., He, Y. (eds.) MSN 2005. LNCS, vol. 3794, pp. 133–142. Springer, Heidelberg (2005)
8. Huang, S.C.-H., Wan, P.-J., Vu, C.T., Li, Y., Yao, F.F.: Nearly constant approximation for data aggregation scheduling in wireless sensor networks. In: INFOCOM, pp. 366–372 (2007)
9. Zhu, J., Hu, X.: Improved algorithm for minimum data aggregation time problem in wireless sensor networks. Journal of Systems Science and Complexity 21(14), 626–636 (2008)
10. Xu, X., Wang, S., Mao, X., Tang, S., Li, X.Y.: An improved approximation algorithm for data aggregation in multi-hop wireless sensor networks. In: Proceedings of the 2nd ACM International Workshop on Foundations of Wireless Ad Hoc and Sensor Networking and Computing, FOWANC 2009, pp. 47–56. ACM (2009)
11. Wan, P.-J., Alzoubi, K.M., Frieder, O.: Distributed construction of connected dominating set in wireless ad hoc networks. MONET 9(2), 141–149 (2004)
12. Wan, P.-J., Yi, C.-W., Jia, X., Kim, D.: Approximation algorithms for conflict-free channel assignment in wireless ad hoc networks: Research articles. Wirel. Commun. Mob. Comput. 6(2), 201–211 (2006)
13. Lutz, M.: Python Pocket Reference, 4th edn. O'Reilly Media (2009)
14. Rossum, G.V., Drake, F.L.: The python language reference, vol. 24(10), p. 47881 (October 2010)

Access to Emergency Services during Overload Traffic Period

Andrey Levakov and Nikolay Sokolov

Saint-Petersburg State University of Telecommunications, Russia
levak66@bk.ru, sokolov@niits.ru

Abstract. Next generation network has a number of peculiarities that must be taken into consideration in the evolution process of telecommunications. One of such peculiarities is the behavior of the network during the dramatic increase of the traffic. This process is important in particular when emergency situations arise. It is caused by two main factors. Firstly, there is a natural requirement for exchange of large volumes of information. Secondly, it is possible that the emergency situation will lead to malfunctioning of a part of a network. In other words, increase of a traffic can be accompanied by decrease of a throughput of the network, which essentially complicates provision of the standardized quality of service indices. In this research the problems of serving dramatically increasing traffic without change of available resources are considered.

1 The Simplest Model of the NGN Node

It is considered that IP-packets flow is best represented by distributions with so called "heavy tails". Duration of service of IP-packet in the node is usually taken as constant value or distribution law with small variation coefficient is applied. Router is considered as single-line queuing system. It means that in Kendall's notation mathematical model of router can be written as $G/G/1$.

Some estimations (usually qualitative) can be obtained by simplification of initial model. Thus for the first stage of researchthe model $M/M/1$ can be chosen. For this model, it is easy to deduce formulas for calculation of two important parameters of quality of service, standardized in ITU-T Recommendation Y.1541 [1]:

- Mean IP-packet transfer delay – *IPTD*;
- IP-packet delay variation – *IPDV*.

The great property of the $M/M/1$ lies in that minimum time of IP-packets delay is zero. It means that *IPDV* equals to p-quantile, usually denoted as t_p in technical literature. Required parameters are determined by the following equations [2]:

$$IPTD = \frac{1}{\mu - \lambda}, \quad IPDV = t_p = \frac{-\ln(1-p)}{\mu - \lambda} \tag{1}$$

where μ is intensity of the service in the queuing system, calculated as b^{-1} (b is the mean service time of IP-packet), λ – intensity of IP-packets flow, it equals to a^{-1} (a is

S. Andreev et al. (Eds.): NEW2AN/ruSMART 2012, LNCS 7469, pp. 424–428, 2012.

the mean time between moments of arrival of IP-packets), p – probability, for which the value of *IPDV* is standardized. In the teletraffic theory for the queuing systems, the value of load is ρ introduced:

$$\rho = \frac{\lambda}{\mu}.$$

For the stable functioning of teletraffic system the condition $\rho < 1$ must be upheld. Let us introduce measure ξ, which equals to difference between *IPDV* and *IPTD*:

$$\xi = \frac{-\ln(1-p)-1}{\mu(1-\rho)} \qquad (2)$$

Measure ξ can be considered as difference of jitter and mean value of IP-packets delay. It is obvious that with $\rho \to 1$ measure ξ tends to infinity. This fact points out the big dispersion of the moments of time, which characterize the studied random process. Apparently, the quality of service decreases significantly. At that numerator of equation is a constant value, which magnitude does not change the essence of studied process. This allows replacement of quantile standardized in ITU-T Y.1541 for $p=0.999$ by some other value. In particular, for the estimations obtained by means of simulation the 95% quantile is usually used.

For the *M/M/*1 model calculation of the new quantile *IPDV*$_2$, specified for probability p_2, is carried out according to norms, determined for the pair *IPDV*$_1$ and p_1:

$$IPDV_2 = IPDV_1 \frac{\ln(1-p_2)}{\ln(1-p_1)} \qquad (3)$$

Quantile t_p can be expressed as:

$$t_p = IPTD + \vartheta\sigma, \qquad (4)$$

where σ is a standard deviation, then coefficient ϑ has the following form:

$$\vartheta = -1 - \ln(1-p).$$

In particular, the standardized in ITU-T Recommendation Y.1541 probability $p=0.999$ yields $\vartheta \approx 5.908$. Value $p=0.95$ gives $\vartheta \approx 1.996$.

Estimation method of *IPDV* parameter via mean value and standard deviation is promising, since it allows significant simplification of the quality of service parameters calculation based on the statistical information. It is usually gathered in the form of histogram in which delay t_i occurs with probability q_i. Then values *IPTD* and σ are calculated by the rules of determining the moments of the random variable:

$$IPTD = \sum_{\{I\}} q_i t_i, \quad \sigma = \sqrt{\sum_{\{I\}} q_i t_i^2 - \left(\sum_{\{I\}} q_i t_i\right)^2}.$$

Equations (1) – (4) are ineffective for the study of NGN with significant increase of traffic. They are useful from the point of view of detecting a number of regularities, which as shown further are valid for a more complicated models used for adequate description of processes of serving packet traffic under the condition of its sharp increase.

2   Elaboration of NGN Node Model

For the analysis of the behaviour of NGN node with dramatic increase of traffic, it is preferable to use $G/M/1$ model. Assumption of an exponential distribution of the packets' service time enables calculation of upper limits for the parameters of IP-packets' delay. For the $G/M/1$ model, there are formulas in [2], which can be used for determination of required parameters of a queuing system. For the IP-packets flow at the model's input, distribution laws with "heavy tails" will be applied. Out of probable distributions only those that have Laplace-Stiltjes transform $\alpha(s)$ of the IP-packets interarrival delay distribution function $A(t)$ are chosen.

For the study of the $G/M/1$ model it is necessary to find value v, which is the only root of the equation

$$v = \alpha(\mu - \mu v), \qquad (5)$$

lying in the range $0 < v < 1$. After solving (5) IPTD and IPDV are determined by formulas:

$$IPTD = \frac{1}{\mu - v\mu}, \quad IPDV = \frac{-\ln(1 - p)}{\mu - v\mu}. \qquad (6)$$

For the $M/M/1$ model solution of the (5) yields: $v = \rho$. It means that expressions (1) and (6) coincide. The formal similarity of equations (1) and (6) affirms the possibility of using formula (3) for simplification of studying of traffic servicing processes. Confirmation of the hypothesis of this kind can be done by means of computation of analyzed values IPDV and IPTD.

Validity of utilizing formula (3) was checked and proved for the studied model with different kinds of distribution $A(t)$. The revealed pattern indicates that the decisive factor for the chosen method of quantile conversion is the distribution of the IP-packet service time $B(t)$. It must be exponential.

Values IPDV and IPTD will increase sharply with increasing load. This is the common property of queuing system regardless of the type of functions $A(t)$ and $B(t)$. From the practical point of view for the studied model $G/M/1$ the relation of indices IPDV and IPTD with load ρ and variation coefficient C_A, determined by the type of function $A(t)$, is especially interesting

As $A(t)$ function the hyper exponential distribution was chosen since. It allows studying the behaviour of the queuing system in a wide range of values C_A. On the figures 1 and 2 the curves are given that depict the nature of the change in values IPDV and IPTD respectively. The graphs are plotted for three values C_A: 2, 6 and 10 in the range $0.8 \leq \rho < 1$. On the same graphs for the point $\rho = 0.99$ values of IPDV and IPTD for the $M/M/1$ model are indicated. On the vertical axis, the values of both studied variables, normalized to the average IP-packets service time b, are denoted.

Behaviour of functions $IPTD = f(\rho, C_A)$ and $IPDV = f(\rho, C_A)$ leads to some interesting and very important practice conclusions. Firstly, behaviour of $IPTD = f(\rho, C_A)$ and $IPDV = f(\rho, C_A)$ curves is identical, that was expected taking into account expression (4). Secondly, values IPTD and IPDV with an increase of traffic significantly exceed

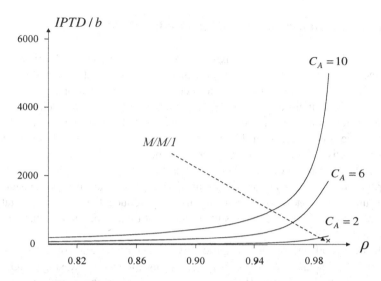

Fig. 1. The dependence of *IPTD* from load and parameter C_A

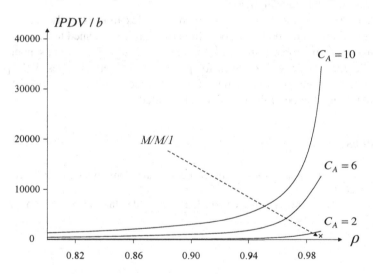

Fig. 2. The dependence of *IPDV* from load and parameter C_A

(by orders of magnitude and not in times) mean IP-packets service time. Thirdly, studied values are significantly affected by the variation coefficient C_A that is expression "behaviour of the teletraffic system during overload" only partially describes arising situation. The kind of input flow for which behaviour of the system under heavy load is studied must also be specified.

This fact must be considered in the development of methods and algorithms of overload prevention. If the $A(t)$ process can be managed in a way that will lead to reduction of variation coefficient then it will be possible to effectively support the maximum achievable level of quality of service for traffic.

3   Conclusion and Direction of Further Research

According to the results of the approximate analysis of the NGN network with a significant increase in traffic the numerical estimates were obtained, which allow making some conclusions. It seems appropriate to focus on the three positions.

First, one of the main conclusions is that frequently used term "behaviour of the teletraffic system during overload" can not be considered exhaustive. In addition to load magnitude it is necessary to know: which distribution law describes the intervals between successive IP-packets entering the NGN node?

Second, the possibility of replacing the quantile, standardized at the 99.9% level of the distribution function by more common 95% quantile, can also be considered as useful result. It becomes possible to significantly reduce the amount of statistics, which is needed to obtain reliable results in terms of quality of service characteristics of traffic in the NGN.

Third, the hypothesis of an asymptotically exponential nature of the distribution function of IP-packets delay is useful for analyzing models of the single NGN node and NGN network as a whole. If this hypothesis is correct, the number of important relations can be obtained analytically.

Further research includes analysis of more complex models, fully taking into account the peculiarities of the NGN node. In particular, it is planned to carry out analysis by simulation of probabilistic and temporal characteristics of NGN node, taking into consideration priority disciplines for processing of certain types of IP-packets. Additionally, it is necessary to take into account the fact that the capacity of the buffer memory at the input node of the NGN node is limited.

References

1. ITU-T Recommendation Y.1541, Network Performance Objectives for IP-Based Services (February 2006)
2. Kleinrock, L.: Queueing Systems, Volume I: Theory. Wiley Interscience, New York (1975)

M2M Applications and Open API: What Could Be Next?

Manfred Sneps-Sneppe[1] and Dmitry Namiot[2]

[1] Ventspils University College
Ventspils International Radioastronomy Centre
Ventspils, Latvia
[2] Lomonosov Moscow State University
Faculty of Computational Mathematics and Cybernetics
Moscow, Russia
{manfreds.sneps,dnamiot}@gmail.com

Abstract. In this paper we describe the current state of open APIs for M2M applications as well as some possible changes and extensions. Our article based on open standards ETSI is going to provide for the rapidly growing M2M market. An open specification, presented as an Application Programming Interface (OpenAPI), provides applications with a rich framework of core network capabilities upon which to build services while encapsulating the underlying communication protocols. OpenAPI is a portable platform for services that may be replicated and ported between different execution environments and hardware platforms. We would like to discuss the possible extensions for ETSI proposals that, by our opinion, let keep telecom development inline with the modern approaches in the web development domain.

Keywords: m2m, open api, etsi, rest, web development, web intents.

1   Introduction

At Machine-to-Machine (M2M) refers to technologies that allow both wireless and wired systems to communicate with other devices of the same ability. M2M uses a device (such as a sensor or meter) to capture an event (such as temperature, inventory level, etc.), which is relayed through a network (wireless, wired or hybrid) to an application (software program), translates the captured event into meaningful information [1].

Considering M2M communications as a central point of Future Internet, European commission creates standardization mandate M/441 [2]. The Standardization mandate M/441, issued on 12th March 2009 by the European Commission to CEN, CENELEC and ETSI, in the field of measuring instruments for the development of an open architecture for utility meters involving communication protocols enabling interoperability, is a major development in shaping the future European standards for smart metering and Advanced Metering Infrastructures. The general objective of the mandate is to ensure European standards that will enable interoperability of utility meters (water, gas, electricity, heat), which can then improve the means by which

S. Andreev et al. (Eds.): NEW2AN/ruSMART 2012, LNCS 7469, pp. 429–439, 2012.

customers' awareness of actual consumption can be raised in order to allow timely adaptation to their demands.

Besides the describing the current state of standards, our goal main here is the proposal for some new additions in M2M APIs architecture. We are going to propose web intents as add-on for the more traditional REST approach in order to simplify the development phases for M2M applications. The key moments in our proposals are: JSON versus XML, asynchronous communications and integrated calls.

Let us start from the basic moments. Right now market players are offering own standards for M2M architecture. We refer to the recent ETSI TC M2M Workshop held on October 26-28, 2011. Figure 1 illustrates the basics of M2M infrastructure (as per ETSI) [3].

As per Cisco [4] the M2M infrastructure includes three primary domains: cloud, network, and edge devices. Each of these domains contains a specific anchor point which conducts the M2M signaling across the infrastructure. The M2M traffic has its own specific characteristics, such as low mobility and offline and online data transmission, which create new challenges for dimensioning the network. Service providers that are trying to customize their networks face the additional challenge of supporting traffic generated from residential and enterprise customer premises equipment (CPE).

And of course, there are several attempts to provide the standard set of software tools for M2M applications. These attempts are well explainable and understandable of course. Because M2M applications are directly linked to hardware devices than the portability of applications, the ability to bring new devices in system etc. becomes the key factors. It is practically the same idea we saw in telecom standards like Parlay for example.

Current customized M2M solutions and platforms tend to assume direct connectivity between the M2M core and devices, with no aggregators. However, linking residential and enterprise M2M gateways to an M2M-ready core opens new business models for service providers. M2M gateways can be bypassed when necessary.

In other words, what we can see now – it is a growing interest to the M2M middleware.

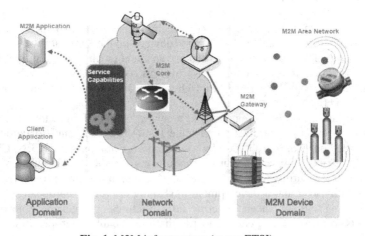

Fig. 1. M2M infrastructure (as per ETSI)

Also M2M middleware helps us with heterogeneity of M2M applications. Heterogeneity of service protocols inhibits the interoperation among smart objects using different service protocols and/or API's. We assume that service protocols and API's are known in advance. This assumption prevents existing works from being applied to situations where a user wants to spontaneously configure her smart objects to interoperate with smart objects found nearby [5]

Alcatel [6], for example proposes the following conceptual view of M2M server (Fig.2).

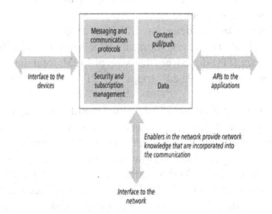

Fig. 2. M2M server architecture [6]

The gateway element should be located on the boundary between a wireless network and the Internet network used by application servers to communicate to a device. So, the M2M server can maintain sessions to application servers on one side, and to devices on the other side. In other words, it acts as a bridge, passing information from the application server to appropriate devices.

The rest of the paper is organized as follows. Section 2 contains an analysis of M2M API standardization activities. In Section 3 we consider Open API for M2M, submitted to ETSI. In Section 4 we offer the never web tool – Web Intents for enhancement of M2M middleware. Sections 5 and 6 are devoted to discussion.

2   On M2M API Standards

ETSI standards are very important of course, but we should mention other participants too [18].

The 3rd Generation Partnership Project maintains and develops technical specifications and reports for mobile communication systems. Mobile networks are also concerned with the integration and support of M2M communications. 3GPP Technical Specifications Group dealing with Service and System Aspects [19], has issued a number of specifications dealing with requirements that M2M services and M2M communication imposes on the mobile network.

The Telecommunications Industry Association is the United States developing industry standards for a wide variety of telecommunication products. The standardization activities are assigned to separate Engineering Committees. The TR-50 Engineering Committee Smart Device Communications [20], has been assigned the task to develop and maintain physical-medium-agnostic interface standards, that will enable the monitoring and bi-directional communication of events and information between smart devices and other devices, applications or networks. It will develop a Smart Device Communications framework that can operate over different types of underlying transport networks (wireless, wired, etc.) and can be adapted to a given transport network by means of an adaptation/convergence layer.

The International Telecommunication Union as a specialized agency of the United Nations is responsible for IT and communication technologies. The Telecommunications Standardization Sector (ITU-T), covers the issue of M2M communication via the special Ubiquitous Sensor Networks-related groups [21]. ITU address the area of networked intelligent sensors.

Open Mobile Alliance (OMA) [22] develops mobile service enabler specifications. OMA drives service enabler architectures and open enabler interfaces that are independent of the underlying wireless networks and platforms. An OMA Enabler is a management object designated for a particular purpose. It is defined in a specification and is published by the Open Mobile Alliance as a set of requirements documents, architecture documents, technical specifications and test specifications. Examples of enablers would be: a download enabler, a browsing enabler, a messaging enabler, a location enabler, etc. Data service enablers from OMA should work across devices, service providers, operators, networks, and geographies.

As there are several OMA standards that map into the ETSI M2M framework, a link has been established between the two standardization bodies in order to provide associations between ETSI M2M Service Capabilities and OMA Supporting Enablers [23]. Specifically, the expertise of OMA in abstract, protocol-independent APIs creation, as well as the creation of APIs protocol bindings (i.e. REST, SOAP) and especially the expertise of OMA in RESTful APIs is expected to complement the standardization activities of ETSI in the field of M2M communications. Additionally, OMA has identified areas where further standardization will enhance support for generic M2M implementations, i.e. device management, network APIs addressing M2M service capabilities, location services for mobile M2M applications [18].

3   Open API from ETSI

This section describes an Open API for M2M, submitted to ETSI. It is probably the most valuable achievement at this moment.

The Open API for M2M applications developed jointly in Eurescom study P1957 [7] and the EU FP7 SENSEI project [8] makes. The OpenAPI has been submitted as a contribution to ETSI TC M2M [9] for standardization.

Actually, in this Open API we can see the big influence of Parlay specification. Parlay Group leads the standard, so called Parlay/OSA API, to open up the networks by defining, establishing, and supporting a common industry-standard APIs. Parlay Group also specifies the Parlay Web services API, also known as Parlay X API, which is much simpler than Parlay/OSA API to enable IT developers to use it without network expertise [10].

The goals are obvious, and they are probably the same as for any unified API. One of the main challenges in order to support easy development of M2M services and applications will be to make M2M network protocols "transparent" to applications. Providing standard interfaces to service and application providers in a network independent way will allow service portability [11].

At the same time an application could provide services via different M2M networks using different technologies as long as the same API is supported and used. This way an API shields applications from the underlying technologies, and reduces efforts involved in service development. Services may be replicated and ported between different execution environments and hardware platforms [12]

This approach also lets services and technology platforms to evolve independently. A standard open M2M API with network support will ensure service interoperability and allow ubiquitous end-to-end service provisioning.

The OpenAPI provide service capabilities that are to be shared by different applications. Service Capabilities may be M2M specific or generic, i.e. providing support to more than one M2M application.

Let us name the main Open API categories and make some remarks:

Grouping
A group here is defined as a common set of attributes (data elements) shared between member elements.   On practice it is about the definition of addressable and exchangeable data sets. Just note, as it is important for our future suggestions, there are no persistence mechanisms for groups

Transactions
Service capability features and their service primitives optionally include a transaction ID in order to allow relevant service capabilities to be part of a transaction. Just for the deploying transactions and presenting some sequences of operations as atomic.

In the terms of transactions management Open API presents the classical 2-phase commit model. By the way, we should note here that this model practically does not work in the large-scale web applications. We think it is very important because without scalability we can think about "billions of connected devices".

Application Interaction
The application interaction part is added in order to support development of simple M2M applications with only minor application specific data definitions: readings, observations and commands.

Application interactions build on the generic messaging and transaction functionality and offer capabilities considered sufficient for most simple application domains.

Messaging
The Message service capability feature offers message delivery with no message duplication. Messages may be unconfirmed, confirmed or transaction controlled. The message modes supported are single Object messaging, Object group messaging, and any object messaging; (it can also be Selective object messaging). Think about this as Message Broker.

Event notification and presence
The notification service capability feature is more generic than handling only presence. It could give notifications on an object entering or leaving a specific group, reaching a certain location area, sensor readings outside a predefined band, an alarm, etc.

It is a generic form. So, for example, geo fencing should fall into this category too.

The subscriber subscribes for events happening at the Target at a Registrar. The Registrar and the Target might be the same object. This configuration offers a publish/subscribe mechanism with no central point of failure.

Compensation
Fair and flexible compensation schemes between cooperating and competing parties are required to correlate resource consumption and cost, e.g. in order to avoid anomalous resource consumption and blocking of incentives for investments. The defined capability feature for micro-payment additionally allows charging for consumed network resources.

It is very similar, by the way, to Parlay's offering for Charging API.

Sessions
In the context of OpenAPI a session shall be understood to represent the state of active communication between Connected Objects

OpenAPI is REST based, so, the endpoints should be presented as some URI's capable to accept (in this implementation) the basic commands GET, POST, PUT, DELETE.

4   Web Intents vs. Open API from ETSI

Let us start from the basic. Users use many different services on the web to handle their day to day tasks, developers use different services for various tasks. In other words – our environment consists of connected applications. And of course, all they expect their applications to be connected and to work together seamlessly.

It is almost impossible for developers to anticipate every new service and to integrate with every existing external service that their users prefer and thus they must choose to integrate with a few select APIs at great expense to the developer.

As per telecom experience we can mention here the various attempts for unified API that started, probably, with Parlay. Despite a lot of efforts, Parlay API's actually increase the time for development. It is, by our opinion, the main reason for the Parlay's failure.

Web Intents solves this problem. Web Intents is a framework for client-side service discovery and inter-application communication. Services register their intention to be able to handle an action on the user's behalf. Applications request to start an action of a certain verb (for example share, edit, view, pick etc.) and the system will find the appropriate services for the user to use based on the user's preference. It is the basic [13].

Intents play the very important role in Android Architecture. Three of the four basic OS component types - activities, services, and broadcast receivers - are activated by an asynchronous message called as intent.

Intents bind individual components to each other at runtime (you can think of them as the messengers that request an action from other components), whether the component belongs to your application or another.

Created intent defines a message to activate either a specific component or a specific type of component - an intent can be either explicit or implicit, respectively.

For activities and services, an intent defines the action to perform (for example, to "view" or "send" something) and may specify the URI of the data to act on (among other things that the component being started might need to know). For example, our intent might convey a request for an activity to show an image or to open a web page. In some cases, you can start an activity to receive a result, in which case, the activity also returns the result in an Intent (for example, you can issue an intent to let the user pick a personal contact and have it returned to you - the return intent includes a URI pointing to the chosen contact) [15].

Going to M2M applications it means that our potential devices will be able to present more integrated data for the measurement visualization for example. The final goal of any M2M based application is to get (collect) measurements and perform some calculations (make some decisions) on the collected dataset. We can go either via low level APIs or use (at least for the majority of use cases) some integrated solutions. The advantages are obvious. We can seriously decrease the time for development.

Web Intents puts the user in control of service integrations and makes the developers life simple.

Here is the modified example for web intents integration for the hypothetical web intents example:

1. Register some intent upon loading our HTML document

```
document.addEventListener("DOMContentLoaded", function() {
    var regBtn = document.getElementById("register");
    regBtn.addEventListener("click", function() {
    window.navigator.register("http://webintents.org/m2m", undefined);
    }, false);
```

2. Start intent's activity

```
    var startButton = document.getElementById("startActivity");
    startButton.addEventListener("click", function() {
    var intent = new Intent();
    intent.action = "http://webintents.org/m2m";
        window.navigator.startActivity(intent);
    }, false);
```

3. Get measurements (note – in JSON rather than XML) and display them in our application

```
window.navigator.onActivity = function(data) {
  var output = document.getElementById("output");
  output.textContent = JSON.stringify(data);
};
}, false);
```

Obviously, that it is much shorter than the long sequence of individual calls as per M2M Open API.

The key point here is onActivity callback, that returns JSON (not XML!) formatted data. As per suggested M2M API we should perform several individual requests, parse XML responses for the each of them and only after that make some visualization. Additionally, web intents based approach is asynchronous by its nature, so, we don need to organize asynchronous calls by our own.

Generally speaking, we expect the initiatives from software companies that will opposite to telecom approach. For example, Paho project [17] (IBM et al.) directly declares the need to provide open source implementations of open and standard messaging protocols that support current and emerging requirements of M2M integration with Web and Enterprise middleware and applications. It will include client implementations for use on embedded platforms along with corresponding server support as determined by the community. This will enable a paradigm shift from legacy point-to-point protocols and the limitations of protocols like SOAP or HTTP into more loosely coupled yet determinable models. It will bridge the SOA, REST, Pub/Sub and other middleware architectures already well understood by Web 2.0 and Enterprise IT shops today, with the embedded and wireless device architectures inherent to M2M.

5   Data Persistence

The next question we would like to discuss relating to the M2M API's is probably more discussion able. Shall we add some persistence API (at least in the form of generic interface)?

The reasons are obvious – save the development time. Again, we should keep in mind that we are talking about the particular domain – M2M. In the most cases our business applications will deal with some metering data. As soon as we admit, that we are dealing with the measurements in the various forms we should make, as seems to us a natural conclusion – we need to save the data somewhere. It is very simple – we need to save data for the future processing.

So, the question is very easy – can we talk about M2M applications without talking about data persistence? Again, the key question is M2M. It is not abstract web API. We are talking about the well-defined domain.

As seems to us, even right now, before the putting some unified API in place, the term M2M almost always coexists with the term "cloud". And as we can see, almost always has been accompanied by the terms like automatic database logging, backup capabilities etc.

So, maybe this question is more for the discussions or it even could be provocative in the some forms, but it is: why there is no reference API for persistence layer in the unified M2M API? It is possible in general to create data gathering API without even mentioning data persistence? Shall we define cloud database API as a part of M2M standard or not?

The use of cloud computing means that data collection, processing, interface, and control can be separated and distributed to the most appropriate resource and device. In contrast, currently M2M implementations tend to combine data collection, processing, interface, and control.

Once transmitted to the cloud, data can be stored, retrieved and processed without having to address many of the underlying computing resources and processes traditionally associated with databases. For M2M applications, this type of virtualized data storage service is ideal [16]

As soon as ETSI standards define the interfaces, the developers we will be able to introduce various implementations. For example, it looks like NoSQL solutions are perfect fit for M2M applications.

These data stores operate by using key-value associations which allows for a flatter non-relational form of association. NoSQL databases can work without fixed table schemes, which makes it easy to store different data formats as well as change and expand formats over time. It is very important for M2M applications (as well as for any type of applications tied with hardware). There are no "unified" devices in the real word. We simply cannot create an efficient schema that will serve all the devices (including new entrants). So M2M stores should schema-less.

NoSQL databases could be easily scaled horizontally. Data is distributed across many servers and disks. Indexing is performed by keys that route the queries to the datastore for the range that serves that key. This means different clusters respond to requests independently from other clusters, what increases throughput and response times. Growth can be accommodated by quickly adding new servers, database instances and disks and changing the ranges of keys.

There are more then enough NoSQL systems on the market, they all have own APIs, so the question for M2M standardization body becomes even more important: shall we include the "unified" interface to data store into standard?

6   New Signaling Demand

Eventually, billions of devices — such as sensors, consumer electronic devices, smart phones, PDAs and computers — will generate billions of M2M transactions. For example: Price information will be pushed to smart meters in a demand-response system. Push notifications will be sent to connected devices, letting a client application know about new information available in the network. The scale of these transactions will go beyond anything today's largest network operators have experienced. Signaling traffic will be the primary bottleneck as M2M communications increase. Alcatel-Lucent Bell Labs traffic modeling studies support this by comparing network capacity against projected traffic demand across multiple dimensions (such as signaling processing load on the radio network controller, air-interface access channel capacity, data volume and memory requirement for maintaining session contexts). The limiting factor is likely to be the number of session set-ups and tear-downs. For the specific traffic model and network deployment considered in the study, it is seen that up to 67 percent of computing resources in the radio network controller is consumed by M2M applications. [6].

How much of the traffic sent is network overhead? As an analysis carried on by A. Sorrevad [14] shows for ZigBee solution, a node is sending at least 40 Mbytes per year with the purpose of maintaining the network and polling for new data. The trigger data traffic for a year is much less - around1-10 Mbytes. Thus we see that the relationship between network and trigger traffic can range between 40:1 to 4:1 in a ZigBee solution that is following the home automation specification.

The traffic sent when maintaining a 6LoWPAN network is application specific. The relationship between network and trigger traffic can then be in the range 2:1 to 5:1.

Why do we think it is a place for traffic talk? Because again, it is not clear completely how can we support transactional APIs (as per ETSI draft [9]), without the dealing with the increased traffic. Simply – in our transactions we need the confirmation that device is alive, that operation has been performed etc. All this is signaling traffic. Actually, this may lead to next provocative questions: do we really need transactional calls for all use cases? For example, the modern large-scale web applications (e.g. social networks) are not transactional internally.

7   Conclusion

In this article we briefly describe the current state for the open unified M2M API from ETSI. We propose some new additions – web intents as add-on for the more traditional REST approach. The main goal for our suggestions is the simplifying the development phases for M2M applications. The key advantages are JSON versus XML, asynchronous communications and integrated   calls. Also we would like to point attention of readers to the couple of important questions that are not covered yet by our opinion: data persistence and signaling traffic.

Acknowledgement. The paper is financed from EDRF's project SATTEH (No. 2010/0189/2DP/2.1.1.2.0./10/APIA/VIAA/019) being implemented in Engineering Research Institute «Ventspils International Radio Astronomy Centre» of Ventspils University College (VIRAC).

References

[1] Machine-to-Machine, http://en.wikipedia.org/wiki/Machine-to-Machine
[2] Standartisation mandate to CEN, CENELEC and ETSI in the field of measuring instruments for the developing of an open architecture for utility meters involving communication protocols enabling interoperability, European Commission, M/441 (2009)
[3] ETSI Machine-to-Machine Communications,
      http://www.etsi.org/website/technologies/m2m.aspx

[4] Managed and Cloud Services Insight Group Machine-to-Machine and Cloud Services, http://www.cisco.com/en/US/solutions/collateral/ns341/ns849/ ns1098/white_paper_c11-663879.html

[5] Park, H., Kim, B., Ko, Y., Lee, D.: InterX: A service interoperability gateway for heterogeneous smart objects. In: 2011 IEEE International Conference on Pervasive Computing and Communications Workshops (PERCOM Workshops), March 21-25, pp. 233–238 (2011)

[6] Viswanathan, H.: Getting Ready for M2M Traffic Growth, http://www2.alcatel-lucent.com/blogs/techzine/2011/ getting-ready-for-m2m-traffic-growth/

[7] EURESCOM project P1957, Open API for M2M applications, http://www.eurescom.de/public/projects/P1900-series/P1957/

[8] Sensei project, http://www.sensei-project.eu/

[9] Draft ETSI TS 102 690 V0.13.3 (2011-07) Technical Specification

[10] Yim, J.-C., Choi, Y.-I., Lee, B.-S.: Third Party Call Control in IMS using. In: The 8th International Conference on Parlay Web Service Gateway Advanced Communication Technology, ICACT 2006, February 20-22, pp. 221–224 (2006)

[11] Grønbæk, I.: Architecture for the Internet of Things (IoT): API and interconnect. In: The Second International Conference on Sensor Technologies and Applications. IEEE (August 2008), doi:10.1109/SENSORCOMM.2008.20, 809

[12] Grønbæk, I., Ostendorf, K.: Open API for M2M applications. In: ETSI M2M Workshop (October 2010)

[13] Web Intents, http://webintents.org/

[14] Sorrevad, A.: M2M Traffic Characteristics, KTH Information and Communication Technology Master of Science Thesis Stockholm, Sweden, TRITA-ICT-EX-2009:212 (2009), http://web.it.kth.se/~maguire/DEGREE-PROJECT-REPORTS/ 091201-Anders_Orrevad-with-cover.pdf

[15] Android Developers, http://developer.android.com/guide/topics/fundamentals.html

[16] Cloud + Machine-to-Machine, http://www.readwriteweb.com/cloud/2011/03/cloud-machine-to-machine-disruptive-innovation-part-1p2.php

[17] Paho project, http://eclipse.org/proposals/technology.paho/

[18] IoT project, http://www.iot-a.eu/public

[19] 3GPP TS 22.368 V11.0.1, 3rd Generation Partnership Project; Technical Specification Group Services and System Aspects; Service requirements for Machine-Type Communications (MTC); Stage 1 (Release 11)

[20] TR-50 standards, http://www.tiaonline.org/standards/committees/ committee.cfm?comm=tr-50

[21] Han, J.-I., Vu, A.-D., Kim, J.-W., Jeon, J.-S., Lee, S.-M., Kim, Y.-M.: The fundamental functions and interfaces for the ITU-T USN middleware components. In: 2010 International Conference on Information and Communication Technology Convergence (ICTC), November 17-19, pp. 226–231 (2010), doi:10.1109/ICTC.2010.5674664, Print ISBN: 978-1-4244-9806-2

[22] OMA, http://www.openmobilealliance.org/

[23] Blum, N., Boldea, I., Magedanz, T., Margaria, T.: Service-oriented access to next generation networks: from service creation to execution. Journal Mobile Networks and Applications Archive 15(3) (June 2010), doi:10.1007/s11036-010-0222-1

Modeling of Hysteretic Signaling Load Control in Next Generation Networks*

Pavel Abaev, Yuliya Gaidamaka, and Konstantin E. Samouylov

Telecommunication Systems Department,
Peoples' Friendship University of Russia,
Ordzhonikidze str. 3, 115419 Moscow, Russia
{pabaev,ygaidamaka,ksam}@sci.pfu.edu.ru
http://www.telesys.pfu.edu.ru

Abstract. In this paper we investigate traffic load control mechanisms for controlling congestion in signaling networks based on three types of thresholds. The goal of the paper is to analyze congestion controlling mechanisms and develop corresponding queuing models of SIP servers. The study is based on hysteretic congestion control, which has been developed for Signaling System 7 (SS7). Models for describing the hysteretic control are developed. The current state and problems of basic overload control mechanism proposed by Internet Engineering Task Force (IETF) for SIP signaling networks are investigated. Approaches to building mathematical models of SIP servers in the form of a queuing system with hysteretic control are proposed.

Keywords: signaling network, SS7, SIP server, hysteretic control, congestion control mechanism, overloading control, threshold control.

1   Introduction

Threshold load control is a key tool in preventing congestion of various types in telecommunication networks [1–4]. To control congestion three types of thresholds are used: congestion onset threshold, congestion abatement threshold and congestion discard threshold. Variations of this mechanism are used for detection of congestion in SS7 networks [1, 4] as well as in networks based on Session Initiation Protocol (SIP) [2, 3, 5–7].

Hysteretic control for SS7 networks was developed by International Telecommunication Union (ITU) on the Recommendations of Q.700 series [1]. Recommendation Q.704 [1] defines three types of congestion: link congestion, route congestion, and signaling point congestion. Processing congestion in SS7 involves two stages: congestion detection and congestion mitigation. The number of messages in transmit or retransmit buffers is controlled in order to detect congestion and prevent it.

---

* This work was supported in part by the Russian Foundation for Basic Research (grants 10-07-00487-a and 12-07-00108), and by Rosobrazovanie (project no. 020619-1-174).

S. Andreev et al. (Eds.): NEW2AN/ruSMART 2012, LNCS 7469, pp. 440–452, 2012.

The main purpose of this paper is to analyze overloading control mechanisms of SIP servers. The basic provisions of the mechanisms are defined in the IETF Standards [2, 3, 6], and in [8–21] it is shown that they are based on the principles of hysteretic control.

Section 2 of this paper shows that in SS7 congestion detection is carried out by introducing thresholds in the transmit buffer, and the action to mitigate congestion is to reduce the load. The respective mechanism is called hysteretic overload control. Section 3 of this paper is devoted to investigation of overload control mechanisms of SIP servers. First, we analyze typical examples of overloading detection and the problems of developing the control mechanisms in accordance with IETF Standards [2, 3]. Second, we introduce the types of overload control and classify the mechanisms of overload control of SIP servers. Note that in the IETF Standards the concept of hysteretic control is not explicitly defined, but these mechanisms that prevent congestion of SIP servers are studied in numerous papers [8–21]. These papers also propose various models for analyzing the parameters and indicators of overload control. Using the concept of hysteretic congestion control imposed for SS7, we design typical models of overload control of SIP servers in the form of a queuing system in section 4. Finally, we study a queuing system, developed for the analysis of overload control of SIP servers.

2   SS7 Signaling Link Congestion Control

SS7 signaling link congestion control developed by the ITU is based on the procedure of hysteretic overload control. One of the first results in this area was published in [22]. The procedure implements the monitoring of the total number of messages in transmit or retransmit buffer and signaling traffic flow control. The signaling traffic flow control procedures are used to divert a given traffic flow (toward one or more destinations) from congested signaling link to an alternative available signaling link.

The mechanism of SS7 signaling link congestion control consists of two stages: to detect congestion and to eliminate or mitigate congestion.

In order to detect congestion the monitoring of the total number of messages in transmit and retransmit buffers is kept on hold. To eliminate congestion limitation of signaling traffic at its source is performed. Limitation (restriction or prohibition) of incoming signalling traffic is needed in case the signalling network is not capable of transferring all signalling traffic offered by the user because of network failures or congestion. The following three types of thresholds are defined in ITU Recommendation Q.704:

- congestion onset threshold H_1 – for detecting the onset of congestion,
- congestion abatement threshold L_1 – for monitoring the abatement of congestion, and
- congestion discard thresholds R_1 – for determining whether, under congestion conditions, a message should be discarded.

In national signalling networks with multiple congestion thresholds up to three separate thresholds are provided for detecting the onset of congestion, and up to three separate thresholds for monitoring the abatement of congestion, respectively. In addition up to two separate thresholds are provided for determining whether, under congestion conditions, a message should be discarded or transmitted using the signalling link.

So there are four congestion detection statuses and three congestion discard statuses:

$$h = \begin{cases} 0, \text{no congestion,} \\ i, i - \text{level of congestion, } i = 1,2,3; \end{cases} \qquad r = \begin{cases} 0, \text{no discard,} \\ i, i - \text{level of discard, } i = 1,2. \end{cases}$$

Criteria for determination of signalling link congestion status are the number of messages in transmit or retransmit buffers, i.e. buffer occupancy. When the buffer occupancy increases and exceeds congestion onset threshold, H_1, congestion is determined. Then the incoming load is reduced to avoid overloading. However, the load does not return to normal load value immediately, but does so after a while when the buffer occupancy decreases and comes to below the congestion abatement threshold, L_1. This technique is called hysteretic overload control. Two-level hysteretic control (onset and abatement) is needed to reduce potential oscillations between control-on and control-off states under certain loading conditions. In national signalling networks using multiple signalling link congestion states, at the onset of congestion when the buffer occupancy is increasing, congestion status is determined by the highest congestion onset threshold exceeded by buffer occupancy. At the abatement of congestion when the buffer occupancy is decreasing, congestion status is determined by the lowest congestion abatement threshold below which the buffer occupancy has dropped.

Fig. 1 shows the process of changing the congestion status depending on the buffer occupancy in the case of $h \in \{0,1\}$.

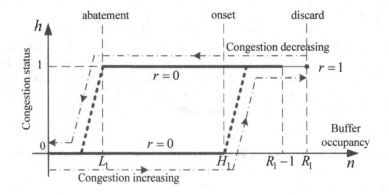

Fig. 1. The congestion detection and discard statuses

Under normal operation when the signalling link is uncongested, the signalling link congestion status is assigned zero value ($h = 0$). When the buffer occupancy increases the congestion status does not change until the predetermined congestion onset threshold, H_1, of the buffer occupancy is crossed. After that the current congestion status is assigned the unit value ($h = 1$). When the buffer occupancy is increasing up to the congestion discard threshold, R_1, or when the buffer occupancy is decreasing up to the congestion abatement threshold L_1, the congestion status has the unit value ($h = 1$). When the buffer occupancy is decreasing and crosses the congestion abatement threshold, L_1, the congestion status is assigned zero value ($h = 0$).

Fig. 2 shows one-level hysteretic overload control where the incoming signalling load $\lambda(h, r, n)$ depends on congestion detection and discard statuses, and buffer occupancy.

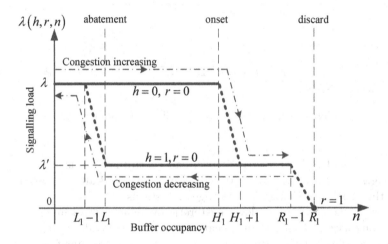

Fig. 2. One-level signalling load hysteretic control

Under normal operation the signalling traffic load has the value, λ. When congestion is detected after the buffer occupancy has crossed the congestion onset threshold H_1, the normal signalling traffic load is restricted by the value, λ'. When the buffer occupancy is increasing and crosses the congestion discard threshold R_1, the signalling traffic load is prohibited, i.e. $\lambda(h, r, n)=0$ for $n \geq R_1$. When the buffer occupancy is decreasing and crosses the congestion discard threshold R_1, the signalling traffic load has value λ', and after the buffer occupancy crosses the congestion abatement threshold L_1, the signalling traffic load has value λ.

In national signalling networks using multiple signalling link congestion states without congestion priority, up to three levels of signalling link congestion status are accommodated in the signalling network (Fig. 3). Let us point out that Fig. 3 shows the case $R_i < L_{i+1}$, but the case $R_i > L_{i+1}$ is not considered in this paper.

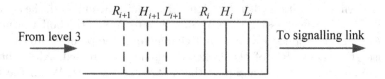

Fig. 3. Signalling link congestion status thresholds

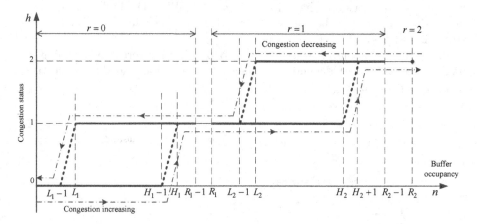

Fig. 4. Two levels of signalling link congestion status

Fig. 4 shows two-level hysteretic overload control for two levels of signalling link congestion status, i.e. $h, r \in \{0, 1, 2\}$. Fig. 4 illustrates all necessary notations about the process of changing the congestion status.

In Fig. 5 one can see two-level hysteretic overload control in the case when congestion onset and congestion discard threshold are equal, i.e. $R_1 = H_1$, $R_2 = H_2$. Let us point out that this particular case is important for further study of SIP server overload control.

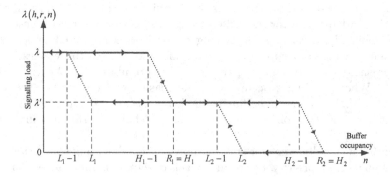

Fig. 5. Two-level signalling load hysteretic control in case of $R_1 = H_1$, $R_2 = H_2$

We consider an example of probability measures of the hysteretic load control [22–27] for two-level hysteretic control shown in Fig. 5. Let triple $(h, r, n) \in X$ be the state of the system, where the set X of all states has a form of the partition: $\mathcal{X} = \mathcal{X}_0 \cup \mathcal{X}_1 \cup \mathcal{X}_2$, where $\mathcal{X}_0$ is the set of normal load states, $\mathcal{X}_1$ is the set of first-level congestion state, $\mathcal{X}_2$ is the set of second-level congestion state:

$$\mathcal{X}_0 = \{(h, r, n) : h = 0, r = 0, 0 \leqslant n < H_1\};$$

$$\mathcal{X}_1 = \mathcal{X}_{11} \cup \mathcal{X}_{12} = \{(h, r, n) : h = 1, r = 0, L_1 \leqslant n < R_1\} \cup$$
$$\cup \{(h, r, n) : h = 1, r = 1, R_1 \leqslant n < H_2\};$$

$$\mathcal{X}_2 = \mathcal{X}_{21} \cup \mathcal{X}_{22} = \{(h, r, n) : h = 2, r = 1, L_2 \leqslant n < R_2\} \cup$$
$$\cup \{(h, r, n) : h = 2, r = 2, n = R_2\}.$$

Now we are ready to define the required quality parameters of hysteretic control — probabilities $P_i = P(\mathcal{X}_i)$, $i = 0, 1, 2$. If the limitations on the quality parameters of the system are given, for example as

$$P(\mathcal{X}_i) \leqslant P_i^*,$$

then the location of the congestion onset threshold H_1, the congestion abatement threshold L_1, and the congestion discard thresholds R_1 could be defined.

So we outlined the general principles of the hysteretic load control in SS7 and introduced the necessary notation to be used below for SIP signaling network analysis.

3   SIP Overload Control

One of the most popular session types that SIP is used for is call session. Fig. 6 shows a scenario of successful call processing [6, 7].

The connection is initiated with a UA-client by sending an INVITE message to a Proxy 1. When processing an INVITE request, a proxy typically responds with a 100 (Trying) response to stop INVITE retransmissions at the previous hop. Transfer of a message with code 180 indicates that channel resources have been allocated. Then the UA server sends a message with code 200, and UA client responds with ACK message. The session is established. When session is over successfully, the connection is disrupted and the allocated network resources are emptied.

With the increasing number of users of services based on SIP protocol, different types of SIP-server overloads arise due to lack of resources for user agent registration and for session establishment and termination. There are two types of overload: a client to server overload ("client-to-server") or a server to server overload ("server-to-server") [3]. "Client-to-server" overload appears in SIP server (Registration server) due to excessive load created by groups of SIP terminals,

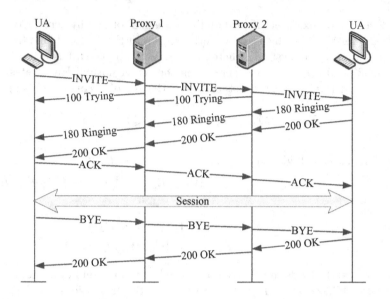

Fig. 6. SIP session establishment

marked in Fig. 6 as UA. This overload can happen when power is restored after a mass power failure in a large metropolitan area, and after the power is restored, a very large number of SIP devices boot up and send out SIP registration requests almost simultaneously, which could easily overload the corresponding SIP registration server. "Server-to-server" overload appears in the SIP server (Proxy server) as a result of some special events, also referred to as flash crowds.

The SIP protocol provides a basic overload control mechanism through the 503 (Service Unavailable) response code [6]. SIP servers that are unable to forward a request due to temporary overload can reject the request with a 503 response. The overloaded server can insert a Retry-After header into the 503 response, which defines the number of seconds during which this server is not available for receiving any further request from the upstream neighbor. A server that receives a 503 response from a downstream neighbor stops forwarding requests to this neighbor for the specified amount of time and starts again after this time is over. Without a Retry-After header, a 503 response only affects the current request and all other requests can still be forwarded to this downstream neighbor. A server that has received a 503 response can try to re-send the request to an alternate server, if one is available. A server does not forward 503 responses toward the UA and converts them to 500 Server Internal Error responses instead.

The list of the problems that arise as a result of overload control mechanism of the SIP server using the 503 (Service Unavailable) message is provided below [2]. Note that in modern IETF Standards these problems are still unsolved.

Load Amplification. The supplementary result of the 503 mechanism is the tendency to significantly amplify the load during periods of overload, thus causing further aggravation of the problem and bringing the collapse of the network closer.

Underutilization. RFC 3261 does not cover how the 503 message recipient should react. In fact, there are some network configurations where it is not possible to clearly identify the sender of the message. So the sender may slow down the load to the entire cluster of servers with no overloaded servers, but not to the actual server in overload.

The Off/On Retry-After Problem. When the sender is balancing requests between a small number of the receivers, the 503 mechanism with Retry-After becomes noneffective because of its all-or-nothing technique. The 503 mechanism with Retry-After tends to cause highly oscillatory behavior under even mild overload.

Ambiguous Usages. The Standards do not clearly determine when the server must send the message with the code 503, and as a result of various implementations the message 503 is used to indicate different states. For example, according to RFC 3398 [28] the signaling gateway sends a message 503 in response to reports of inability to handle the request, which does not necessarily mean that the gateway is overloaded.

Depending on the method by which the sender determines the state of the receiver and manages the load, the congestion control mechanisms can be divided into explicit overload control mechanisms, that are in fact feedback based, and implicit overload control mechanisms, that are self-limiting.

According to RFC 6357 [3], overload control can be implemented end-to-end, hop-by-hop or locally as shown in Fig. 7.

The main ideas of the five explicit overload control feedback mechanisms are formulated as follows [3, 29].

Rate-based Overload Control. The mechanism consists of reducing the transmission rate of the sender based on the current value of the receiver performance.

Loss-based Overload Control. The mechanism allows the receiver to request the sender to reduce the load on a given number of percentages, which is calculated by the recipient taking into account its current load.

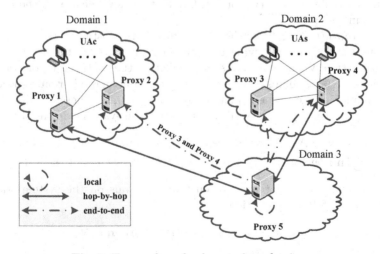

Fig. 7. Types of overload control mechanisms

Window-based Overload Control. The mechanism consists of controlling the size of the window and is inspired by TCP. The mechanism allows the sender to transmit a certain number of messages before it needs to receive a confirmation for the messages in transit.

Signal-based Overload Control. According to this mechanism the sender uses 503 Service Unavailable message from the receiver as a signal of overloading of the receiver.

On-/Off Overload Control. The mechanism uses the 503 (Service Unavailable) response with a Retry-After header and enables the receiver to turn the traffic it is receiving either on or off.

There are a number of papers [9–13, 15–18, 20] with approaches to building models with a threshold overload control. As the authors know, neither in the IETF Standards nor in other sources available there are any analytical models for SIP-server overload control. So in the next section we construct a model of SIP-server on the principles of the SS7 hysteretic load control introduced in section 2.

4   Queuing Model of SIP Servers Overload Control

We consider a single-server queuing system [30], with working vacations and hysteretic congestion control and denote it as $M|M|1|\langle L \rangle |B|WV$. A Poisson customer flow arrives at the system and customers are queued and take service in accordance with congestion control algorithm. The server operates in two modes: normal and congestion. To detect an overload we introduce two thresholds. When the queue becomes full the system recognizes detecting congestion and newly arrived customers are discarded. After that, when the queue length becomes less than L, the system recognizes that congestion is removed and starts putting new customers into the queue. The server takes a working vacation at times when the system is empty. During the vacation period newly arrived customers are stored in the buffer. The server only takes another new vacation if there are no customers in the queue.

We denote $n_1(t) \in \{1, 2\}$ as the server state in the instant t, where state "1" means the server is busy serving a customer and state "2" the server takes a working vacation. We need also to specify whether the control is on or not. Thus we let $n_2(t) \in \{0, 1\}$ equals 0 or 1 to indicate the control is off or on respectively. The occupancy of the queue is denoted by $n_3(t) = \overline{0, B}$. Therefore the Markov process $\mathbf{N}(t) = (n_1, n_2, n_3)$ describes completely the system over the state space

$$\mathcal{N} = \mathcal{N}_0 \cup \mathcal{N}_1, \, \mathcal{N}_0 = \{\mathbf{n} : n_1 = 1, 2; \, n_2 = 0; \, 0 \leq n_3 \leq B - 1\},$$
$$\mathcal{N}_1 = \{\mathbf{n} : (n_1 = 1, 2; \, n_2 = 0; 0 \leq n_3 \leq B - 1) \vee (n_2 = 1; \, n_3 = B)\}.$$

The dependence of the intensity of customer arrivals on the system states is specified by the following relation

$$\lambda(\mathbf{n}) = \lambda \cdot u\left((1 - n_2) \cdot (B - n_3)\right), \mathbf{n} \in \mathcal{N}, \tag{1}$$

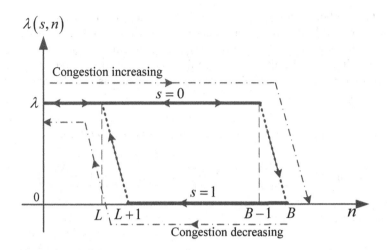

$\lambda(s,n)$

Congestion increasing

$s = 0$

λ

$s = 1$

0

$L \quad L+1$ $B-1 \quad B$ n

Congestion decreasing

Fig. 8. Hysteretic congestion control mechanism

where $u(\cdot)$ is the Heaviside function. The qualitative interpretation of the dependence is presented in Fig. 8.

We assume the customer service time and the vacation durations to be exponentially distributed with parameters μ_1 and μ_2, respectively. We consider that stationary probabilities $p_{n_1 n_2 n_3} = \lim_{t \to \infty} P\{\mathbf{N}(t) = \mathbf{n}\}$, $\mathbf{n} \in \mathcal{N}$ exist, and satisfy the system of equilibrium equations

$$\lambda p_{200} = \mu_1 p_{100}, \tag{2}$$

$$(\lambda + \mu_2)\, p_{20i} = \lambda p_{20i-1}, \; i = \overline{1, B-1}, \tag{3}$$

$$\mu_2 p_{21B} = \lambda p_{20B-1}, \tag{4}$$

$$(\lambda + \mu_1)\, p_{10i} = \lambda u(i)\, p_{10i-1} + \mu_1 p_{10i+1} + \mu_2 p_{20i+1}, i = \overline{0, B-2}, \, i \neq L, \tag{5}$$

$$(\lambda + \mu_1)\, p_{10L} = \lambda p_{10L-1} + \mu_1 p_{10L+1} + \mu_2 p_{20L+1} + \mu_1 p_{11L+1}, \tag{6}$$

$$(\lambda + \mu_1)\, p_{10B-1} = \lambda p_{10B-2}, \tag{7}$$

$$\mu_1 p_{11B} = \lambda p_{10B-1}, \tag{8}$$

$$p_{11i} = p_{11i+1}, \; i = \overline{L+1, B-2}, \tag{9}$$

$$\mu_1 p_{11B-1} = \mu_1 p_{11B} + \mu_2 p_{21B}. \tag{10}$$

In fact the number of states is so large that problem of performance evaluation becomes intractable. We derive a recursive algorithm for efficient calculation of the probabilities $p_{n_1 n_2 n_3}$. We express $p_{n_1 n_2 n_3}$ in terms of p_{200}, i.e., $p_{n_1 n_2 n_3} = x_{n_1 n_2 n_3} p_{200}$. The coefficients $x_{n_1 n_2 n_3}$ fulfill conditions formulated in the lemma below. In order to find the probability p_{200}, we make use of the normalization condition $\sum_{\mathbf{n} \in \mathcal{N}} p_{n_1 n_2 n_3} = 1$.

Lemma 1. *The coefficients $x_{n_1 n_2 n_3}$ obey the following relations*

$$x_{200} = 1, \ x_{20i} = \left(\frac{\lambda}{\lambda + \mu_2} \right)^i, \ i = \overline{1, B-1}, \ x_{21B} = \frac{\lambda}{\mu_2} \cdot \left(\frac{\lambda}{\lambda + \mu_2} \right)^{B-1}, \quad (11)$$

$$x_{100} = \frac{\lambda}{\mu_1}, \quad (12)$$

$$x_{10i+1} = \mu_1^{-1} \left[(\lambda + \mu_1) x_{10i} - \lambda u(i) x_{10i-1} - \mu_2 x_{20i+1} \right], \ i = \overline{0, L-1},$$

$$x_{11L+i} = \frac{\lambda a_{B-L-2} + \mu_2 x_{21B}}{\mu_1 + \lambda b_{B-L-2}}, \ i = \overline{1, B-L-1}, \quad (13)$$

$$x_{10L+i} = a_{i-1} - b_{i-1} x_{11L+1}, \ i = \overline{1, B-L-1}, \quad (14)$$

$$x_{11B} = \frac{\lambda}{\mu_1} x_{10B-1}, \quad (15)$$

where a_i and b_i are given by

$$a_0 = \mu_1^{-1} \left(\lambda x_{20L} + \lambda x_{10L} - \mu_2 x_{20L+1} \right),$$
$$a_i = \mu_1^{-1} \left(\lambda x_{20L+i} + \lambda a_{i-1} - \mu_2 x_{20L+i+1} \right), \ i = \overline{1, B-L-2}, \quad (16)$$

$$b_0 = 1, \ b_i = 1 + \frac{\lambda}{\mu_1} b_{i-1}, \ i = \overline{1, B-L-2}. \quad (17)$$

The proof of Lemma 1 is presented in [30].

5   Conclusion

Major standards organizations 3GPP, ITU and ETSI have all adopted SIP as a key signalling protocol for Next Generation Networks based on the Internet Multimedia Subsystem (IMS). Nowadays, IETF is holding intensive standardization work on SIP readiness of handling overload. The adopted standards and drafts as well as a number of scientific papers still do not give answers to the problems of SIP server overload control mechanism and its modeling. Our possible work for the next step may include the development of queuing models with hysteretic load control as well as mechanisms based on SIP processor occupancy.

References

1. ITU-T Recommendation Q.704: Signalling System No.7 – Message Transfer Part, Signalling network functions and messages (1996)
2. Rosenberg, J.: Requirements for Management of Overload in the Session Initiation Protocol. RFC5390 (2008)
3. Hilt, V., Noel, E., Shen, C., Abdelal, A.: Design Considerations for Session Initiation Protocol (SIP) Overload Control. RFC6357 (2011)
4. Samouilov, K., Zharkov, M., Gaidamaka, Y.: Inconsistency between Q.706 and E.733 and queuing delay calculations in Q.706. COM11-D1479. ITU-O SG11, Geneva, Switzerland (November-December 1999)
5. Abaev, P., Gaidamaka, Y., Samouylov, K.: Load Control Technique with Hysteresis in SIP Signalling Server. In: XXIX International Seminar on Stability Problems for Stochastic Models, the Autumn Session of the V International Seminar on Applied Problems of Probability Theory and Mathematical Statistics related to Modeling of Information Systems, Svetlogorsk, Russia, October 10-16, pp. 67–69 (2011)
6. Rosenberg, J., Schulzrinne, H., Camarillo, G., et al.: SIP: Session Initiation Protocol. RFC3261 (2002)
7. Johnston, A., Donovan, S., Sparks, R., et al.: Session Initiation Protocol (SIP) Basic Call Flow Examples. RFC3665 (2003)
8. Kasera, S., Pinheiro, J., Loader, C., et al.: Fast and robust signaling overload control. In: Ninth International Conference on Network Protocols, pp. 323–331 (2001)
9. Ohta, M.: Overload Protection in a SIP Signaling Network. In: International Conference on Internet Surveillance and Protection, pp. 205–210 (2006)
10. Ohta, M.: Overload Control in a SIP Signaling Network. International Journal of Electrical and Electronics Engineering, 87–92 (2009)
11. Hilt, V., Widjaja, I.: Controlling Overload in Networks of SIP Servers. In: IEEE International Conference on Network Protocols, pp. 83–93 (2008)
12. Shen, C., Schulzrinne, H., Nahum, E.: Session Initiation Protocol (SIP) Server Overload Control: Design and Evaluation. In: Schulzrinne, H., State, R., Niccolini, S. (eds.) IPTComm 2008. LNCS, vol. 5310, pp. 149–173. Springer, Heidelberg (2008)
13. Montagna, S., Pignolo, M.: Performance Evaluation of Load Control Techniques in SIP Signaling Servers. In: Proceedings of Third International Conference on Systems (ICONS), pp. 51–56 (2008)
14. Yang, J., Huang, F., Gou, S.: An optimized algorithm for overload control of SIP signaling network. In: Proceedings of the 5th International Conference on Wireless Communications, Networking and Mobile Computing, WiCOM (2009)
15. Garroppo, R.G., Giordano, S., Spagna, S., Niccolini, S.: Queueing Strategies for Local Overload Control in SIP Server. In: IEEE Global Telecommunications Conference, pp. 1–6 (2009)
16. Montagna, S., Pignolo, M.: Load Control techniques in SIP signaling servers using multiple thresholds. In: 13th International Telecommunications Network Strategy and Planning Symposium, NETWORKS, pp. 1–17 (2008)
17. Garroppo, R.G., Giordano, S., Spagna, S., Niccolini, S.: IEEE Transactions on Network and Service Management 8(1), 39–51 (2011)
18. Homayouni, M., Nemati, H., Azhari, V., Akbari, A.: Controlling Overload in SIP Proxies: An Adaptive Window Based Approach Using No Explicit Feedback. In: 2010 IEEE Global Telecommunications Conference, GLOBECOM 2010, pp. 1–5 (2010)

19. Abdelal, A., Matragi, W.: Signal-Based Overload Control for SIP Servers. In: 7th IEEE Consumer Communications and Networking Conference (CCNC), pp. 1–7 (2010)
20. Montagna, S., Pignolo, M.: Comparison between two approaches to overload control in a Real Server: "local" or "hybrid" solutions? In: 15th IEEE Mediterranean Electrotechnical Conference, MELECON 2010, pp. 845–849 (2010)
21. Hong, Y., Huang, C., Yan, J.: Mitigating SIP Overload Using a Control-Theoretic Approach. In: IEEE Global Telecommunications Conference, GLOBECOM 2010, pp. 1–5 (2010)
22. Gebhart, R.F.: A queuing process with bilevel hysteretic service-rate control. Naval Research Logistics Quarterly 14(1), 55–68 (1967)
23. Brown, P., Chemouil, P., Delosme, B.: A Congestion Control Policy for Signalling Networks. In: Proceedings of 7th IeCC, pp. 717–724 (1984)
24. Filipiak, J.: Modelling and Control of Dynamic Flows in Communication Networks, 202 pages. Springer, New York (1988)
25. Roughan, M., Pearce, C.: A martingale analysis of hysteretic overload control. Advances in Performance Analysis: A Journal of Teletraffic Theory and Performance Analysis of Communication Systems and Networks 3, 1–30 (2000)
26. Takagi, H.: Analysis of a Finite-Capacity M/G/1 Queue with a Resume Level. Performance Evaluation 5, 197–203 (1985)
27. Yum, T.P., Yen, H.M.: Design algorithm for a hysteresis buffer congestion control strategy. In: Proceedings of the IEEE International Conference on Communications, pp. 499–503 (1983)
28. Camarillo, G., Roach, A., Peterson, J., Ong, L.: Integrated Services Digital Network (ISDN) User Part (ISUP) to Session Initiation Protocol (SIP) Mapping. RFC3398 (2002)
29. Gurbani, V., Hilt, V., Schulzrinne, H.: Session Initiation Protocol (SIP) Overload Control. draft-ietf-soc-overload-control-08 (2012)
30. Abaev, P.O.: Algorithm for Computing Steady-State Probabilities of the Queuing System with Hysteretic Congestion Control and Working Vacations. Bulletin of Peoples' Friendship University of Russia (3), 58–62 (2011)

Modeling the Positioning Algorithms Based on RSS Characteristics in IEEE 802.11g Networks

Vladimir Sukhov[1], Mstislav Sivers[2], and Sergey Makarov[1]

[1] St. Petersburg State Polytechnic University, Saint-Petersburg, Russia
sukhov@smtp.ru
[2] The Bonch-Bruevich Saint-Petersburg State University of Telecommunications,
Saint-Petersburg, Russia

Abstract. Using different methods of navigation, as well as consideration of various sources of signal level determine the resultant value of error in estimating location of the mobile station. Modeling variables of the received signal strength is based on known models of channel structure and service packs (standard IEEE 802.11), with taking into account statistical characteristics. Recursive algorithms use to estimate the constant component of the stochastic processes for solving navigational tasks. Accuracy of the location of subscriber devices is assessed with simulation.

Keywords: indoor positioning, wireless channel model, navigation algorithm, location estimation, modeling.

1 Introduction

Despite the fairly high interest and a lot of papers and articles devoted to various aspects of navigation in the WLAN there are few research works about the ultimate accuracy of the coordinates of the Subscriber Unit (SU). In this article we would like to consider all modeling stages of the positioning system – from obtaining raw data to an estimating of mobile station coordinates. It is necessary thought to bring the range of potential accuracy of coordinates, depending on specific conditions.

Positioning technologies are classified according to the parameter signal that is measured for estimating the distance up to the base station with known coordinates (Locator Point). In WLAN networks, the implementation of positioning services most frequently use algorithms based on the measurement of received signal strength RSS (Received Signal Strength). Locating a mobile station (SU) is particularly difficult indoor due to unpredictable propagation of electromagnetic waves. It is one of the reasons of considerable errors. Algorithm for estimating the location of the SU is based on calculating the distance to several LP's and solving of navigation algorithm at the next stage.

LP sends signals with management frames (beacons) at intervals of about 100 ms in accordance with the requirements of IEEE 802.11. Having received these signals SU estimates its level. Been known the power level, we determine the amount of signal attenuation in the path of LP-SU. Further in accordance with the selected model the distance can be estimate.

S. Andreev et al. (Eds.): NEW2AN/ruSMART 2012, LNCS 7469, pp. 453–461, 2012.

It should be noted that there are fluctuations of the receive power level due to RSSI parameter is an integral characteristic of the signal.

2   Modeling RSS Characteristics

Let us consider the simulation of the RSS-performance from two points of view: in terms of creation and in terms of evaluation. In forming characteristics should be taken into account the properties of the channel distortion and intruded noise, as well as the permissible width of the spectrum of the transmitted signal.

In propagation process, environment parameters are been changed over time in a random manner. Therefore, the channel is a linear system with random parameters. The transmission coefficient of such a system for each frequency is a random process. Deterministic signal, after passing through the channel, acquires the characteristics of a random process.

The signal at the receiver is a set of components. In a generalized model as a parameters there are the number of pathways and propagation delay time for each path. Strictly speaking, the number of pathways N and time delay of propagation τ_n are time-varying values. For most channels it is permissible to assume the number of components and values of propagation delays for each path are changed very slowly, so they can be taken to be constant. Given this, the complex envelope of the impulse response of the channel and multipath signal can be described by (1) and (2), respectively [1].

$$\tilde{c}(\tau,t) = \sum_{n=1}^{N} \tilde{a}_n(t)\delta(\tau-\tau_n),\qquad(1)$$

where $\tilde{a}_n(t)$ - complex envelope of the damping factor components of the signal emanating from the delay τ by the n-th path.

$$\tilde{y}(t) = \sum_{n=1}^{N} \tilde{a}_n(t)\tilde{s}(t-\tau_n)\qquad(2)$$

Block-diagram of a multipath signal is shown in fig.1.

To simplify the task of modeling and to limit bandwidth, let make the transition to a TDL with the same delay interval T_S. In accordance with Kotelnikov's theorem the signal $\tilde{s}(t)$, limited range of the highest frequency by ω_M, can be represented by

$$\tilde{s}(t) = \sum_{k=-\infty}^{\infty} \tilde{s}(kT_S)\text{sinc}(\omega_M(t-kT_S)).\qquad(3)$$

At the same time taking into account (2) $\tilde{y}(t)$ is defined as follows

$$\tilde{y}(t) = \sum_{k=-K}^{K} \tilde{s}(kT_S)\tilde{g}_k(t),\qquad(4)$$

where

$$\tilde{g}_k(t) = \sum_{n=1}^{N} \tilde{a}_n(t)\text{sinc}(\omega_M(t-kT_S-\tau_n)) = \sum_{n=1}^{N} \tilde{a}_n(t)\alpha(n,k)\qquad(5)$$

and K is chosen so that for $k > K$ $\left| g_k(t) \right| \to 0$. Function $\alpha(n,k)$ is decreasing rapidly, so the number of delay elements for a model with limited bandwidth fairly small. Thus, a transition to the TDL with equal values of the delay T_s, while the input discrete signal samples $\tilde{s}\left[t_k \right]$ are led to the input line. Complex envelope of the signal attenuation coefficient is given by (5) [1].

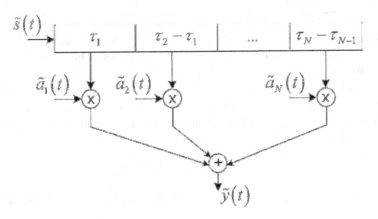

Fig. 1. TDL Channel Model with Variable Spacing

Rayleigh and Rice channel models describe the phenomena of fading with high accuracy. They describe the effect of multipath propagation, the Doppler effect, which can be invoked as the relative motion of transmitter and receiver, as well as movement of objects in the space around. In this case the spectrum of each process $g_k(t)$ will practically have the same form $S(f)$.

For simulation we have used the library simulation environment MATLAB Communication Toolbox. A classification of typical SOHO areas is proposed in [2]. It depends on their sizes, RMS propagation delay values σ_τ and number of clusters. Six models were allocated in accordance with the selected classification. In the simulation we use the models C and D with the parameters given in the appendix C to [1]. The signal assessment has been fulfilled by using function "stdchan(ts,fd,'802.11g',t_rms)" with the values of the delay [1e-9 2e-9]. The purpose of modeling is to create RSS features and to evaluate there options. Subsequently, these parameters will be used in the formation of signals from the LP during assessing the accuracy of positioning. In all cases the distribution of the noise component may be approximated as the centered normal distribution. The simulation results (dispersion of the distribution) are shown in Table 1.One should pay attention to the required estimating frequency of the RSS level. The assessment of RSS level was carried out with frequencies of 1, 2, 5 and 10 Hz; according the interval between administering packages. At higher frequencies f_d, dispersion of the RSS is higher. This can be explained that on higher frequencies the contribution of each ray component is more considerable.

Table 1. Distribution Fitting Results

Channel model	$f_d = 10$ Hz	$f_d = 5$ Hz	$f_d = 2$ Hz	$f_d = 1$ Hz
stdchan802.11g, $\sigma_\tau = 100$ ns	4.79	2.26	0.93	0.51
stdchan802.11g', $\sigma_\tau = 200$ ns	3.01	1.49	0.59	0.32
Model C (NLOS) $\sigma_\tau = 30$ ns	13.98	7.3	2.48	1.41
Model D (LOS) $\sigma_\tau = 50$ ns	9.17	4.49	1.89	1.14

To estimate the distance between the SU and LP by the power of received signal, it is necessary use the model that reflects their interrelation. Most of the models are logarithmic. The most commonly used ones are (6) and (7). As the parameter in equation (6) is being used the damping at the distance d_0, in (7) – the characteristics of the receiving and transmitting antennas. In both cases, there is a coefficient n characterizing the properties of signal attenuation at particular cases, as well as a random variable X_α with normal distribution.

$$P_L(d) = P_L(d_0) + 10n\log(d/d_0) + X_\alpha,\tag{6}$$

where d is the distance between SU and LP, P_L is the magnitude of loss depending on the distance.

$$P_r = P_t + G_t + G_r + 20\log\lambda - 20\log 4\pi - 10n\log d - X_\alpha,\tag{7}$$

where P_t is the magnitude of transmitter output (dBm), G_t, G_r are mean values of transmitter and receiver antenna gain (dBi), λ is the wavelength.

It is important to be note that the construction of the whole model for the entire area or a building is impossible. Changing the propagating properties caused by radiation pattern of antennas, changing the amplitude-time characteristics of the channel, etc. have a significant impact on the resulting power of the received signal.

We assume to create a set of monotonic parameters of attenuation of the signal for each specific area at the stage of system training. It would be useful during positioning systems design. Linearization of (6) and (7) is allowed without a significant loss in accuracy in cases of small range of distances d.

3   Modeling Navigation Algorithms

The problem of determining the coordinates can be formulated in two points of view: a) as a problem of determining original parameters and b) as step-by-step clarifying their values by proposing the amendments.

In addition to solving the navigation problem by the minimum volume of measurements, widely used methods based on the involvement of excessive number of measurements are applied. All ones use techniques of statistical analysis. In statistical processing, random (weakly correlated) components of the measurement error are being smoothed. These components have an impact on a limited number of

LP. Thus excess of the information is to improve the accuracy of coordinate determination of SU.

Finite and iterative methods of solution can be used in order to construct navigation algorithms based on the minimal measurements volume.

3.1   Finite Method

The solution of the navigation problem is reduced to solving systems of nonlinear equations. To determine the spatial coordinates of the object it is sufficient to make distance measurements up to three LP.

Let's impose on the measurement object the Cartesian coordinate system. The coordinates of SU (x, y) can be found by solving the systems of nonlinear equations

$$\begin{cases} r_1^2 = (x_{01} - x)^2 + (y_{01} - y)^2 \\ r_2^2 = (x_{02} - x)^2 + (y_{02} - y)^2 \\ r_3^2 = (x_{03} - x)^2 + (y_{03} - y)^2 \end{cases} \tag{8}$$

where x_{0i}, y_{0j} are coordinates of the j-th LP, r_i is the measured distance between the SU and the LP. Introducing the notation

$$\rho^2 = x^2 + y^2 \quad a_{31} = \frac{\rho_{03}^2 - \rho_{01}^2 + r_1^2 - r_3^2}{2}$$

$$\rho_{0i}^2 = x_{0i}^2 + y_{0i}^2 \quad a_{21} = \frac{\rho_{02}^2 - \rho_{01}^2 + r_1^2 - r_2^2}{2} \tag{9}$$

we'll solve the system (8) and finally obtain the solution

$$\begin{cases} x = \dfrac{a_{31} \cdot (y_{02} - y_{01}) - a_{21} \cdot (y_{03} - y_{01})}{(x_{03} - x_{01}) \cdot (y_{02} - y_{01}) - (x_{02} - x_{01}) \cdot (y_{03} - y_{01})} \\ y = \dfrac{a_{21} \cdot (x_{03} - x_{01}) - a_{31} \cdot (x_{02} - x_{01})}{(x_{03} - x_{01}) \cdot (y_{02} - y_{01}) - (x_{02} - x_{01}) \cdot (y_{03} - y_{01})} \end{cases} \tag{10}$$

3.2   Iteration Method of Solving by Minimum Measurements Volume

The system of equations (8) by Newton's method is a process of iterations by the formula [3]

$$\mathbf{q}_k = \mathbf{q}_{k-1} + \mathbf{C}_{k-1}^{-1} \mathbf{R}_{k-1}, \tag{11}$$

where $\mathbf{R}_{k-1} = \mathbf{R}_{0(k-1)} - \mathbf{R}_M$ is the difference vector of calculated $\mathbf{R}_{0(k-1)}$ and measured $\mathbf{R}_M$ values – vector of residuals and $\mathbf{C}_{k-1}$ is the matrix of observations

$$\mathbf{C}_{k-1} = \begin{bmatrix} \mathbf{C}_{1(k-1)} \\ \vdots \\ \mathbf{C}_{m(k-1)} \end{bmatrix} = \begin{bmatrix} \dfrac{\partial R_1}{\partial q_1} & \dfrac{\partial R_2}{\partial q_1} \\ \vdots & \vdots \\ \dfrac{\partial R_1}{\partial q_m} & \dfrac{\partial R_2}{\partial q_m} \end{bmatrix} = \begin{bmatrix} \dfrac{x_{ci}-x_{k-1}}{r_{01(k-1)}} & \dfrac{y_{ci}-y_{k-1}}{r_{01(k-1)}} \\ \vdots & \vdots \\ \dfrac{x_{ci}-x_{k-1}}{r_{0m(k-1)}} & \dfrac{y_{ci}-y_{k-1}}{r_{0m(k-1)}} \end{bmatrix}. \tag{12}$$

Matrix $\mathbf{C}_{k-1}$ and the difference vector $\mathbf{R}_{k-1}$ in the first iteration are calculated on the basis of a priori data and on subsequent iterations - on the basis of the data obtained in previous iterations. Iterative cycles are repeated until $\delta q = |q_{k-1} - q_k| < \varepsilon$ where ε is the given error value.

3.3   Iteration Method of Solving by Excess Measurements Volume

Most common statistical methods are based on the least square method (LMS). The solution of the navigation problem (8) using the LMS can be written as

$$\mathbf{q}_k = \mathbf{q}_{k-1} + \left(\mathbf{C}_{k-1}^T \mathbf{P} \mathbf{C}_{k-1}\right)^{-1} \mathbf{C}_{k-1}^T \mathbf{P} \mathbf{R}_{k-1}, \tag{13}$$

where $\mathbf{q}_k$ is the estimation of SU position on k-th step. $\mathbf{P}$ is the symmetrical non-negative definite matrix of weights

$$\mathbf{P} = \begin{bmatrix} \sigma_1^2 & \cdots & r_{1,n}\sigma_1\sigma_n \\ \vdots & \ddots & \vdots \\ r_{1,n}\sigma_1\sigma_n & \cdots & \sigma_n^2 \end{bmatrix}^{n \times n}, \tag{14}$$

where r_{ij} is the correlation coefficient of the distance error measurement of i-th and j-th LP. The matrix $\mathbf{K}_q = \mathbf{C}^{-1}\mathbf{P}^{-1}\left(\mathbf{C}^{-1}\right)^T$ is the correlation matrix of the errors of the unknown parameters. The resulting error of positioning is expressed in trace of the correlation matrix $\sigma_q = \sqrt{\mathrm{Sp}(\mathbf{K}_q)}$. To correct linearization errors it's possible to organize additional iterative process constructed on the Newton's scheme. At each measurement of RSS there are six additional iterations. This number of iterations $k_{\mathit{\Pi}}$ is chosen in the modeling process as the optimum in terms of the lowest computational complexity, taking into account the minimum number of measurements of RSS and the lowest residual measurement error.

3.4   Kalman Filter Euations

To solve the problem of estimating the constant component of the received signal strength it is convenient to use Kalman filter (algorithm). The algorithm is defined by the following set of expressions [4]. A priori estimate of the state vector of SU is determined as

$$\mathbf{q}_k^{(-)} = \mathbf{F}_{k-1}\mathbf{q}_{k-1}^{(+)}, \tag{15}$$

where $\mathbf{q}^{(-)}$ is a priori estimate of the state vector $\mathbf{q}$, and $\mathbf{q}^{(+)}$ is posteriori value of $\mathbf{q}$. On the first stage $\mathbf{q}_0^{(+)}$ is defined as expectation of $\mathbf{q}_k$, i.e. $E(\mathbf{q}) = \mathbf{q}_0^{(+)}$. Correlation matrix of errors $\mathbf{P}_k$ characterizes the accuracy of the state vector estimate $\mathbf{q}_k^{(-)}$. Its priori estimate on the k-th step is defined as

$$\mathbf{P}_k^{(-)} = \mathbf{F}_{k-1}\mathbf{P}_{k-1}^{(+)}\mathbf{F}_{k-1}^T + \mathbf{W}_{k-1}, \tag{16}$$

where $\mathbf{P}_k^{(-)}$ is a priori estimate of the error correlation matrix, $\mathbf{P}_{k-1}^{(+)}$ is posteriori value of error correlation matrix got on the $(k-1)$ step, $\mathbf{W}_{k-1}$ covariance matrix of the dynamic system model. Posteriori value of error correlation matrix on the k-th step is defined as

$$\mathbf{P}_k^{(+)} = [\mathbf{I} - \mathbf{K}_k\mathbf{H}_k]\mathbf{P}_k^{(-)}, \tag{17}$$

where $\mathbf{K}_k$ is the Kalman filter gain

$$\mathbf{K}_k = \mathbf{P}_k^{(-)}\mathbf{H}_k^T\left[\mathbf{H}_k\mathbf{P}_k^{(-)}\mathbf{H}_k^T + \mathbf{V}_k\right]^{-1}, \tag{18}$$

where $\mathbf{V}_k$ is the covariance matrix of the measurement model. Posteriori value of the estimated parameters of SU is defined as

$$\mathbf{q}_k^{(+)} = \mathbf{q}_k^{(-)} + \mathbf{K}_k\left[\mathbf{R}_k - \mathbf{H}\mathbf{q}_k^{(-)}\right]. \tag{19}$$

4   Experimental Results

It is obvious that to solve the navigation problem on the first stage it's necessary to evaluate the unknown parameter, and on the second stage to determine the coordinates of SU. The overall block diagram is shown in fig.2. Permanent part of RSS, and the value of the distance between the LP and SU can be assessed with using KF. In the first case it is possible to use a set of parameters $\mathbf{H} = 1$, $\mathbf{V} = 20$, $\mathbf{F} = 1$, $\mathbf{W} = 1$ with subsequent transformation of the model (7) in the value of the distance and then solving of navigation problem. In the second case, the measurement sensitivity matrix it is necessary to determine by linearization expression (7) in the area of the point $q = q_0$ and the subsequent decision of navigation problem.

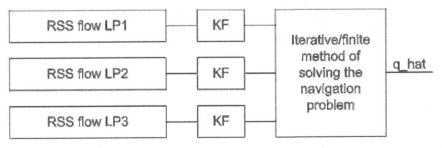

Fig. 2. Structure Model of Navigation Algorithm

Assessment of the potential accuracy of navigation has been carried out by simulation. In Fig.3 dashed line shows the trajectory of the SU. Large dots indicate LP position. Table 2 shows the confidence intervals of maximum navigation accuracy depending on the confidence probability α and various sampling frequencies of RSS f_d. These values have been obtained for the velocity of SU 0.7 m/s with channel model D [2]. Despite larger dispersion on higher frequencies f_d, more accurate positioning can be reached in the interval [2; 5] Hz depending on the positioning algorithm.

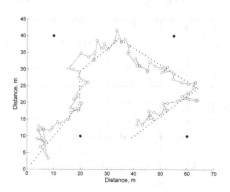

Fig. 3. Modeling of Navigation for Moving SU

Table 2. Maximum Distance Range Error for Specific Confidence Probability α

	Using KF for distance estimation			Using KF for RSS estimation		
	$\alpha = 0.5$	$\alpha = 0.8$	$\alpha = 0.9$	$\alpha = 0.5$	$\alpha = 0.8$	$\alpha = 0.9$
$f_d = 1\text{Hz}$	16.1-17.7	15.9-17.9	15.8-18.0	6.1-6.7	6.0-6.8	6.0-6.8
$f_d = 2\text{Hz}$	12.9-14.7	12.7-14.9	12.6-15.0	3.8-4.6	3.8-4.6	3.7-4.7
$f_d = 5\text{Hz}$	10.8-12.6	10.6-12.8	10.5-12.9	5.6-6.6	5.5-6.7	5.4-6.8
$f_d = 10\text{Hz}$	11.4-13.0	11.2-13.2	11.2-13.2	11.3-12.7	11.2-12.8	11.1-12.9

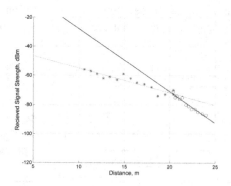

Fig. 4. Linearization of $P_r(d)$

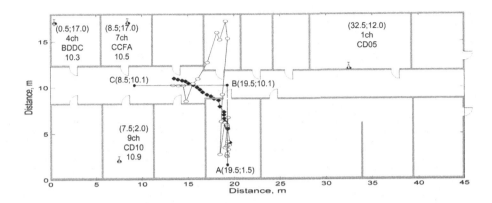

Fig. 5. Position Fixes

The experiment has been fulfilled on the territory of one a company in Saint-Petersburg to confirm values in Table 2. Floor plan, the trajectory of the SU and the coordinates of LP are shown in Fig.5. It was initially assessed the relationship in a distance between SU – LP and the RSS level. The trajectory of the SU is divided into two sections AB and BC. Parameters of $P_r(d)$ were estimated by using LSM (on the assumption of linearity) for each LP at each part (AB and BC). Fig. 4 shows the linearization of $P_r(d)$ to LP #1 BDBC and experimental values. Markers "o" denote the values for AB part. Markers "*" denote the values for BC part. Results of the solution are shown in Fig. 5. The solid line with markers "o" denotes solutions obtained by finite method. The solid line with markers «diamond» denotes the solutions obtained by the iterative method.

References

1. Jeruchim, M.C., Balaban, P., Sam Shanmugan, K.: Simulation of Communication Systems. Modelling, Methodology, and Techniques, 2nd edn., 937 p. Kluwer Academic Publishes (2002) ISBN 0-306-46971-5
2. Erceg, V., et al.: TGn Channel Models. IEEE 802.11. document 11-03/0940r4
3. Shebshaevich, V.S., Dmitriev, P.P., Ivantsevich, N.V., et al.: Network Satellite Radio Navigation Systems, 408 p. Radio i Svyaz Publishers, Moscow (1993) ISBN 5-256-00174-4 (in Russian)
4. Grewal, M., Andrews, A.: Kalman filtering: Theory and practice using MATLAB. John-Wiey&Sons, Inc. (2001) ISBN 0-471-39254-5

Author Index